PERGAMON INTERNATIONAL LIBRARY
of Science, Technology, Engineering and Social Studies
*The 1000-volume original paperback library in aid of education,
industrial training and the enjoyment of leisure*
Publisher: Robert Maxwell, M.C.

The Dynamical Behaviour of Structures

THE PERGAMON TEXTBOOK
INSPECTION COPY SERVICE

An inspection copy of any book published in the Pergamon International Library
will gladly be sent to academic staff without obligation for their consideration for
course adoption or recommendation. Copies may be retained for a period of 60 days
from receipt and returned if not suitable. When a particular title is adopted or
recommended for adoption for class use and the recommendation results in a sale
of 12 or more copies, the inspection copy may be retained with our compliments.
If after examination the lecturer decides that the book is not suitable for adoption
but would like to retain it for his personal library, then a discount of 10% is
allowed on the invoiced price. The Publishers will be pleased to receive suggestions
for revised editions and new titles to be published in this important International
Library.

STRUCTURES AND SOLID BODY MECHANICS SERIES
Series Editor: P R O F E S S O R B. G. N E A L

Other Titles in the Series

A K R O Y D
Concrete: Its Properties and Manufacture

A L L E N
Analysis and Design of Structural Sandwich Panels

B R O W N E
Basic Theory of Structures

C H A R L T O N
Model Analysis of Plane Structures

C H E U N G
Finite Strip Method in Structural Analysis

D U G D A L E
Elements of Elasticity

D U T T O N
A Student's Guide to Model Making

D Y M
Introduction to the Theory of Shells

H E N D R Y
Elements of Experimental Stress Analysis

H E Y M A N
Beams and Framed Structures, 2nd Edition

H O R N E & M E R C H A N T
The Stability of Frames

J A E G E R
Cartesian Tensors in Engineering Science

L E N C Z N E R
Elements of Loadbearing Brickwork

The terms of our inspection copy service apply to all the above
books. Full details of all books listed will gladly be sent upon
request.

The Dynamical Behaviour of Structures

SECOND EDITION

G. B. WARBURTON

Professor of Applied Mechanics, University of Nottingham

PERGAMON PRESS

OXFORD · NEW YORK · TORONTO
SYDNEY · PARIS · FRANKFURT

U.K.	Pergamon Press Ltd., Headington Hill Hall, Oxford OX3 0BW, England
U.S.A.	Pergamon Press Inc., Maxwell House, Fairview Park, Elmsford, New York 10523, U.S.A.
CANADA	Pergamon of Canada Ltd., P.O. Box 9600, Don Mills M3C 2T9, Ontario, Canada
AUSTRALIA	Pergamon Press (Aust.) Pty. Ltd., 19a Boundary Street, Rushcutters Bay, N.S.W. 2011, Australia
FRANCE	Pergamon Press SARL, 24 rue des Ecoles, 75240 Paris, Cedex 05, France
WEST GERMANY	Pergamon Press GmbH, 6242 Kronberg/Taunus, Pferdstrasse 1, Frankfurt-am-Main, West Germany

First Edition 1964

Second Edition 1976

Library of Congress Cataloging in Publication Data

Warburton, Geoffrey Barratt, 1924–
The dynamical behaviour of structures.

(Structures and solid body mechanics)
(Pergamon international library of science, technology, engineering, and social studies)
Includes bibliographical references and index.
1. Structural dynamics. I. Title.
TA654.W3 1976   624′.171   76-13551
ISBN 0-08-020364-7
ISBN 0-08-020363-9 pbk.

Printed in Great Britain by A. Wheaton & Co., Exeter

Contents

271413

Preface to the Second Edition

DURING the twelve years that have elapsed since the preparation of the first edition there has been a considerable increase of interest in, and knowledge of, structural dynamics. Practical reasons for this increase include: the growth in size of structures of various types with a consequential increase in the dynamic forces in existence; the requirement to be able to predict the response of structures to possible earthquakes; and the growth of interest in off-shore structures. The development of computer programs that will solve practical problems in structural dynamics is a vital contributory factor; the numerical methods upon which these programs are based are of obvious significance to structural dynamics.

Consideration of the above points has led to the following major inclusions and changes: an introduction to the random vibration of single- and multi-degree-of-freedom systems; the finite element method is presented as the basic approximate method for determining natural frequencies and response of the various structures considered in the book, although the Rayleigh–Ritz method is included also as it is simpler in concept but less versatile; inclusion of selected numerical integration methods for determining response; more emphasis on the determination of response with allowance for damping and the use of the Lagrange equation when determining response by approximate methods. (The power of this equation, together with its contribution to a unified approach to the subject, outweighs the disadvantages mentioned in the preface to the first edition.) Application of the chosen approximate methods effectively replaces the actual elastic structure by an approximate multi-degree-of-freedom system and the general matrix analysis of Chapter 2 shows how natural frequencies and the response of such systems can be determined. However, the order of presentation—single-degree-of-freedom systems, multi-degree-of-freedom systems, elastic bodies of increasing complexity—has been maintained and is governed by increasing conceptual difficulties in dynamics. Thus methods of generating approximate multi-degree-of-freedom models for elastic structures are presented in later chapters rather than the basic methods of solving dynamic problems for these systems.

SI units, which have many advantages for dynamic problems over any gravitational system of units, are used in this edition, although the use of units is limited to some numerical examples.

Sections of the manuscript have been used when teaching undergraduate and postgraduate courses at the University of Nottingham; the complete manuscript was used as a teaching text in a special course in the Engineering Acoustics Graduate Program of Pennsylvania State University. I appreciate greatly the comments that I have received from these groups. I wish to thank particularly Professor J. C. Snowdon and Dr. J. J. Webster for their helpful and stimulating discussions of many aspects of structural dynamics and Mrs. L. Mlejnecky and Mrs. E. O. Wigginton for typing the manuscript.

CHAPTER 1

Systems with One Degree of Freedom

1.1. Introduction

Although the dynamic response of a practical structure will be complex, it is necessary to begin a study of the dynamical behaviour of structures by considering the fundamentals of vibrations of simple systems. A rough guide to the complexity of a dynamical system is the number of *degrees of freedom* possessed by the system. This number is equal to the number of independent coordinates required to specify completely the displacement of the system. For instance, a rigid body constrained to move in the XY-plane requires three coordinates to specify its position completely—namely the linear displacements in the X- and Y-directions and the angular rotation about the Z-axis (perpendicular to the plane XY); thus this body has three degrees of freedom. The displacement of an elastic body, e.g., a beam, has to be specified at each point by using a continuous equation, so that an elastic body has an infinite number of degrees of freedom. In a dynamical problem the number of modes of vibration in which a structure can respond is equal to the number of degrees of freedom; thus the simplest structure has only one degree of freedom.

Figure 1.1 shows the conventional representation of a system with one degree of freedom; it consists of a mass m constrained to move in the X-direction by frictionless guides and restrained by the spring of stiffness k. It is assumed that the mass of the spring is negligible compared with m. Thus the displacement of the system is specified completely by x, the displacement of the mass, and the system has one degree of freedom. For the purpose of analysing their dynamic response it is possible to treat some simple structures as systems with one degree of freedom. In the simple frame of Fig. 1.2 it is assumed that the

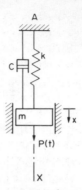

FIG. 1.1. Single-degree-of-freedom system.

horizontal member *BC* is rigid, that the vertical members *AB* and *CD* have negligible mass compared with that of *BC*, and that in any swaying motion *BC* remains horizontal. Then the motion of the system is given by the horizontal displacement of *BC*, *x*, and the frame can be treated as a system with one degree of freedom. Equations and results derived for the system of Fig. 1.1 will be applicable to that of Fig. 1.2.

There are several types of excitation to which structures may be subjected. In Fig. 1.1 a force $P(t)$ is shown applied to the mass; this force is a function of time. There are three main types of exciting force $P(t)$: (i) Harmonic forces, such as $P(t) = P_0 \sin \omega t$ or $P(t) = C\omega^2 \sin \omega t$. (The latter is typical of a component of the force produced by out of balance in a rotating machine.) A force which is periodic but

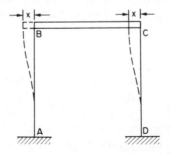

FIG. 1.2. Simple frame with one degree of freedom.

not harmonic can be expressed as a sum of harmonic terms, using Fourier series, and for a linear system the total response can be obtained by superposing the individual responses from each harmonic component of the force. Thus forces which are periodic but not harmonic will not be considered further. (ii) Transient or aperiodic forces: usually these are forces which are applied suddenly or for a short interval of time; simple examples, illustrating the two types, are shown in Fig. 1.3 a and b. (iii) Random forces: the force $P(t)$ cannot be specified as a known function of time, but can be described only in statistical terms; forces due to gusts of wind form an example of this type of excitation.

For (i) the steady-state response of the mass to the harmonic force is required. For (ii) the transient response is required; usually the maximum displacement of the mass or the maximum extension of the spring (the stress in the elastic member of the system is proportional to this extension), occurring during the period of application of the force or in the motion immediately following this period, will be of greatest interest. For (iii) the response can only be determined statistically. Methods of determining the response for the above types of exciting force will be developed later in this chapter.

Vibrations may also be excited by impressed motion of the support, that is by vertical motion of the support A in Fig. 1.1 or by horizontal motion of the foundation AD in Fig. 1.2. As in the case of the exciting force applied to the mass, the impressed motion of the support may be harmonic, transient or random. The response of structures to vibra-

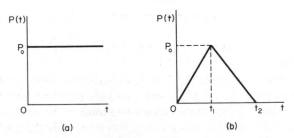

Fig. 1.3. Examples of transient force excitation.

tions transmitted through the ground by earthquakes, traffic, pile-drivers, hammers, explosions, etc., is an important practical subject. If this ground motion (displacement, velocity or acceleration) has been recorded for a past event, this may be treated as a transient response problem. However, if for future events the ground motion can be described only in statistical terms, we have a random vibration problem. Although these different forms of excitation have been illustrated by referring to the simple system with one degree of freedom, the response of more complex structures to these forms of excitation will be considered later.

1.2. General Equation of Motion

The general equation of motion is derived by considering the forces acting on the mass m of Fig. 1.1 at any time t. If the displacement of the mass, x, is measured from the position of static equilibrium, the gravity force mg need not be included in the equation as it is balanced by the restoring force in the spring, $k\delta_{st}$ where δ_{st} is the static deflection of m and k is the stiffness of the spring or the force required to produce unit deflection in the spring; it is assumed that the spring is linear, i.e. k is a constant.

In any real system there will be some damping; this may take various forms, but here *viscous damping* will be assumed, and thus the damping force is proportional to the velocity $\dot{x}$ and opposes the motion. (A dot over a symbol indicates differentiation with respect to time; thus velocity $dx/dt \equiv \dot{x}$ and acceleration $d^2x/dt^2 \equiv \ddot{x}$.) Conventionally viscous damping is represented by the dashpot, shown in parallel with the spring in Fig. 1.1; in practice the damping force is caused by internal friction in the spring, etc., and thus is collinear with the spring force.

Newton's second law of motion is applied to the system; this can be expressed as the product of the mass and the resulting acceleration in the X-direction is equal to the net applied force in the X-direction. For this system the latter has three components, namely the applied force $P(t)$, the restoring or spring force $(-kx)$ and the damping force $(-c\dot{x})$.

Thus the equation of motion is

$$m\ddot{x} = P(t) - kx - c\dot{x}$$

or

$$m\ddot{x} + c\dot{x} + kx = P(t). \tag{1.1}$$

The solution of equation (1.1) gives the response of the mass to the applied force $P(t)$.

For numerical problems SI units are used throughout the book. Thus the relevant fundamental units are: mass in kilograms, displacement in metres and time in seconds. As each term in equation (1.1) represents a force, the units of this equation are Newtons. (A force of 1 Newton applied to a mass of 1 kg produces an acceleration of 1 m/s^2.)

1.3. Free Vibration

Equation (1.1) gives the equation of motion of a single-degree-of-freedom system subject to a disturbing force. If no disturbing force is applied to the system, but the mass is subjected to an initial displacement or velocity, free vibrations will occur; these vibrations will gradually die out due to the damping present in the system.

Thus the equation of free vibration is

$$m\ddot{x} + c\dot{x} + kx = 0. \tag{1.2}$$

By substitution it is seen that equation (1.2) is satisfied by a solution of the form:

$$x = A \exp(\lambda t) \tag{1.3}$$

provided that

$$m\lambda^2 + c\lambda + k = 0,$$

i.e.

$$\lambda = -\frac{c}{2m} \pm \left[\left(\frac{c}{2m} \right)^2 - \frac{k}{m} \right]^{1/2}. \tag{1.4}$$

Case (i). If $(c/2m)^2 > k/m$, the two roots of equation (1.4) are real and negative. Thus the complete solution of equation (1.2) is of the form

$$x = A_1 \exp(\lambda_1 t) + A_2 \exp(\lambda_2 t)$$

and with λ_1 and λ_2 real and negative this represents a gradual creeping back of the mass m towards the equilibrium position. For this to occur the damping must be considerably greater than that existing in practice (with the exception of a few special applications that are not of interest here), so that this case does not require further consideration.

Case (ii). If $(c/2m)^2 < (k/m)$, the roots of equation (1.4) are complex and can be written:

$$\lambda = -\frac{c}{2m} \pm i\left[\frac{k}{m} - \left(\frac{c}{2m}\right)^2\right]^{1/2}$$

where

$$i = (-1)^{1/2}.$$

As

$$\exp(i\theta) = \cos\theta + i\sin\theta$$

and

$$\exp(-i\theta) = \cos\theta - i\sin\theta,$$

the solution of equation (1.2) can be written, after some manipulation, as

$$x = \exp\left(-\frac{c}{2m}t\right)(A_1 \sin\omega^1 t + A_2 \cos\omega^1 t) \tag{1.5}$$

where

$$\omega^1 = \left[\frac{k}{m} - \left(\frac{c}{2m}\right)^2\right]^{1/2}.$$

The constants A_1 and A_2 are determined from the initial conditions, usually the values of x and $\dot{x}$ at $t = 0$. Before considering the solution (1.5) in greater detail, two special cases will be considered.

Special case (a): free undamped vibration, i.e. $c = 0$. When damping is neglected, the solution from (1.5) becomes

$$x = A_1 \sin \omega_n t + A_2 \cos \omega_n t \qquad (1.6)$$

where

$$\omega_n^2 = k/m. \qquad (1.7)$$

Equation (1.6) can be rewritten:

$$x = A \sin (\omega_n t + \alpha)$$

where

$$A = (A_1^2 + A_2^2)^{1/2} \quad \text{and} \quad \tan \alpha = A_2/A_1. \qquad (1.8)$$

Equation (1.8), representing free undamped vibrations, is plotted in Fig. 1.4. It is seen that the amplitude, defined as the maximum

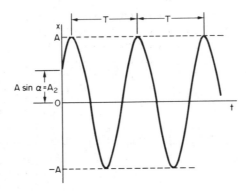

FIG. 1.4. Free undamped vibration.

displacement from the mean position, is A; the period T is the time required for one complete cycle of vibrations, and the frequency f is the number of vibrations in unit time. From equation (1.8) and the above definitions

$$\omega_n T = 2\pi \quad \text{and} \quad f = 1/T,$$

i.e.

$$f = \omega_n/2\pi = 1/T. \qquad (1.9)$$

Thus ω_n is the circular natural frequency of the system, ω_n being measured in radians per second if the natural frequency f is in Hertz. The angle α in equation (1.8) is a phase angle.

The constants A and α in equation (1.8) (or A_1 and A_2 in equation (1.6)) are determined from the initial conditions. For example, suppose that at time $t = 0$ the mass m is given a displacement x_0 and a velocity V. Then substituting in equation (1.6)

$$x_0 = A_2 .$$

Differentiating equation (1.6) with respect to time

$$\dot{x} = A_1 \omega_n \cos \omega_n t - A_2 \omega_n \sin \omega_n t$$

and substituting

$$\dot{x} = V \quad \text{at} \quad t = 0,$$

$$V = A_1 \omega_n .$$

Thus

$$x = \frac{V}{\omega_n} \sin \omega_n t + x_0 \cos \omega_n t$$

or

$$x = \left[\left(\frac{V}{\omega_n} \right)^2 + x_0^2 \right]^{1/2} \sin \left(\omega_n t + \alpha \right)$$

with

$$\tan \alpha = \omega_n x_0 / V.$$

Special case (b): critical damping c_c. Critical damping is the value of the damping coefficient at the change-over from the "creeping" motion of (i) to the damped vibrations of (ii). (It will be used when defining non-dimensional parameters.)

Thus

$$(c_c/2m)^2 = k/m$$

or

$$c_c^2 = 4km,$$

the condition for equal roots of the quadratic equation in λ, i.e.

$$c_c = 2(km)^{1/2} = 2k/\omega_n = 2m\omega_n \qquad (1.10)$$

using equation (1.7).

Case (ii) *continued*: *free damped vibration*. The damping ratio is defined as $\gamma = c/c_c$, i.e. ratio of the actual damping constant to the critical damping value. Thus from equation (1.10) $c/2m = \gamma\omega_n$ and the solution for free damped vibration [equation (1.5)] can be written as:

$$x = A \exp(-\gamma\omega_n t) \sin(\omega^1 t + \alpha). \qquad (1.11)$$

Equation (1.11) is plotted in Fig. 1.5; the displacement, represented by

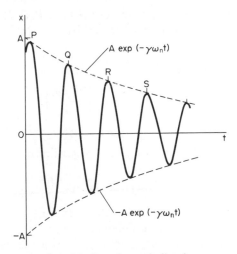

FIG. 1.5. Free damped vibration.

the solid line, lies between the two envelope curves $\pm A \exp(-\gamma\omega_n t)$. Now

$$\omega^1 = [k/m - (c/2m)^2]^{1/2}$$
$$= \omega_n(1 - \gamma^2)^{1/2}. \qquad (1.12)$$

In practice, $\omega^1 \simeq \omega_n$, as except for special applications where heavy damping has been added to a system, γ is small (<0.2). In structures the value of γ depends on the material and on the type of connections at the joints (any looseness increasing the damping); a typical value for a steel structure is $\gamma = 0.03$. Thus for practical values of damping the frequency of free damped vibrations is approximately equal to the natural frequency of free vibrations of the system. In Fig. 1.5 the points P, Q, R and S are successive maximum displacements in the positive direction. At these points $dx/dt = 0$ and x is positive; from equation (1.11) these conditions lead to $\sin(\omega^1 t + \alpha) = (1 - \gamma^2)^{1/2}$. These points are identified easily on an experimental record of free damped vibrations. [The points at which the curve for displacement touches the envelope curve $A \exp(-\gamma\omega_n t)$ are given by $\sin(\omega^1 t + \alpha) = 1$. For small damping each of these points lies close to, but is not coincident with, the corresponding point P, Q, R or S.] For P, Q, R and S the time interval between adjacent points is given by $t^1 = 2\pi/\omega^1$. Thus

$$\frac{x_P}{x_Q} = \frac{x_Q}{x_R} = \frac{x_R}{x_S} = \cdots = \exp(\gamma\omega_n t^1)$$

$$= \exp\left(\frac{2\pi\gamma\omega_n}{\omega^1}\right)$$

$$= \exp[2\pi\gamma(1 - \gamma^2)^{-1/2}]$$

using equation (1.12). Defining the logarithmic decrement δ as $\ln(x_P/x_Q)$

$$\delta = 2\pi\gamma(1 - \gamma^2)^{-1/2}. \tag{1.13}$$

In practice, for reasons given above, $\delta \simeq 2\pi\gamma$. Equation (1.13) gives the relation between logarithmic decrement, which is easily determined from experimental records of free vibration, and the damping factor, which as will be shown is an important parameter in the response to harmonic excitation.

1.4. Response to Harmonic Excitation

The response of the system with one degree of freedom (Fig. 1.1) to a harmonic applied force will now be considered, i.e. $P(t) = P_0 \cos \omega t$ in Fig. 1.1, P_0 is a constant and $\omega/2\pi$ is the frequency of the applied force. From equation (1.1) the relevant equation of motion is

$$m\ddot{x} + c\dot{x} + kx = P_0 \cos \omega t. \tag{1.14}$$

The complete solution of equation (1.14) consists of the complementary function and particular integral; the former is the solution of equation (1.2) which has been obtained in Section 1.3. There are several methods of determining the particular integral; the method to be given was chosen because of its applicability to the multi-degree-of-freedom systems of Chapter 2. On the right-hand side of equation (1.14) cos ωt is replaced by exp $(i\omega t)$, where $i = (-1)^{1/2}$, the modified equation is solved to give a complex expression for x, and the real part of this expression is the required solution. As

$$\exp(i\omega t) = \cos \omega t + i \sin \omega t, \tag{1.15}$$

this procedure is based upon cos ωt being the real part of exp $(i\omega t)$. (If the applied force in equation (1.14) is $P_0 \sin \omega t$, the procedure is similar, except that the imaginary part of the complex expression for x is the required solution.) Thus the modified equation is

$$m\ddot{x} + c\dot{x} + kx = P_0 \exp(i\omega t). \tag{1.16}$$

The particular integral is of the form $x = X \exp(i\omega t)$; thus $\dot{x} = i\omega x$ and $\ddot{x} = -\omega^2 x$. Substituting in equation (1.16)

$$(k - m\omega^2 + ic\omega)x = P_0 \exp(i\omega t),$$

i.e.

$$x = \frac{P_0(k - m\omega^2 - ic\omega) \exp(i\omega t)}{(k - m\omega^2)^2 + c^2\omega^2}. \tag{1.17}$$

This can be expressed as

$$x = H(\omega)P_0 \exp(i\omega t)$$

where the complex frequency response or receptance

$$H(\omega) = \frac{(k - m\omega^2 - ic\omega)}{(k - m\omega^2)^2 + c^2\omega^2}. \tag{1.18}$$

(Equation (1.18) will be used in Section 1.9.) Taking the real part of the right-hand side of equation (1.17),

$$x = \frac{P_0[(k - m\omega^2) \cos \omega t + c\omega \sin \omega t]}{(k - m\omega^2)^2 + c^2\omega^2}$$

$$= \frac{P_0 \cos (\omega t - \alpha)}{[(k - m\omega^2)^2 + c^2\omega^2]^{1/2}} \tag{1.19}$$

with

$$\tan \alpha = \frac{c\omega}{k - m\omega^2}. \tag{1.20}$$

The complete solution is the sum of expressions (1.5) and (1.19) and can be written:

$$x = \exp (-\gamma\omega_n t)(A_1 \sin \omega^1 t + A_2 \cos \omega^1 t)$$

$$+ \frac{P_0 \cos (\omega t - \alpha)}{[(k - m\omega^2)^2 + c^2\omega^2]^{1/2}}. \tag{1.21}$$

As before, the constants A_1 and A_2 are determined from the initial conditions. Physically, the complete response is the sum of the starting transient (the complementary function), which decreases exponentially with time, and the steady-state response (the particular integral). If the vibrations during the first few cycles are of interest, equation (1.21) must be investigated, but in most problems only the response after the starting transient has died away is required; thus further consideration will be given to the steady-state response of equation (1.19).

If X is the amplitude of steady forced vibration, i.e. $x = X \cos (\omega t - \alpha)$, and $X_{st} = P_0/k =$ the displacement for a static force P_0, then

$$\frac{X}{X_{st}} = \frac{1}{[(1 - \omega^2/\omega_n^2)^2 + (c\omega/k)^2]^{1/2}}.$$

Using $c_c = 2k/\omega_n$ and $\gamma = c/c_c$ and defining $r = \omega/\omega_n =$ ratio of the frequency of the applied force to the natural frequency of the system

$$\frac{X}{X_{st}} = \frac{1}{[(1 - r^2)^2 + (2\gamma r)^2]^{1/2}}. \tag{1.22}$$

Also

$$\tan \alpha = \frac{2\gamma r}{1 - r^2}. \tag{1.23}$$

The angle α is the phase angle by which the response lags behind the applied force. The ratio X/X_{st} is called the (dynamic) *magnification factor* or *gain*. Equations (1.22) and (1.23) express the magnification factor and the phase angle in terms of the frequency ratio r and the damping factor γ. They are interpreted graphically in Figs. 1.6 and 1.7,

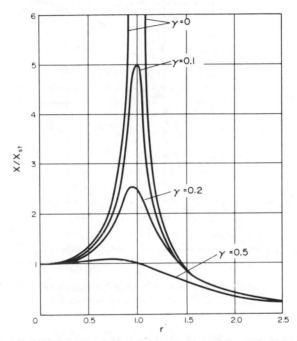

FIG. 1.6. Response to forced vibration: variation of dynamic magnification factor X/X_{st} with frequency ratio r ($=\omega/\omega_n$) for values of the damping factor γ ($=c/c_c$).

where magnification factor and phase angle are plotted against frequency ratio for various values of the damping factor. These figures illustrate the phenomenon of *resonance*—that when the frequency of the applied force equals the natural frequency the amplitude of forced vibration is large for practical values of damping. The value of r at which X/X_{st} is a maximum for a given value of γ is found after evaluating: $[d(X/X_{st})/dr] = 0$ to be given by

$$r^2 = 1 - 2\gamma^2. \tag{1.24}$$

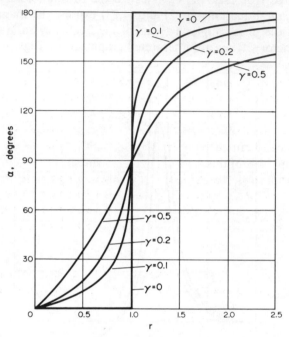

FIG. 1.7. Variation of phase lag α with frequency ratio r for values of γ.

Thus for practical values of damping the magnification factor is a maximum when the frequency ratio is slightly less than unity. (For $\gamma = 0.2$, the maximum gain occurs when $r = 0.96$.) When $r = 1$,

$$X/X_{st} = 1/2\gamma. \tag{1.25}$$

For small damping the maximum amplitude of forced vibration is given approximately by equation (1.25). (It can be shown that the error in this approximation, i.e. in assuming that the maximum value occurs when $r = 1$, rather than when r is given by equation (1.24), is ≤ 2 per cent for $\gamma \leq 0.2$.) Thus if the frequency of the applied force equals the natural frequency, the amplitude tends to a large value, which is inversely proportional to the damping factor. Obviously, if large amplitudes are undesirable, the frequency of the applied force must not be near to the natural frequency of the system. Irrespective of the value of the damping, the amplitude of forced vibration approaches the static displacement when $\omega \ll \omega_n$ and the amplitude of forced vibration becomes very small for $\omega \gg \omega_n$.

For practical damping the phase angle in the neighbourhood of the resonant condition is very sensitive to the value of r. For all values of damping it is equal to $\pi/2$ when $r = 1$.

It has been mentioned previously that in many practical systems the applied force is caused by out of balance in a rotating machine and that a component of this force will be $C\omega^2 \sin \omega t$. The response to this type of applied force is found by solving equation (1.14) with $C\omega^2 \sin \omega t$ replacing $P_0 \cos \omega t$ on the right-hand side. The details are left to the reader, but the amplitude of vibration X is given by

$$\frac{m}{C} X = \frac{r^2}{[(1 - r^2)^2 + (2\gamma r)^2]^{1/2}} . \tag{1.26}$$

For practical values of damping the maximum magnification factor (mX/C in this case) occurs when r is slightly greater than unity, and is given approximately by the value for $r = 1$, namely $mX/C = 1/2\gamma$.

1.5. Energy Expressions

For a vibrating system, energy, equal to the work done by the applied force, is supplied to the system, energy is dissipated by the damping mechanism and energy is stored as *kinetic energy*—the energy of the moving masses—and as *potential energy* consisting of the potential energy of deformation of the elastic members (the *strain energy*) and the potential energy of the masses by virtue of their positions. From the principle of conservation of energy,

(work done by the applied force) − (energy dissipated by damping) =

increase in kinetic energy and potential energy.   (1.27)

(The equation of motion of a single-degree-of-freedom system, (1.1), can be derived from the above energy equation.)

For free undamped vibration, there is no applied force and no energy is dissipated by damping; thus the sum of the kinetic energy and potential energy is a constant. For a single-degree-of-freedom system the kinetic energy is $\frac{1}{2}m\dot{x}^2$; the strain energy is $\frac{1}{2}k$(extension of spring)^2, i.e. $\frac{1}{2}k(x + \delta_{st})^2$, where x is measured from the position of static equilibrium and δ_{st} is the static deflection; and the potential energy of position is $-mg(x + \delta_{st})$ (negative as x is positive downwards, Fig. 1.1).

Thus, $\frac{1}{2}m\dot{x}^2 + \frac{1}{2}k(x + \delta_{st})^2 - mg(x + \delta_{st}) = $ a constant.

Taking constant terms to the right-hand side of the equation and noting that $k\delta_{st} = mg$,

$$\tfrac{1}{2}m\dot{x}^2 + \tfrac{1}{2}kx^2 = C$$

where C is another constant. If $x = A \sin(\omega_n t + \alpha)$,

$$\dot{x} = A\omega_n \cos(\omega_n t + \alpha)$$

and the total energy is a constant, only if the coefficients of the terms in $\sin^2(\omega_n t + \alpha)$ and $\cos^2(\omega_n t + \alpha)$ are equal; i.e.

$$\tfrac{1}{2}mA^2\omega_n^2 = \tfrac{1}{2}kA^2$$

or

$$\omega_n^2 = k/m,$$

agreeing with equation (1.7). It follows that the kinetic energy in the mean position $(\frac{1}{2}mA^2\omega_n^2)$ is equal to the strain energy corresponding to a displacement from the mean position $x = 0$ to the position of maximum displacement $x = A$. This principle is used in the Rayleigh method of determining natural frequencies (Section 3.7).

Applying the energy principle to one complete cycle of steady-state forced vibrations, the work done by the applied force, $P_0 \cos \omega t$, is $\int P_0 \cos \omega t \, dx$. The steady-state response $x = X \cos(\omega t - \alpha)$; thus $dx = -\omega X \sin(\omega t - \alpha) \, dt$. The limits of integration, corresponding to one complete cycle, are $t = 0$ and $t = 2\pi/\omega$. Thus the work done by the applied force per cycle

$$= -\omega P_0 X \int_0^{2\pi/\omega} \cos \omega t \sin(\omega t - \alpha) \, dt$$

$$= \pi P_0 X \sin \alpha. \tag{1.28}$$

The energy dissipated by damping per cycle

$$= \int c\dot{x}\, dx$$

$$= \int_0^{2\pi/\omega} c\dot{x}^2\, dt$$

$$= \int_0^{2\pi/\omega} cX^2\omega^2 \sin^2(\omega t - \alpha)\, dt$$

$$= \pi cX^2\omega. \tag{1.29}$$

From the energy expression (1.27), as there is no net change in kinetic and potential energy over a complete cycle, the work done by the applied force equals the energy dissipated by damping. Using equations (1.19) and (1.20) it can be seen that equations (1.28) and (1.29) are equivalent.

1.6. Hysteretic Damping

From equation (1.29) the energy dissipated per cycle by a viscous damper increases linearly with the frequency of vibration when the amplitude of vibration is constant. In later chapters we want to introduce mathematical models for the dynamic behaviour of elastic bodies and to allow for the imperfections in elasticity that must exist in real bodies. Although damping mechanisms are complex and more knowledge is required, a mechanism that leads to the energy dissipation per cycle being independent of the frequency is a better approximation than viscous damping.

Thus we replace the viscous damper, where the damping force is $-c\dot{x}$ as in Fig. 1.1, by a hysteretic damper, where the damping force is $-h\dot{x}/\omega$; h is the hysteretic damping constant and ω the frequency of harmonic vibration. Thus the equation of motion of the mass for the system with a hysteretic damper, shown in Fig. 1.8, is

$$m\ddot{x} + h\dot{x}/\omega + kx = P_0 \cos \omega t. \tag{1.30}$$

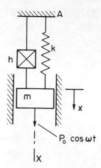

FIG. 1.8. Single-degree-of-freedom system with hysteretic damping.

For steady-state vibration we can obtain results for hysteretic damping from comparable expressions for viscous damping by replacing c by h/ω. Thus the energy dissipated per cycle by hysteretic damping is $\pi h X^2$. The steady-state response is

$$x = X \cos (\omega t - \alpha)$$

with

$$\frac{X}{X_{st}} = \frac{1}{[(1 - r^2)^2 + \mu^2]^{1/2}} \tag{1.31}$$

and

$$\tan \alpha = \frac{\mu}{1 - r^2}$$

where $\mu = h/k$, $r = \omega/\omega_n$, $\omega_n^2 = k/m$ and $X_{st} = P_0/k$. From equation (1.31) the maximum amplitude occurs when $r = 1$ for any value of μ and is given by

$$\left(\frac{X}{X_{st}}\right)_{max} = \frac{1}{\mu}.$$

For small damping the response curves are similar to those shown for viscous damping in Fig. 1.6, but unlike viscous damping, as the non-dimensional hysteretic damping parameter μ increases, maximum amplitudes occur when the excitation frequency is exactly equal to the natural frequency of the system. Hysteretic damping is included when the response of multi-degree-of-freedom systems and beams to harmonic excitation is considered in Chapters 2 and 3 respectively.

1.7. Transient Response

In this section the response of a system with one degree of freedom to transient disturbances will be considered. If a force is applied to the mass for a short interval of time and then removed, the subsequent motion of the mass will be free vibrations, which will decrease gradually in amplitude due to the damping present in the system. Usually the maximum displacement of the mass is of interest (or the related maximum stress in the spring); this will occur during the time of application of the force or in the first cycle of free vibration after removal of the force. Thus considering transient disturbances of short duration, only a few cycles of vibration of the system are of interest. If the damping is small, its effect on the vibration during the first few cycles will be small.

Thus damping is often neglected, as this simplifies the analysis considerably and actual maximum amplitudes are only slightly less than computed values neglecting damping. However, viscous damping is included in the general analysis of this section for completeness.

Two types of transient disturbance will be considered; in the first a force $P(t)$ is applied to the mass m (Fig. 1.9a); the equation of motion is

$$m\ddot{x} + c\dot{x} + kx = P(t). \tag{1.32}$$

In the second the support or frame (point A in Fig. 1.9a) has a motion imposed upon it of $x_0(t)$. (The force $P(t)$ shown in Fig. 1.9a is assumed to be no longer acting.) Thus the extension of the spring is $(x - x_0)$ and the restoring force in the spring is $k(x - x_0)$; similarly the damping force is $c(\dot{x} - \dot{x}_0)$. The equation of motion is

$$m\ddot{x} = -k(x - x_0) - c(\dot{x} - \dot{x}_0)$$

or

$$m\ddot{x} + c\dot{x} + kx = kx_0 + c\dot{x}_0 . \tag{1.33}$$

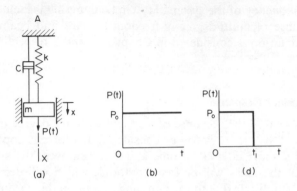

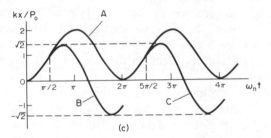

FIG. 1.9. Transient response. (a) System. (b) and (d) Transient forces. (c) Response with $c = 0$ for: applied force (b)—curve A, applied force (d) and $\omega_n t_1 = \pi/2$—curve B, applied force (d) with $\omega_n t_1 = 5\pi/2$—curve C.

Writing equation (1.33) in terms of $x_r = x - x_0$, the displacement of the mass relative to the support,

$$m\ddot{x}_r + c\dot{x}_r + kx_r = -m\ddot{x}_0 . \qquad (1.34)$$

Equations (1.32), (1.33) and (1.34) are mathematically similar so that the following discussion, leading to a solution for equation (1.32) and to results for particular cases, may be applied to equations (1.33) and (1.34) with appropriate changes of nomenclature.

If the mass in Fig. 1.9a is initially at rest and is then subjected at time $t = 0$ to a suddenly applied force P_0 (Fig. 1.9b), the resulting vibrations of the mass are shown in Fig. 1.9c (curve A) and are given by

$$kx/P_0 = 1 - \cos \omega_n t$$

if damping is neglected.

If a rectangular pulse of amplitude P_0 and duration t_1 (Fig. 1.9d) is applied to the mass when at rest at $t = 0$, the initial motion of the mass will be represented by curve A in Fig. 1.9c; at time t_1 the curve for a rectangular pulse will diverge from curve A and for $t > t_1$ the system will perform free vibrations about the mean position, $x = 0$. In Fig. 1.9c curve B gives the displacement of m if $\omega_n t_1 = \pi/2$ and curve C the displacement if $\omega_n t_1 = 5\pi/2$. From curve B for $\omega_n t_1 = \pi/2$, the maximum displacement during the free vibration $(t > t_1)$ is given by $kx/P_0 = 2^{1/2}$ and is *greater* than the maximum displacement during the application of the pulse (given by $kx/P_0 = 1$ when $t = t_1$). From curve C for $\omega_n t_1 = 5\pi/2$, the maximum displacement during the free vibration is given by $kx/P_0 = 2^{1/2}$ and is *smaller* than the maximum during the application of the pulse, given by $kx/P_0 = 2$ when $\omega_n t = \pi$. It can be inferred from the above considerations that whether the maximum transient response occurs during or after the period of application of the disturbing force depends on the ratio t_1/T, i.e. the pulse length/period of the system.

The response of the system of Fig. 1.9a to a general force $P(t)$, which is applied to the mass m, will be considered now with damping included in the analysis. (The results can be used to derive the curves of Fig. 1.9c and comparable curves for other pulse shapes.) If the mass is at rest in its equilibrium position at time $t = 0$ and an impulse $\mathfrak{I}$ is applied to the mass, then from conservation of momentum the initial velocity V of the mass is given by $\mathfrak{I} = mV$.

For $t > 0$ (i.e. after application of the impulse), free damped vibrations of the mass occur. From equation (1.11) this displacement is given by

$$x = \exp(-\gamma\omega_n t)(A \sin \omega^1 t + B \cos \omega^1 t)$$

where $\gamma = c/c_c$, $\omega^1 = \omega_n(1 - \gamma^2)^{1/2}$. The constants A and B have values which satisfy the initial conditions of $x = 0$ and $\dot{x} = \mathfrak{I}/m$ at $t = 0$. Thus from the displacement condition:

$$B = 0$$

and from the velocity condition:

$$\Im/m = \omega^1 A.$$

Hence,

$$x = \frac{\Im}{m\omega^1} \exp\left(-\gamma\omega_n t\right) \sin \omega^1 t.$$

By similar reasoning, if the mass had been at rest until time $t = \tau$ and an impulse $\Im_1$ had been applied at $t = \tau$, the resulting motion would be

$$x = \frac{\Im_1}{m\omega^1} \exp\left[-\gamma\omega_n(t - \tau)\right] \sin \omega^1(t - \tau), \qquad t \geq \tau.$$

This can be expressed as

$$x = \Im_1 h(t - \tau)$$

where

$$h(t - \tau) = \exp\left[-\gamma\omega_n(t - \tau)\right] \sin \omega^1(t - \tau)/m\omega^1 \qquad (1.35)$$

and $h(t - \tau)$ is the response function at time t for a unit impulse applied at time τ. (See Appendix 1 for the relation between the impulse response function and the complex frequency response $H(\omega)$ of equation (1.18).)

The general force $P(t)$ of Fig. 1.10 can be considered as a series of impulses; the shaded area $\Im_1 = P(\tau) \Delta\tau$ is a typical impulse occurring

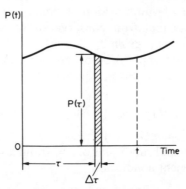

FIG. 1.10. General transient force.

at $t = \tau$. The increment in displacement Δx at time $t(t > \tau)$ due to the impulse $\mathfrak{I}_1$ is

$$\Delta x = \frac{P(\tau)\,\Delta\tau}{m\omega^1} \exp\left[-\gamma\omega_n(t - \tau)\right] \sin \omega^1(t - \tau).$$

Summing for all the impulses into which the curve of $P(t)$ is divided between $t = 0$ and $t = t$, the response at time t is:

$$x = \frac{1}{m\omega^1} \int_0^t P(\tau) \exp\left[-\gamma\omega_n(t - \tau)\right] \sin \omega^1(t - \tau)\, d\tau. \quad (1.36)$$

The Duhamel integral, equation (1.36), assumes that at $t = 0$ the displacement and velocity of the mass m are zero. If these conditions are not satisfied, the free vibration term $A \exp\left(-\gamma\omega_n t\right) \sin\left(\omega^1 t + \alpha\right)$ must be added, where A and α are determined from the initial conditions. If damping is neglected, equation (1.36) is simplified and becomes

$$x = \frac{1}{m\omega_n} \int_0^t P(\tau) \sin \omega_n(t - \tau)\, d\tau. \quad (1.37)$$

Equations (1.36) and (1.37) can be expressed alternatively in terms of the appropriate impulse response function as

$$x = \int_0^t P(\tau)h(t - \tau)\, d\tau. \quad (1.38)$$

As an example, the response of the system to the rectangular pulse of Fig. 1.9d will be determined with damping neglected.

For $0 \le t \le t_1$, $P = P_0$; hence substituting in equation (1.37) and integrating

$$x = \frac{P_0}{m\omega_n^2}(1 - \cos \omega_n t)$$

or

$$kx/P_0 = 1 - \cos \omega_n t. \quad (a)$$

The motion for $t > t_1$ will be a free vibration which can be represented by

$$kx/P_0 = A_1 \sin \omega_n(t - t_1) + A_2 \cos \omega_n(t - t_1). \quad (b)$$

The constants A_1 and A_2 are found by equating values of displacement and velocity from equations (a) and (b) at $t = t_1$, i.e.

$$1 - \cos \omega_n t_1 = A_2$$

and

$$\omega_n \sin \omega_n t_1 = \omega_n A_1.$$

Substituting in equation (b)

$$kx/P_0 = 2 \sin \tfrac{1}{2}\omega_n t_1 [\sin \omega_n(t - t_1) \cos \tfrac{1}{2}\omega_n t_1$$

$$+ \cos \omega_n(t - t_1) \sin \tfrac{1}{2}\omega_n t_1]$$

$$= 2 \sin \tfrac{1}{2}\omega_n t_1 \sin \omega_n(t - \tfrac{1}{2}t_1). \tag{c}$$

From equation (a) the maximum displacement x_m is given by

$$kx_m/P_0 = 2, \tag{d}$$

provided $\omega_n t_1 \geq \pi$, i.e. $t_1/T \geq \tfrac{1}{2}$ where T is the period. If $t_1/T < \tfrac{1}{2}$, the maximum displacement during the pulse is given by

$$kx_m/P_0 = 1 - \cos \omega_n t_1$$

$$= 2 \sin^2 \tfrac{1}{2}\omega_n t_1.$$

From equation (c) the amplitude of free vibration $(t > t_1)$ is

$$kx_m/P_0 = 2 \sin \tfrac{1}{2}\omega_n t_1. \tag{e}$$

Comparing the above values, the maximum displacement occurs during the pulse if $t_1/T \geq \tfrac{1}{2}$ and is given by equation (d); it occurs after the end of the pulse if $t_1/T < \tfrac{1}{2}$ and is given by equation (e).

If the force P_0 is applied statically to the system, the static displacement $x_{st} = P_0/k$. Thus the left-hand sides of equations (d) and (e) are the ratio of the maximum dynamic displacement to the static displacement, or the *dynamic magnification factor*. In Fig. 1.11 the results of

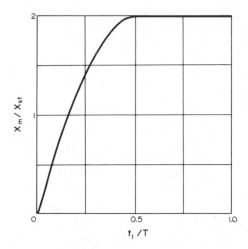

FIG. 1.11. Variation of dynamic magnification factor x_m/x_{st} with pulse time t_1/period T for the force of Fig. 1.9d.

equations (d) and (e) are combined, and the dynamic magnification factor is plotted against the ratio t_1/T.

The above example is relatively simple to evaluate. Pulses of other shapes are of practical interest, but the determination of the dynamic response may involve considerable algebra; Fig. 1.12 shows some of these pulse shapes; triangular, half-sine (an approximation for the force due to an impact) and exponential (an approximation for blast loading). Jacobsen and Ayre[41]* give graphs of dynamic

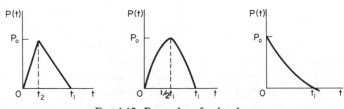

FIG. 1.12. Examples of pulse shape.

* An alphabetical list of references is given on pages 337–340.

magnification factor for many pulse shapes; these show the variation with t_1/T of the maximum displacement of the mass for a force of given pulse shape applied to the mass and also the maximum extension of the spring for a support motion of given pulse shape. The difference between these two maxima with other conditions unchanged is illustrated in Problem 10.

Other methods exist for evaluating the transient response. Readers who are familiar with Laplace transforms may prefer to use them to evaluate the response directly from equation (1.32). Numerical integration of Duhamel's integral is another method, suitable when the form of $P(t)$ makes direct integration difficult or when $P(t)$ is given only in graphical form (e.g. from experimental results). Numerical integration of the equation of motion (1.32) is possible, if some assumption is made about the variation of the displacement or acceleration over a short time interval. Two of the many methods of numerical integration, which can be used to determine the response of multi-degree-of-freedom systems, will be described in Section 2.8; their application to single-degree-of-freedom systems is straightforward. The method given in this section can be used only when the spring or restoring force on the mass is linear, but it should be noted that the methods of numerical integration of Section 2.8 can be used with systems having non-linear restoring forces.

1.8. Random Processes

In previous sections the response of single-degree-of-freedom systems to various types of excitation has been considered; these excitations are all deterministic, i.e. they are specified functions of time, and thus the response is completely determined as a function of time. In practice many forms of excitation, e.g. due to wind and earthquakes, can be described only by their statistical properties and then the corresponding response is described also statistically. Thus in these cases we can predict the probability that the stress or displacement in the structure exceeds a specified level. Essential concepts for dealing with random processes are described briefly in this section. Their application to the determination of the response of single- and multi-degree-of-freedom systems is considered in Sections 1.9 and 2.9, respectively.

More detailed discussion of these concepts will be found in texts on random vibration[24, 66]. It is assumed that the excitation is a stationary ergodic random process and has a Gaussian or normal probability distribution. The meaning of these assumptions will be illustrated later in the section.

We define $\Pr(x(t) > x_0)$ as the probability that a continuously random variable $x(t)$ exceeds the specified value x_0. In a given situation the probability will decrease as x_0 increases. Considering vibrations, where large positive and negative displacements are equally undesirable, the probability of the modulus of $x(t)$ exceeding x_0 may be required; this is expressed as $\Pr(|x(t)| > x_0)$. The probability density function $p_d(x)$ of variable x is defined by

$$\Pr(x_a < x < x_b) = \int_{x_a}^{x_b} p_d(x) \, dx, \qquad (1.39)$$

or

$$\frac{d\Pr(x)}{dx} = p_d(x).$$

It is noted that $p_d(x) > 0$ and $\int_{-\infty}^{\infty} p_d(x) \, dx = 1$. A set of records $x_1(t)$, $x_2(t)$, $x_3(t)$ ... of a random variable x, e.g. earthquake acceleration records or records of wind pressure on a structure, constitutes a random process. If the mean value of x from all the records (when the number is large) at a particular time t_1 is independent of this time t_1, the process is stationary. If also the time average of a record (over a long time) does not depend upon the particular record chosen from the group, the process is ergodic.

It is assumed that the random processes to be considered are stationary and ergodic. (Lin[53] considers non-stationary processes.) The mean value of any function is assumed to be zero. (Functions with a non-zero mean can be treated simply by shifting the origin.) With this simplification the variance σ_x^2 of the variable x, defined as the mean square deviation of x from the mean, equals the mean square value of x. Thus

$$\sigma_x^2 = \overline{x^2(t)}$$

$$= \int_{-\infty}^{\infty} x^2 p_d(x)\, dx. \tag{1.40}$$

In this section, and also in Sections 1.9 and 2.9, the use of a bar over a function denotes the mean value. Thus $\overline{x(t)y(t + \tau)}$ is the mean of the product of x at time t and y at time $t + \tau$, averaged over the time interval $-T$ to $+T$ as $T \to \infty$.

The Gaussian probability distribution represents closely actual random loadings. Also if the excitation has a Gaussian distribution, the response of a linear system also has a Gaussian distribution. For a Gaussian probability distribution:

$$p_d(x) = \frac{1}{(2\pi)^{1/2}\sigma_x} \exp\left[-x^2/2\sigma_x^2\right]. \tag{1.41}$$

From the distribution (1.41) and equation (1.39) the probability that $|x|$ exceeds $n \times$ (root mean square value, σ_x) is 31·7, 4·6, 0·3 and 0·006 per cent for $n = 1, 2, 3$ and 4, respectively.

The autocorrelation function $R_x(\tau)$ of a variable x is defined as

$$R_x(\tau) = \overline{x(t)\,.\,x(t + \tau)}, \tag{1.42}$$

i.e. the mean of the product of the variable at time t and its value at time $(t + \tau)$; the averaging procedure is applied with respect to t over the interval $-T$ to T with $T \to \infty$. From the stationary property $R_x(\tau)$ depends only upon τ. Next the spectral density $S_x(\omega)$ is defined so that $2R_x(\tau)$ and $S_x(\omega)$ form a Fourier transform pair; thus from Appendix 1,

$$R_x(\tau) = \frac{1}{4\pi} \int_{-\infty}^{\infty} S_x(\omega) \exp\left(i\omega t\right) d\omega \tag{1.43}$$

and

$$S_x(\omega) = 2 \int_{-\infty}^{\infty} R_x(\tau) \exp\left(-i\omega\tau\right) d\tau. \tag{1.44}$$

From equation (1.42) $R_x(\tau) = R_x(-\tau)$, so the autocorrelation function is an even function, and it follows from equation (1.44) that $S_x(\omega)$ is also an even function. From equations (1.42) and (1.43)

$$\overline{x^2(t)} = R_x(0)$$

$$= \frac{1}{4\pi} \int_{-\infty}^{\infty} S_x(\omega) \, d\omega$$

$$= \frac{1}{2\pi} \int_{0}^{\infty} S_x(\omega) \, d\omega. \qquad (1.45)$$

From equation (1.45) the spectral density shows the distribution of the harmonic content of the variable x over the frequency range from zero to infinity.

1.9. Random Vibrations

In the previous section a random variable $x(t)$, which may represent a displacement, an acceleration, a pressure, etc., was considered and, assuming a stationary ergodic random process and a Gaussian probability distribution, various statistical relationships introduced. In this section firstly the spectral densities of response and excitation for a vibrating system must be related; this will be derived in general terms so that it is applicable to the multi-degree-of-freedom systems of Chapter 2, as well as to the simpler structures of this chapter.

For the displacement $x_s(t)$ the spectral density and autocorrelation functions are defined as $S_x(\omega)$ and $R_x(t)$, respectively; for the random excitation force $P_j(t)$ the spectral density and autocorrelation functions are defined as $S_p(\omega)$ and $R_p(t)$, respectively. For single-degree-of-freedom systems $x_s(t)$ is the displacement of the mass m of Fig. 1.1 and the mass is subjected to the force $P_j(t)$. For multi-degree-of-freedom systems $x_s(t)$ is the displacement associated with degree of freedom s and $P_j(t)$ is the excitation force applied at degree of freedom j; the integers j and s can take any values between 1 and n, where n is the number of degrees of freedom of the structure. From definition (1.42)

$$R_x(\tau) = \lim_{T \to \infty} \frac{1}{2T} \int_{-T}^{T} x_s(t) x_s(t + \tau) \, dt. \qquad (1.46)$$

From equation (A1.11)

$$x_s(t) = \int_{0}^{\infty} P_j(t - \tau_1) h_{sj}(\tau_1) \, d\tau_1 \qquad (1.47)$$

where $h_{sj}(\tau_1)$ is the response function for a unit impulse, defined in Appendix 1. Similarly

$$x_s(t + \tau) = \int_0^\infty P_j(t + \tau - \tau_2)h_{sj}(\tau_2)\, d\tau_2 . \tag{1.48}$$

Substituting from equations (1.47) and (1.48) in equation (1.46) and rearranging the order of integration,

$$R_x(\tau) = \int_0^\infty h_{sj}(\tau_1) \int_0^\infty h_{sj}(\tau_2) \lim_{T \to \infty} \frac{1}{2T}$$

$$\times \int_{-T}^T P_j(t - \tau_1)P_j(t + \tau - \tau_2)\, dt\, d\tau_2\, d\tau_1$$

$$= \int_0^\infty h_{sj}(\tau_1) \int_0^\infty h_{sj}(\tau_2)R_p(\tau - \tau_2 + \tau_1)\, d\tau_2\, d\tau_1 \tag{1.49}$$

using definition (1.46) for the autocorrelation function $R_p(\tau - \tau_2 + \tau_1)$. Equation (1.44) defines $S_x(\omega)$ in terms of $R_x(\tau)$ and substituting from equation (1.49) in the former equation

$$S_x(\omega) = 2 \int_{-\infty}^\infty \exp\,(-i\omega\tau) \int_0^\infty h_{sj}(\tau_1)$$

$$\times \int_0^\infty h_{sj}(\tau_2)R_p(\tau - \tau_2 + \tau_1)\, d\tau_2\, d\tau_1\, d\tau. \tag{1.50}$$

As $h_{sj}(t) = 0$ for $t < 0$, the lower limits of the inner integrals can be changed from zero to $-\infty$. After some rearrangement equation (1.50) becomes:

$$S_x(\omega) = 2 \int_{-\infty}^\infty h_{sj}(\tau_1) \exp\,(i\omega\tau_1) \int_{-\infty}^\infty h_{sj}(\tau_2) \exp\,(-i\omega\tau_2)$$

$$\times \int_{-\infty}^\infty R_p(\tau - \tau_2 + \tau_1) \exp\,[-i\omega(\tau + \tau_1 - \tau_2)]\, d\tau\, d\tau_2\, d\tau_1$$

$$= \int_{-\infty}^\infty h_{sj}(\tau_1) \exp\,(i\omega\tau_1) \int_{-\infty}^\infty h_{sj}(\tau_2) \exp\,(-i\omega\tau_2)$$

$$\times 2 \int_{-\infty}^{\infty} R_p(\tau_3) \exp\left(-i\omega\tau_3\right) d\tau_3 \, d\tau_2 \, d\tau_1$$

where $\tau_3 = \tau - \tau_2 + \tau_1$. Substituting from equation (A1.14) for the first and second integrals and from equation (1.44) for the third integral,

$$S_x(\omega) = H_{sj}(-\omega) \cdot H_{sj}(\omega) \cdot S_p(\omega)$$

where $H_{sj}(\omega)$ is the complex frequency response and is the Fourier transform of $h_{sj}(t)$. Now $H_{sj}(\omega)$ and $H_{sj}(-\omega)$ are complex conjugates [see expressions (1.18), (2.126) and (2.127) for $H_{sj}(\omega)$]; thus $H_{sj}(\omega)H_{sj}(-\omega) = |H_{sj}(\omega)|^2$ and

$$S_x(\omega) = |H_{sj}(\omega)|^2 S_p(\omega). \tag{1.51}$$

Thus the spectral density of the displacement at frequency ω equals the spectral density of the excitation at ω multiplied by the square of the modulus of the complex frequency response at ω. From equations (1.45) and (1.51)

$$\overline{x_s^2(t)} = \frac{1}{2\pi} \int_0^{\infty} |H_{sj}(\omega)|^2 S_p(\omega) \, d\omega. \tag{1.52}$$

We consider now the single-degree-of-freedom system with viscous damping of Fig. 1.1 with the mass subjected to a random force $P(t)$ of spectral density $S_p(\omega)$. Equation (1.51) relates the spectral density of the displacement $S_x(\omega)$ to $S_p(\omega)$. For this system the complex frequency response is obtained from equation (1.18) and thus

$$|H(\omega)|^2 = \frac{1}{k^2[(1 - \omega^2/\omega_n^2)^2 + (2\gamma\omega/\omega_n)^2]}. \tag{1.53}$$

The subscripts s and j in equations (1.46) to (1.52) can be omitted for this case. Figure 1.13 shows three curves for the spectral density $S_p(\omega)$: (a) $S_p(\omega) = S_0$, i.e. a spectrum that is uniform over the complete frequency range from 0 to ∞, usually called white noise. This is useful for analytical purposes, although it predicts an infinite mean square value for $P(t)$. (b) $S_p(\omega)$ is uniform up to a cut-off frequency ω_c, which is well above the natural frequency ω_n. (c) A curve for $S_p(\omega)$ which varies fairly slowly with ω. Figure 1.14 shows the variation of $|H(\omega)|^2$

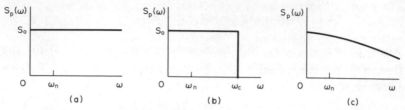

FIG. 1.13. Examples of spectral density $S_p(\omega)$ for an applied force $P(t)$.

with ω for a low value of the damping ratio γ. From equation (1.51) the spectral density of the response $S_x(\omega)$ will be: the same shape as Fig. 1.14 for white noise excitation; the same shape as Fig. 1.14 up to $\omega = \omega_c$ and zero for $\omega > \omega_c$ for (b); and a slightly modulated form of Fig. 1.14 for (c).

For white noise excitation (Fig. 1.13a) the mean square response is found from equations (1.52) and (1.53) as

$$\overline{x^2(t)} = \frac{1}{2\pi} \int_0^\infty \frac{S_0 \, d\omega}{k^2[(1 - \omega^2/\omega_n^2)^2 + (2\gamma\omega/\omega_n)^2]}. \qquad (1.54)$$

Equation (1.54) can be integrated by the calculus of residues to give

$$\overline{x^2(t)} = \frac{S_0 \omega_n}{8\gamma k^2}. \qquad (1.55)$$

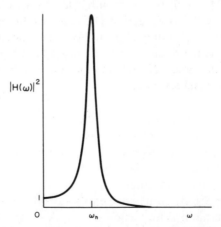

FIG. 1.14. Variation of $|H(\omega)|^2$ with ω for a low value of the damping ratio γ.

The mean square response for excitation (b) is given by equation (1.54) with the upper limit in the integral changed to ω_c. The integral (1.54) with limits ω_c and ∞ can be evaluated using an asymptotic series expression for the integrand, valid for large values of ω. Subtracting this integral from the previous result (1.55) gives the required mean square response as

$$\overline{x^2(t)} = \frac{S_0 \omega_n}{8\gamma k^2} \left[1 - \frac{4}{3\pi} \frac{\gamma}{(\omega_c/\omega_n)^3} - \frac{8(1 - 2\gamma^2)\gamma}{5\pi(\omega_c/\omega_n)^5} \cdots \right]. \qquad (1.56)$$

Even for ω_c/ω_n as low as 2 and γ as high as 0·1 the square bracket in equation (1.56) differs from unity by less than 1 per cent. Thus equation (1.55) can be used to give the mean square response when the excitation spectrum is uniform up to a cut-off frequency well beyond the natural frequency.

If $S_p(\omega)$ varies slowly in the vicinity of ω_n, as in Fig. 1.13c, the mean square value for the response is often approximated as

$$\overline{x^2(t)} \simeq \frac{S_p(\omega_n)\omega_n}{8\gamma k^2} \qquad (1.57)$$

where $S_p(\omega_n)$ is the value of the spectral density of the excitation at the natural frequency. This is justified because the dominant contribution to $\overline{x^2(t)}$ comes from frequencies in the vicinity of ω_n. If the curve of Fig. 1.13c is

$$S_p(\omega) = S_0 - S_2(\omega/\omega_n)^2 \qquad (1.58)$$

substitution of $S_p(\omega)$ and $|H(\omega)|^2$ in equation (1.52) and integration by the calculus of residues gives

$$\overline{x^2(t)} = \frac{(S_0 - S_2)\omega_n}{8\gamma k^2}$$

$$= \frac{S_p(\omega_n)\omega_n}{8\gamma k^2}.$$

Thus for this special case there is no approximation involved in equation (1.57).

If hysteretic damping with non-dimensional damping factor μ replaces the viscous damping, $|H(\omega)|^2$ can be found from equation (1.53) by replacing c by h/ω; as $h/k = \mu$, $2\gamma\omega/\omega_n$ is replaced by μ. For the white noise excitation of Fig. 1.13a the mean square value of response

$$\overline{x^2(t)} = \frac{1}{2\pi} \int_0^\infty \frac{S_0 \, d\omega}{k^2[(1 - \omega^2/\omega_n^2)^2 + \mu^2]}$$

$$\simeq \frac{S_0 \omega_n}{4\mu k^2} \qquad (1.59)$$

by the calculus of residues. This expression is valid provided that $\mu^2 \ll 1$. For reasons given above $\overline{x^2(t)}$ is often approximated for a general spectral density excitation $S_p(\omega)$, which varies slowly in the vicinity of ω_n, by

$$\overline{x^2(t)} \simeq \frac{S_p(\omega_n)\omega_n}{4\mu k^2}. \qquad (1.60)$$

For the excitation of Fig. 1.13c with $S_p(\omega)$ given by equation (1.58), equation (1.60) is valid, provided that $\mu^2 \ll 1$.

Noting that the variance $\sigma_x^2 = \overline{x^2(t)}$ when the mean value is zero, expressions (1.39) and (1.41) for probability and probability density function for a Gaussian distribution can be used to predict the probability that the displacement x exceeds any specified value when the mean square value $\overline{x^2(t)}$ has been determined.

In this section it has been assumed that a random force $P(t)$ is applied to the mass m of Fig. 1.9a. In practice, random vibrations are often caused by random motion of the support A in Fig. 1.9a. When considering transient vibration in Section 1.7, the relevant equations, (1.32) to (1.34), for force and support excitation have been given. Specifically, if a random acceleration $\ddot{x}_0(t)$ is applied to the support and the relative displacement of the mass x_r, where $x_r = x - x_0$, is required, the input spectral density $S_p(\omega)$ relates to the acceleration $\ddot{x}_0(t)$ and, comparing equations (1.32) and (1.34), the mean square response $\overline{x_r^2(t)}$ is obtained by replacing k^2 by ω_n^4 in the denominators of equation (1.53) and subsequent equations.

The analysis given here is for stationary processes; thus the statistical properties of a random variable, shown in a set of records, are independent of the sampling time. From this it is inferred that the records are theoretically of infinite duration, but in practice records of finite duration of specific events must be used. Provided that the duration of the record t_1 /the period of the system $> (0.4/\gamma)$ approximately, there is only a slight error in using the analysis for stationary processes[53]. (An analogous situation occurs when the starting transient is neglected in comparison with the steady-state amplitude in the response to harmonic excitation (Section 1.4); this is valid provided that the time at which the amplitude is required, t_1, satisfies an inequality, similar to that given above.)

Problems

1. A frame consists of a rigid horizontal member BC of mass 2000 kg, supported by two light elastic vertical members, AB and DC, each of stiffness in flexure 1·0 MN/m (Fig. 1.2). Find the natural frequency of horizontal vibrations of BC.
2. The member BC of the frame of Problem 1 is given a horizontal displacement of 20 mm at time $t = 0$ and released. Neglecting damping, obtain an expression for the resulting displacement of BC.
3. The free oscillations of BC in Problem 1 are recorded; it is found that each maximum displacement in the positive direction is 0·9 × (the preceding maximum displacement in that direction). Determine the coefficient of viscous damping, the damping ratio γ and the logarithmic decrement.
4. The member BC in Problem 1 is subjected to a horizontal applied force 200 sin ωt N. Assuming the existence of the viscous damping defined in Problem 3, plot the amplitude of steady forced vibration, and also the phase angle between displacement and disturbing force, against ω. Find the maximum amplitude and the corresponding values of ω and the phase angle.
5. The natural frequency of a single-degree-of-freedom system is $20/2\pi$ Hz. The stiffness of the spring is 8·0 kN/m. The coefficient of viscous damping is 200 N s/m. (a) Find the amplitude of steady forced vibration if a force 100 sin 40t N is applied to the mass. (b) If the above force is applied at time $t = 0$ when the system is at rest, obtain an expression for the resulting displacement of the mass and sketch the variation of this displacement with time.
6. The mass of a single-degree-of-freedom system is subjected to a force $C\omega^2$ sin ωt. Allowing for viscous damping, obtain an expression for the amplitude of steady forced vibration. Determine the value of ω, at which the amplitude is a maximum, in terms of the damping ratio γ.
7. In the system of Fig. 1.1 the mass m is 40 kg, the stiffness of the spring k is 32 kN/m, the viscous damping coefficient c is 300 N s/m, and the excitation force $P(t) = $ 0·04 ω^2 sin ωt N. Find the natural frequency of free undamped vibrations of the system. Find the frequency of the exciting force for which the amplitude of steady forced vibrations is a maximum and this maximum amplitude.
   Determine the frequency ranges for which the amplitude will be less than $\frac{1}{2}$ (the maximum amplitude found above).
8. In the system of Fig. 1.1 the support A is given a displacement B sin ωt. If (a) the

steady-state amplitude of the mass $\leq 0.1B$ and damping is neglected, (b) the amplitude $\leq 0.1B$ and the damping factor $\gamma = 0.2$, (c) the amplitude $\leq 0.01B$ and damping is neglected, and (d) the amplitude $\leq 0.01B$ and $\gamma = 0.2$, show that the maximum allowable stiffness is given by $k = \beta m \omega^2$ and obtain the numerical value of β in each case.

9. The mass of a single-degree-of-freedom system of period T is subjected to a transient force:

$$P(t) = P_0 \sin \pi t/T, \qquad 0 \leq t \leq T.$$

$$P(t) = 0, \qquad t > T.$$

Find the maximum value of the displacement of the mass, and the time at which it occurs, in terms of P_0, T and the spring stiffness k. Neglect damping.

10. The support of a single-degree-of-freedom system of period T is subjected to a displacement x_0 in the form of a rectangular pulse, i.e. $x_0 = a$, $0 \leq t \leq t_1$; $x_0 = 0$, $t > t_1$. Investigate the variation with t_1/T of (a) the maximum extension of the spring, and (b) the maximum displacement of the mass.

11. The base AD of the frame, described in Problem 1, is subjected to a random acceleration $\ddot{x}_0(t)$, which has a spectral density

$$S_0(\omega) = 1 \ \text{m}^2/\text{s}, \qquad 0 \leq \omega \leq 50 \ \text{rad/s}$$

$$= 0, \qquad \omega > 50 \ \text{rad/s}.$$

The viscous damping ratio γ is 0.02. Sketch the spectral density for the displacement of the mass relative to the base, x_r.

Determine the mean square and root mean square values of the force in the vertical member AB, (a) using the above spectral density for the input acceleration, and (b) assuming white noise for $S_0(\omega)$. What is the probability that the magnitude of the force in AB exceeds 40 kN?

If the random acceleration is replaced by a harmonic acceleration, i.e. $\ddot{x}_0 = \ddot{X}_0 \sin \omega t$, and $\ddot{X}_0$ is chosen so that the root mean square values of the harmonic and random forms for $\ddot{x}_0(t)$ are equal, find the maximum amplitude of the force in AB.

CHAPTER 2

Systems with Several Degrees of Freedom

THE main topic of this chapter is the vibration analysis of multi-degree-of-freedom systems with emphasis on determining their response for various types of excitation (Sections 2.4 to 2.9). From the figures used to illustrate them these systems may appear to represent a limited class of practical structures. However, as shown in subsequent chapters the dynamic behaviour of complex elastic structures can be analysed by approximate methods (e.g. the Rayleigh–Ritz and finite element methods), and these analyses effectively reduce the continuous structure to a multi-degree-of-freedom system. Thus the methods of Sections 2.4 to 2.9 have wide practical applications.

The general analysis of multi-degree-of-freedom systems can be handled conveniently only in matrix form. Before considering this general analysis some new vibration concepts are introduced by analysing systems with two degrees of freedom, where matrix algebra is not essential. In Sections 2.1, 2.2 and 2.3, respectively, free undamped vibrations, the steady-state response to harmonic excitation and the response to a general exciting force of systems with two degrees of freedom are considered. The second topic may be regarded as a special case of the last one, but is treated separately because of its practical importance and the existence of special methods of determining the response for harmonic excitation.

2.1. Free Vibrations (Two-degree-of-freedom Systems)

A system has two degrees of freedom if its configuration at any time can be represented by two independent coordinates. Three examples are shown in Figs. 2.1, 2.2a and 2.3. In Fig. 2.1 the masses m_1 and m_2

are constrained to move in the direction XX' by frictionless guides; the three springs have stiffnesses k_1, k_2 and k_3 and the upper end of spring k_1 and the lower end of spring k_3 are anchored to rigid immovable blocks. The motion of the system will be represented by the two independent coordinates x_1 and x_2, the absolute displacements of the masses m_1 and m_2 from their positions of static equilibrium. (It will be seen that although there are only two independent coordinates, they can be chosen in different ways; e.g. the absolute displacement of mass m_1 and the displacement of mass m_2 relative to mass m_1 could be chosen instead of those defined above.)

Two equations of motion for the free undamped vibrations of the system of Fig. 2.1 can be written down, by considering the equilibrium

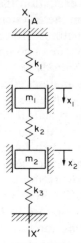

FIG. 2.1. Two-degree-of-freedom system.

of each mass under the applied and inertia forces. For mass m_1:

$$m_1 \ddot{x}_1 = -k_1 x_1 + k_2(x_2 - x_1).$$

For mass m_2:

$$m_2 \ddot{x}_2 = -k_3 x_2 - k_2(x_2 - x_1).$$

On rearrangement the two equations become

$$m_1 \ddot{x}_1 + (k_1 + k_2)x_1 - k_2 x_2 = 0$$

and

$$m_2 \ddot{x}_2 + (k_2 + k_3)x_2 - k_2 x_1 = 0. \tag{2.1}$$

Before obtaining the solution of equations (2.1) the equations of motion for the systems of Figs. 2.2a and 2.3 will be considered.

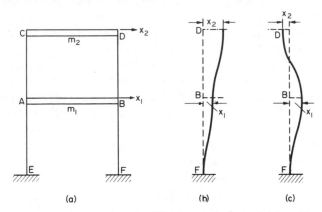

(a)   (b)   (c)

FIG. 2.2. Frame with two degrees of freedom: only flexural deformation of the vertical members in the plane of the frame permitted. (a) Details of frame. (b) and (c) First and second modes of vibration.

Fig. 2.2a represents a two-storey frame, consisting of the two rigid horizontal members *AB* and *CD* and the four light, elastic vertical members *AE*, *BF*, *CA* and *DB*. This frame is an example of a *shear building* with two storeys. This concept of a shear building is very useful in simplifying the analysis of vibrations of buildings; for the general case of a shear building with *n* storeys the following assumptions are made: all the floors are rigid so that no joint can rotate; all the columns are inextensible; and the mass of the building is concentrated at the floor levels, so that vibration of the *n*-storey building is reduced to the vibration of a system with *n* degrees of freedom. (Frames in which *all* members are elastic and have distributed mass can be analysed as systems of beams (Section 4.5) or investigated approximately using the finite element method (Section 3.9).) Only swaying of the frame in its plane, *ABCDEF*, will be considered; the absolute displacements of the members *AB* and *CD* in the horizontal direction are defined as x_1 and x_2 respectively. Any displacement x_1 of member AB will be resisted by elastic restoring forces in the four stanchions. If k_1 is the resultant stiffness of AE and BF, there will be a force on AB of $-k_1 x_1$ for a displacement x_1; if k_2 is the resultant stiffness of CA and DB, there will be forces on AB and CD of $-k_2(x_1 - x_2)$ and $+k_2(x_1 - x_2)$, respectively. Thus the equations of motion become

$$m_1 \ddot{x}_1 + (k_1 + k_2)x_1 - k_2 x_2 = 0$$

and

$$m_2 \ddot{x}_2 + k_2 x_2 - k_2 x_1 = 0. \tag{2.2}$$

As AB and CD are assumed to be rigid members, which remain horizontal, a possible deformation of the stanchions BF and DB is as shown in Fig. 2.2b. If the stanchions are treated as beams of uniform cross-section, standard methods of analysing beams with built-in ends when one end sinks relative to the other show that the stiffnesses k_1 and k_2 are[17]

$$k_1 = 24B_1/l_1^3,$$
$$k_2 = 24B_2/l_2^3, \tag{2.3}$$

where B_1 and l_1 are the flexural rigidity and length respectively of AE and BF, and B_2 and l_2 are the flexural rigidity and length respectively of CA and DB.

Figure 2.3 shows an unsymmetrical single-storey frame, consisting of a rigid horizontal member AB, supported on the two stanchions AC

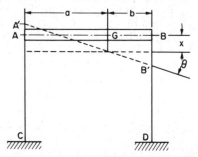

FIG. 2.3. Frame with two degrees of freedom: only extensional deformation of the vertical members permitted.

and BD. In this case it will be assumed that the point A is constrained to move in the direction AC and B in the direction BD; thus any displacement of the member AB to a position $A'B'$ can be considered to be a combination of a vertical displacement of AB plus a rotation of AB about a horizontal axis through the centre of gravity G. This is resisted by the elastic restoring forces, caused by axial compression or tension in the bars AC and BD. If the stiffnesses of the stanchions AC and BD are k_1 and k_2, respectively, and their cross-sectional areas are η_1 and η_2, respectively, then $k_1 = \eta_1 E/l$ and $k_2 = \eta_2 E/l$, where $l =$ the length AC. Defining the independent coordinates as $x =$ the vertical displacement of G (positive downwards) and $\theta =$ the rotation of AB about the horizontal axis through G (positive clockwise), $AA' = x - a\theta$ and $BB' = x + b\theta$ (Fig. 2.3). Two equations of motion are obtained by resolving in the vertical direction for the forces acting on AB including the inertia force $m\ddot{x}$ and by taking moments about G including the inertia moment $I\ddot{\theta}$, where $I =$ the moment of inertia of AB about the horizontal axis through G. That is,

$$m\ddot{x} = -k_1(x - a\theta) - k_2(x + b\theta)$$

and

$$I\ddot{\theta} = k_1(x - a\theta)a - k_2(x + b\theta)b.$$

These equations can be rewritten as:

$$m\ddot{x} + (k_1 + k_2)x + (k_2 b - k_1 a)\theta = 0,$$
$$I\ddot{\theta} + (k_1 a^2 + k_2 b^2)\theta + (k_2 b - k_1 a)x = 0. \tag{2.4}$$

If in the system of Fig. 2.3, $k_1 a = k_2 b$, equations (2.4) reduce to two independent equations, one in x and the other in θ. In this special case (which includes the symmetrical frame where $a = b$ and $k_1 = k_2$), the motion is *uncoupled*; from Chapter 1, the natural frequencies of vertical translation and rocking are given by $\omega_1^2 = (k_1 + k_2)/m$ and $\omega_2^2 = (k_1 a^2 + k_2 b^2)/I$, respectively. In the general case, where $k_1 a \neq k_2 b$, coupling exists between vertical translation and rocking and equations (2.4) are of similar form to equations (2.1) and (2.2).

The equations of motion of any two-degree-of-freedom system can be written as

$$m_1 \ddot{y}_1 + k_{11} y_1 + k_{12} y_2 = 0,$$

$$m_2 \ddot{y}_2 + k_{21} y_1 + k_{22} y_2 = 0. \qquad (2.5)$$

The systems of Figs. 2.1, 2.2a and 2.3 are particular examples; the relations between the parameters in equations (2.5) and those in equations (2.1), (2.2) and (2.4) are easily established and are given in Table 2.1.

TABLE 2.1

Symbol in equations (2.5)	Equivalent expression in		
	equations (2.1), Fig. 2.1	equations (2.2), Fig. 2.2	equations (2.4), Fig. 2.3
y_1	x_1	x_1	x
y_2	x_2	x_2	θ
m_1	m_1	m_1	m
m_2	m_2	m_2	I
k_{11}	$k_1 + k_2$	$k_1 + k_2$	$k_1 + k_2$
$k_{12} = k_{21}$	$-k_2$	$-k_2$	$k_2 b - k_1 a$
k_{22}	$k_2 + k_3$	k_2	$k_1 a^2 + k_2 b^2$

Solutions of equations (2.5) are sought in the form

$$y_1 = Y_1 \sin(\omega t + \alpha), \qquad y_2 = Y_2 \sin(\omega t + \alpha). \qquad (2.6)$$

Equations (2.6) are solutions, provided that

$$(k_{11} - m_1 \omega^2) Y_1 + k_{12} Y_2 = 0$$

and

$$k_{21} Y_1 + (k_{22} - m_2 \omega^2) Y_2 = 0. \qquad (2.7)$$

Elimination of Y_1 / Y_2 leads to the frequency equation:

$$m_1 m_2 \omega^4 - (k_{11} m_2 + k_{22} m_1) \omega^2 + (k_{11} k_{22} - k_{12}^2) = 0, \qquad (2.8)$$

noting that $k_{12} = k_{21}$ in an elastic system by the reciprocal theorem. The solution of equation (2.8), a quadratic equation in ω^2, is:

$$\omega^2 = \frac{1}{2}\left(\frac{k_{11}}{m_1} + \frac{k_{22}}{m_2}\right) \pm \frac{1}{2}\left[\left(\frac{k_{11}}{m_1} - \frac{k_{22}}{m_2}\right)^2 + \frac{4k_{12}^2}{m_1 m_2}\right]^{1/2}. \qquad (2.9)$$

Equation (2.9) gives two real roots for ω^2 (i.e. ω_1^2 and ω_2^2) as the term in the square bracket is always positive. Comparing equation (2.8) with the equation

$$(\omega^2 - \omega_1^2)(\omega^2 - \omega_2^2) = 0,$$

$$\omega_1^2 + \omega_2^2 = (k_{11} m_2 + k_{22} m_1)/(m_1 m_2)$$

and

$$\omega_1^2 \omega_2^2 = (k_{11} k_{22} - k_{12}^2)/(m_1 m_2). \tag{2.10}$$

Thus provided that $k_{11} k_{22} > k_{12}^2$, ω_1^2 and ω_2^2 will be positive. It can be shown that this inequality must be satisfied for the strain energy of the system to be positive always. Hence equation (2.9) gives the two natural frequencies of the system, ω_1 and ω_2; if $\omega_1 < \omega_2$, ω_1 corresponds to the minus sign in equation (2.9). The amplitude ratios Y_2/Y_1 are obtained from either of equations (2.7) after substituting the appropriate value of ω^2. Thus if β_1 and β_2 are the values of Y_2/Y_1 in the first and second modes respectively,

$$\beta_1 = (m_1 \omega_1^2 - k_{11})/k_{12}$$

and

$$\beta_2 = (m_1 \omega_2^2 - k_{11})/k_{12}. \tag{2.11}$$

From equations (2.11), after substituting from equations (2.10) for the sum and product of the roots, ω_1^2 and ω_2^2,

$$\beta_1 \beta_2 = -m_1/m_2. \tag{2.12}$$

Thus one of the ratios β is positive and the other negative. For the examples in Figs. 2.1 and 2.2, noting that k_{12} is negative, β_1 is positive and β_2 is negative. That is, the masses are vibrating in phase in the lower mode at the natural frequency ω_1 and they are vibrating with a phase difference of 180° in the higher mode at the frequency ω_2. This is illustrated in Fig. 2.2 b and c.

If A_1 and A_2 are the amplitudes of the mass m_1 in the first and second modes respectively, the *normal* modes are:

First or lower mode:

$$y_1 = A_1 \sin (\omega_1 t + \alpha_1); \qquad y_2 = \beta_1 A_1 \sin (\omega_1 t + \alpha_1).$$

Second or higher mode:

$$y_1 = A_2 \sin (\omega_2 t + \alpha_2); \qquad y_2 = \beta_2 A_2 \sin (\omega_2 t + \alpha_2). \quad (2.13)$$

The general solution of equations (2.5), obtained by superposing the normal modes, is

$$y_1 = A_1 \sin (\omega_1 t + \alpha_1) + A_2 \sin (\omega_2 t + \alpha_2),$$
$$y_2 = \beta_1 A_1 \sin (\omega_1 t + \alpha_1) + \beta_2 A_2 \sin (\omega_2 t + \alpha_2). \quad (2.14)$$

In equations (2.14) ω_1, ω_2, β_1 and β_2 are given by equations (2.9) and (2.11); the unknown amplitudes A_1 and A_2 and phase angles α_1 and α_2 are determined from the initial conditions, i.e. the values of y_1, y_2, $\dot{y}_1$ and $\dot{y}_2$ at time $t = 0$. In general, the initial conditions will be such that the resulting motion has components of both normal modes. In practice, damping must be present in the system, causing a slow decay in the vibrations; in general, the components of the two normal modes will decay at different rates.

The *orthogonal* property of the normal modes,

$$m_1 A_1 A_2 + m_2(\beta_1 A_1)(\beta_2 A_2) = 0, \quad (2.15)$$

follows from equation (2.12). The modes are *normalized* if A_1 and A_2, the amplitudes of the mass m_1 in the first and second modes, satisfy

$$m_1 A_1^2 + m_2(\beta_1 A_1)^2 = 1,$$
$$m_1 A_2^2 + m_2(\beta_2 A_2)^2 = 1. \quad (2.16)$$

By a change of coordinates it is possible to derive two uncoupled equations. The new coordinates, q_1 and q_2, which are functions of time, are defined by the equations

$$y_1 = A_1 q_1 + A_2 q_2,$$
$$y_2 = \beta_1 A_1 q_1 + \beta_2 A_2 q_2 \quad (2.17)$$

with A_1 and A_2 satisfying the normalizing conditions (2.16). In terms of q_1 and q_2 the equations of motion (2.5) become

$$m_1 A_1 \ddot{q}_1 + m_1 A_2 \ddot{q}_2 + k_{11} A_1 q_1 + k_{11} A_2 q_2$$
$$+ k_{12} \beta_1 A_1 q_1 + k_{12} \beta_2 A_2 q_2 = 0 \quad (2.18)$$

and

$$m_2\beta_1 A_1 \ddot{q}_1 + m_2\beta_2 A_2 \ddot{q}_2 + k_{21}A_1 q_1 + k_{21}A_2 q_2$$
$$+ k_{22}\beta_1 A_1 q_1 + k_{22}\beta_2 A_2 q_2 = 0.$$

Multiplying the first and second equations by A_1 and $(\beta_1 A_1)$, respectively, and adding the resulting equations,

$$\ddot{q}_1(m_1 A_1^2 + m_2\beta_1^2 A_1^2) + \ddot{q}_2(m_1 A_2 A_1 + m_2\beta_2 A_2 \beta_1 A_1)$$
$$+ q_1[(k_{11}A_1 + k_{12}\beta_1 A_1)A_1 + \beta_1 A_1(k_{21}A_1 + k_{22}\beta_1 A_1)]$$
$$+ q_2[(k_{11}A_2 + k_{12}\beta_2 A_2)A_1 + (k_{21}A_2 + k_{22}\beta_2 A_2)\beta_1 A_1] = 0.$$

(2.19)

Equations (2.7), rewritten in terms of the normalized modes (2.13), are

$$k_{11}A_1 + k_{12}(\beta_1 A_1) = m_1\omega_1^2 A_1$$

and

$$k_{21}A_1 + k_{22}(\beta_1 A_1) = m_2\omega_1^2(\beta_1 A_1) \qquad (2.20)$$

for the first mode, and

$$k_{11}A_2 + k_{12}(\beta_2 A_2) = m_1\omega_2^2 A_2$$

and

$$k_{21}A_2 + k_{22}(\beta_2 A_2) = m_2\omega_2^2(\beta_2 A_2) \qquad (2.21)$$

for the second mode. Applying equations (2.16), (2.15), (2.20) and (2.21) to the coefficients of $\ddot{q}_1$, $\ddot{q}_2$, q_1 and q_2, respectively, equation (2.19) reduces to

$$\ddot{q}_1 + q_1(m_1\omega_1^2 A_1^2 + m_2\omega_1^2 \beta_1^2 A_1^2)$$
$$+ q_2(m_1\omega_2^2 A_1 A_2 + m_2\omega_2^2 \beta_1 \beta_2 A_1 A_2) = 0. \qquad (2.22)$$

Using the normalizing condition (2.16) and the orthogonal property (2.15)

$$\ddot{q}_1 + \omega_1^2 q_1 = 0. \qquad (2.23)$$

If the first and second of equations (2.18) are multiplied by A_2 and $(\beta_2 A_2)$, respectively, and summed, the resulting equation reduces to

$$\ddot{q}_2 + \omega_2^2 q_2 = 0. \qquad (2.24)$$

Equations (2.23) and (2.24) are uncoupled; that is, their solutions

$$q_1 = \sin(\omega_1 t + \alpha_1), \qquad q_2 = \sin(\omega_2 t + \alpha_2) \qquad (2.25)$$

are independent of each other. It will be noted that the general solution for free vibrations (2.14) is obtained again by substituting from equation (2.25) in (2.17). Uncoupled equations will be used in later sections when evaluating the dynamic response.

The initial conditions required if the resulting motion must consist of only the first normal mode are that q_1 and/or $\dot{q}_1$ are non-zero and q_2 and $\dot{q}_2$ are zero. Thus, for example, the motion will be harmonic of frequency ω_1, if the initial conditions are $y_1 = C$, $y_2 = \beta_1 C$, $\dot{y}_1 = \dot{y}_2 = 0$ [from equation (2.17)]; it will be harmonic of frequency ω_2, if the initial conditions are $y_1 = C$, $y_2 = \beta_2 C$, $\dot{y}_1 = \dot{y}_2 = 0$.

2.2. Response to Harmonic Excitation (Two-degree-of-freedom Systems)

In this section it is assumed that a harmonic force $P \sin \omega t$ is applied to the mass m_1 of equations (2.5). (In the equivalent systems of Figs. 2.1, 2.2a and 2.3 the corresponding forces are $P \sin \omega t$ acting: on m_1 in the direction XX' in Fig. 2.1; on m_1 in the direction AB in Fig. 2.2; and through G in the X-direction in Fig. 2.3.)

It is essential to include damping if the response at resonance is required. For example, in Fig. 2.1 viscous dampers of constants c_1, c_2 and c_3 are introduced in parallel with the springs of stiffness k_1, k_2 and k_3, respectively. Thus on masses m_1 and m_2 a damping force $c_2(\dot{x}_1 - \dot{x}_2)$ acts, in addition to the spring force $k_2(x_1 - x_2)$. Introducing damping in all the systems of Figs. 2.1 to 2.3 the equations of motion for a general two-degree-of-freedom system can be written

$$m_1 \ddot{y}_1 + c_{11}\dot{y}_1 + c_{12}\dot{y}_2 + k_{11}y_1 + k_{12}y_2 = P \sin \omega t,$$

$$m_2 \ddot{y}_2 + c_{12}\dot{y}_1 + c_{22}\dot{y}_2 + k_{12}y_1 + k_{22}y_2 = 0. \qquad (2.26)$$

The relations between c_{rs} in equations (2.26) and actual viscous damping parameters in the systems of Figs. 2.1 to 2.3 can be inferred from the equivalent expressions for stiffness given in Table 2.1. In equations (2.26) a harmonic force is applied to mass m_1 only. When the response has been obtained from equations (2.26), the response to a force $P_2 \sin \omega t$, acting on mass m_2, can be found by appropriate changes of subscripts.

Alternatively, hysteretic damping can be included in the analysis. In this case the damping force between the masses m_1 and m_2 of Fig. 2.1 is $h_2 (\dot{x}_1 - \dot{x}_2)/\omega$, where h_2 is the hysteretic damping constant. The general equations with hysteretic damping included, corresponding to equations (2.26) for viscous damping, are

$$m_1 \ddot{y}_1 + h_{11} \dot{y}_1/\omega + h_{12} \dot{y}_2/\omega + k_{11} y_1 + k_{12} y_2 = P \sin \omega t,$$

$$m_2 \ddot{y}_2 + h_{12} \dot{y}_1/\omega + h_{22} \dot{y}_2/\omega + k_{12} y_1 + k_{22} y_2 = 0. \tag{2.27}$$

The complete solution of equations (2.26) consists of a damped vibration, which decays with time, and the steady-state response of frequency ω. We consider only the latter, which for either equations (2.26) or (2.27) is of the form

$$y_1 = Y_1 \sin (\omega t + \alpha_1), \qquad y_2 = Y_2 \sin (\omega t + \alpha_2). \tag{2.28}$$

As in Section 1.4, $\sin \omega t$ on the right-hand side of equations (2.26) and (2.27) is replaced by $\exp (i\omega t)$; the equations are solved, giving complex expressions for y_1 and y_2; and the imaginary part of the solution taken, as $\sin \omega t$ is the imaginary part of $\exp (i\omega t)$. As the complex solutions are of the form $y_r = Y_r \exp (i\omega t)$, $\dot{y}_r = i\omega y_r$, and $\ddot{y}_r = -\omega^2 y_r$, $r = 1, 2$. Thus equations (2.26) can be written

$$(k_{11} - m_1 \omega^2 + i\omega c_{11}) y_1 + (k_{12} + i\omega c_{12}) y_2 = P \exp (i\omega t),$$

$$(k_{12} + i\omega c_{12}) y_1 + (k_{22} - m_2 \omega^2 + i\omega c_{22}) y_2 = 0. \tag{2.29}$$

The solution of equations (2.29) is

$$y_1 = (k_{22} - m_2 \omega^2 + i\omega c_{22}) P \exp (i\omega t)/f(\omega),$$

$$y_2 = -(k_{12} + i\omega c_{12}) P \exp (i\omega t)/f(\omega) \tag{2.30}$$

where

$$f(\omega) = (k_{11} - m_1 \omega^2 + i\omega c_{11})(k_{22} - m_2 \omega^2 + i\omega c_{22}) - (k_{12} + i\omega c_{12})^2.$$

This solution can be written in the form

$$y_r = (C_r + iD_r) P (\cos \omega t + i \sin \omega t)/(C_0 + iD_0), \qquad r = 1, 2$$

where the constants C_r and D_r ($r = 0$, 1 and 2) can be expressed in terms of the system parameters and ω. Rationalizing and taking the imaginary part of the complex expression for y_r,

$$y_r = Y_r \sin(\omega t + \alpha_r), \qquad r = 1, 2$$

where the amplitude

$$Y_r = P\left(\frac{C_r^2 + D_r^2}{C_0^2 + D_0^2}\right)^{1/2}, \qquad r = 1, 2$$

and the phase angle α_r is given by

$$\tan \alpha_r = \frac{D_r C_0 - C_r D_0}{C_r C_0 + D_r D_0}, \qquad r = 1, 2.$$

Substituting for the constants C_r and D_r ($r = 0$, 1 and 2) from equations (2.30)

$$Y_1 = P\left[\frac{(k_{22} - m_2\omega^2)^2 + \omega^2 c_{22}^2}{F(\omega)}\right]^{1/2}$$

and

$$Y_2 = P\left[\frac{k_{12}^2 + \omega^2 c_{12}^2}{F(\omega)}\right]^{1/2} \tag{2.31}$$

where

$$F(\omega) = [(k_{11} - m_1\omega^2)(k_{22} - m_2\omega^2) - k_{12}^2 - \omega^2(c_{11}c_{22} - c_{12}^2)]^2$$
$$+ \omega^2[c_{11}(k_{22} - m_2\omega^2) + c_{22}(k_{11} - m_1\omega^2) - 2c_{12}k_{12}]^2. \tag{2.32}$$

If damping is neglected ($c_{rs} = 0$, r, $s = 1, 2$), the amplitudes Y_1 and Y_2 approach infinity as the excitation frequency ω approaches either of the two frequencies for which $F(\omega) = 0$; these frequencies are given by

$$(k_{11} - m_1\omega^2)(k_{22} - m_2\omega^2) - k_{12}^2 = 0. \tag{2.33}$$

It has been shown, equation (2.8), that equation (2.33) is the frequency equation for the general two-degree-of-freedom system and yields the two natural frequencies ω_1 and ω_2. Thus, in the absence of damping the amplitudes are infinite if the excitation frequency is equal to either of the natural frequencies. With damping included in the analysis (and real systems must contain some damping) $F(\omega)$ is positive for all values of ω. For light damping the response curves (i.e. graphs of the amplitudes Y_r plotted against the excitation frequency ω) have, in general, two sharp peaks, which occur approximately when $\omega = \omega_1$ and $\omega = \omega_2$. This is the phenomenon of *resonance* – when the excitation frequency equals either natural frequency, the amplitude of forced vibration is large for practical values of damping. For heavy damping the response curves have less sharp peaks and these maxima occur at frequencies differing from ω_1 and ω_2; these frequencies can be obtained from the solutions of $\partial Y_r / \partial \omega = 0$, $r = 1, 2$.

For hysteretic damping, the response of the system, governed by equations (2.27), can be found from equations (2.31) and (2.32) by replacing ωc_{rs} by h_{rs}, $r, s = 1, 2$.

Example. The lower mass AB of the frame of Fig. 2.2a is subjected to a horizontal force $P \sin \omega t$. Each of the masses AB and CD is equal to m. The combined stiffnesses in flexure of the stanchions AE and BF and also of AC and BD are each equal to k. In parallel with each restoring force $k(x_r - x_s)$ there is a viscous damping force $c(\dot{x}_r - \dot{x}_s)$.

Determine the amplitudes of vibration of the two masses and show how these depend upon the excitation frequency ω, if the damping parameter γ, equal to $c(mk)^{-1/2}$, is (a) 0·02 and (b) 0·2.

In equations (2.29) $k_{11} = 2k$, $k_{12} = -k$, $k_{22} = k$, $c_{11} = 2c$, $c_{12} = -c$, $c_{22} = c$ and $m_1 = m_2 = m$. If the amplitudes of masses AB and CD are X_1 and X_2, respectively, equations (2.31) and (2.32) can be rewritten as

$$\frac{kX_1}{P} = \left[\frac{(1 - r^2)^2 + \gamma^2 r^2}{F(r)} \right]^{1/2},$$

$$\frac{kX_2}{P} = \left[\frac{1 + \gamma^2 r^2}{F(r)} \right]^{1/2} \tag{a}$$

where

$$F(r) = [(2 - r^2)(1 - r^2) - 1 - \gamma^2 r^2]^2 + \gamma^2 r^2 (2 - 3r^2)^2,$$

$$r = \omega/\omega_0 \quad \text{and} \quad \omega_0^2 = k/m.$$

Equations (a) give the dynamic magnification factors for the two masses as functions of the frequency ratio r and the damping

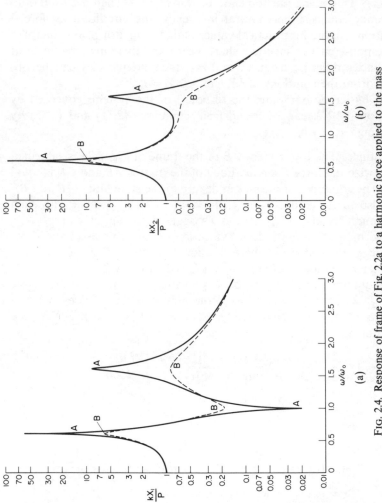

FIG. 2.4. Response of frame of Fig. 2.2a to a harmonic force applied to the mass AB. Curve A—damping parameter $\gamma = 0.02$. Curve B—damping parameter $\gamma = 0.2$. (a) Response of mass AB. (b) Response of mass CD.

parameter γ. In Fig. 2.4 kX_1/P and kX_2/P are plotted against r for $\gamma = 0.02$ and $\gamma = 0.2$. For the lower damping value the two resonances are clearly shown; the dynamic magnification factor for $\omega = \omega_1$ is considerably larger than that for $\omega = \omega_2$; the curve for X_1 has a marked minimum or anti-resonance at $r = 1$. Damping affects the response appreciably only in the vicinity of the two resonances and for X_1 in the vicinity of the anti-resonance. For $\gamma = 0.2$ the peak for X_2 associated with $\omega = \omega_2$ is almost completely suppressed.

2.3. Transient Response
(Two-degree-of-freedom Systems)

It is assumed that a general force $P_1(t)$ acts upon the mass m_1 of equations (2.5). In order to avoid excessive complexity damping is neglected. (Damping will be included in the matrix analysis for multi-degree-of-freedom systems in Section 2.6.) Thus the equations of motion are:

$$m_1\ddot{y}_1 + k_{11}y_1 + k_{12}y_2 = P_1(t),$$

$$m_2\ddot{y}_2 + k_{12}y_1 + k_{22}y_2 = 0. \tag{2.34}$$

The force $P_1(t)$ may be any prescribed function of time, e.g. the suddenly applied force of Fig. 1.3a or one of the transient forces of Fig. 1.12. The response can be determined by an extension of the method of Section 1.7. The general solution for free vibrations of the undamped system is given by equations (2.14). If the mass m_1 is subjected to an impulse $\Im$ at time $t = 0$, its initial velocity V_1 is given by $\Im = m_1 V_1$. The system then undergoes free vibrations subject to the conditions that at $t = 0$: $y_1 = 0 = y_2 = \dot{y}_2$, $\dot{y}_1 = V_1$. Substituting these conditions in equations (2.14) it is found that

$$y_1 = \frac{\Im \sin \omega_1 t}{m_1\omega_1(1 - \beta_1/\beta_2)} + \frac{\Im \sin \omega_2 t}{m_1\omega_2(1 - \beta_2/\beta_1)}$$

and

$$y_2 = \frac{\beta_1 \Im \sin \omega_1 t}{m_1\omega_1(1 - \beta_1/\beta_2)} + \frac{\beta_2 \Im \sin \omega_2 t}{m_1\omega_2(1 - \beta_2/\beta_1)}$$

with β_1 and β_2 given by equations (2.11). Considering the force $P_1(t)$ to be divided into a large number of impulses, as in Section 1.7, the dynamic response of the masses m_1 and m_2 is

$$y_1 = \frac{1}{m_1(\beta_2 - \beta_1)} \int_0^t P_1(\tau) \left[\frac{\beta_2}{\omega_1} \sin \omega_1(t - \tau) - \frac{\beta_1}{\omega_2} \sin \omega_2(t - \tau) \right] d\tau$$

and

$$y_2 = \frac{\beta_1 \beta_2}{m_1(\beta_2 - \beta_1)} \int_0^t P_1(\tau) \left[\frac{1}{\omega_1} \sin \omega_1(t - \tau) - \frac{1}{\omega_2} \sin \omega_2(t - \tau) \right] dt.$$

$$(2.35)$$

Equations (2.35) may be used to obtain the response of any two-degree-of-freedom system to any disturbing force. It is assumed that until the disturbing force is applied at $t = 0$ the system is at rest, as in equation (1.37). The response of the system to disturbing forces $P_1(t)$ and $P_2(t)$ applied to masses m_1 and m_2, respectively, can be formed from equations (2.35) by manipulation of subscripts and superposition.

Equations (2.35) may be used also to determine the response to prescribed support motion. If the upper support A in Fig. 2.1 is subject to motion in the direction XX', $x_0(t)$, the resulting equations are given by equations (2.5) with the addition of the term $k_1 x_0(t)$ to the right-hand side of the first equation. Similarly if the support EF in Fig. 2.2a undergoes a horizontal displacement $x_0(t)$, the term $k_1 x_0(t)$ must be added to the right-hand side of the first of equations (2.2). Comparing these modified equations with equations (2.34), it is apparent that equations (2.35) with $P_1(\tau)$ replaced by $k_1 x_0(\tau)$ give the dynamic response of the system to support motion. If the support CD in Fig. 2.3 is given a vertical displacement $x_0(t)$, terms $(k_1 + k_2)x_0(t)$ and $(k_2 b - k_1 a)x_0(t)$ must be added to the right-hand sides of the first and second of equations (2.4). This is equivalent to adding terms to the right-hand sides of both of equations (2.34) and thus the response can be found using equations (2.35), as outlined above.

The dynamic response of the system to a force $P_1(t)$ applied to mass m_1 [equations (2.34)] can be derived alternatively by considering the appropriate uncoupled equations. Comparing equations (2.5) and (2.34) the only difference is the addition of the term $P_1(t)$ to the right-hand side of the first of equations (2.34). Introducing the change of coordinates, equation (2.17), into (2.34), and following the analysis of equations (2.18) to (2.24), the uncoupled equations, corresponding to (2.34), are

$$\ddot{q}_1 + \omega_1^2 q_1 = A_1 P_1(t),$$

$$\ddot{q}_2 + \omega_2^2 q_2 = A_2 P_1(t) \tag{2.36}$$

with A_1 and A_2 satisfying the normalizing conditions (2.16).

The solutions of equations (2.36) are given in terms of Duhamel's integral as

$$q_1 = \frac{A_1}{\omega_1} \int_0^t P_1(\tau) \sin \omega_1(t - \tau) \, d\tau,$$

$$q_2 = \frac{A_2}{\omega_2} \int_0^t P_1(\tau) \sin \omega_2(t - \tau) \, d\tau. \tag{2.37}$$

Returning to the original coordinates, y_1 and y_2, the dynamic response of masses m_1 and m_2 is [from equations (2.17) and (2.37)]

$$y_1 = \frac{A_1^2}{\omega_1} \int_0^t P_1(\tau) \sin \omega_1(t - \tau) \, d\tau + \frac{A_2^2}{\omega_2} \int_0^t P_1(\tau) \sin \omega_2(t - \tau) \, d\tau,$$

$$y_2 = \frac{\beta_1 A_1^2}{\omega_1} \int_0^t P_1(\tau) \sin \omega_1(t - \tau) \, d\tau + \frac{\beta_2 A_2^2}{\omega_2} \int_0^t P_1(\tau) \sin \omega_2(t - \tau) \, d\tau.$$

$$\tag{2.38}$$

Now

$$A_1^2 = \frac{1}{m_1[1 + (m_2/m_1)\beta_1^2]}$$

and

$$A_2^2 = \frac{1}{m_1[1 + (m_2/m_1)\beta_2^2]}$$

from equation (2.16). Substituting $m_2/m_1 = -1/\beta_1\beta_2$ from equation (2.12),

$$A_1^2 = \frac{1}{m_1(1 - \beta_1/\beta_2)}$$

and

$$A_2^2 = \frac{1}{m_1(1 - \beta_2/\beta_1)}.$$

Thus the expressions for the response (2.35) and (2.38) are equivalent.

In this section two methods of determining the response of systems with two degrees of freedom have been given; the second method, in which uncoupled equations are obtained by a change to principal coordinates, is an introduction to the normal mode method of determining response. The latter will be applied to multi-degree-of-freedom systems in Section 2.6, where matrix notation is used, in order to avoid many complicated equations and for ease of manipulation, and damping is included.

2.4. General Equations

For a system with n degrees of freedom there are, by definition, n independent or generalized coordinates and n equations of motion are obtained. For illustrative purposes typical equations for the n-storey frame of Fig. 2.5 will be derived, but by incorporating these into a more general matrix form, and giving analysis in later sections for this general equation, most linear multi-degree-of-freedom systems can be treated.

Figure 2.5 shows part of an n-storey frame; as in the two-degree-of-freedom system of Fig. 2.2a the assumptions pertaining to a shear building are made; i.e. each horizontal member is assumed to be rigid and capable of motion only in the horizontal direction in the plane of

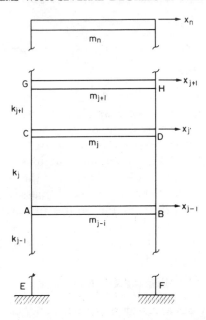

FIG. 2.5. Frame with n degrees of freedom: only flexural deformation of the vertical members in the plane of the frame permitted.

the diagram; each vertical member is assumed to be massless and elastic. The combined stiffness of the stanchions AC and BD in flexure is k_j. The horizontal displacements x_j of the masses m_j, $j = 1, 2, \ldots, n$, are the generalized coordinates. The system is subjected to applied forces in the horizontal direction; typically the force $P_j(t)$ is applied to member CD. Viscous damping, resisting relative motion between adjacent horizontal members, is assumed to exist (not shown on Fig. 2.5). Thus on the member CD there are damping forces $-c_j(\dot{x}_j - \dot{x}_{j-1})$ and $-c_{j+1}(\dot{x}_j - \dot{x}_{j+1})$, where c_j is the viscous damping coefficient for the stanchions AC and BD and c_{j+1} is the coefficient for stanchions CG and DH. The equation of motion for member CD is

$$m_j\ddot{x}_j = P_j(t) - k_j(x_j - x_{j-1}) + k_{j+1}(x_{j+1} - x_j) - c_j(\dot{x}_j - \dot{x}_{j-1})$$
$$+ c_{j+1}(\dot{x}_{j+1} - \dot{x}_j),$$

i.e.

$$m_j \ddot{x}_j + (c_j + c_{j+1})\dot{x}_j - c_j \dot{x}_{j-1} - c_{j+1}\dot{x}_{j+1} + (k_j + k_{j+1})x_j - k_j x_{j-1}$$
$$- k_{j+1}x_{j+1} = P_j(t). \tag{2.39}$$

Similar equations can be written down for the other masses.

The frame of Fig. 2.5 is one example of a system with n degrees of freedom. For a multi-storey frame of the type shown in Fig. 2.3, two equations of motion are obtained for each mass by considering vertical translation and rocking. Each equation contains six independent coordinates, namely the vertical displacements of the centres of gravity of the relevant mass and of the two adjacent masses and the rotations about horizontal axes through the centres of gravity of these three masses. In order to accommodate these two types of structure and other multi-degree-of-freedom systems the following general matrix equation will be considered:

$$\mathbf{M}\ddot{x} + \mathbf{C}\dot{x} + \mathbf{K}x = \mathbf{p}(t). \tag{2.40}$$

In equation (2.40) $\mathbf{x}$ is a vector of the independent coordinates, i.e.

$$\mathbf{x} = \begin{bmatrix} x_1 \\ x_2 \\ x_3 \\ \cdot \\ \cdot \\ x_n \end{bmatrix}. \tag{2.41}$$

$\dot{x}$ and $\ddot{x}$ are the corresponding velocity and acceleration vectors. (The vector $\mathbf{x}$ may contain linear and angular displacements; this would occur for a multi-storey frame, similar to Fig. 2.3.) The vector $\mathbf{p}(t)$ lists the applied forces, i.e.

$$\mathbf{p}(t) = \begin{bmatrix} P_1(t) \\ P_2(t) \\ \cdot \\ \cdot \\ P_n(t) \end{bmatrix}. \tag{2.42}$$

A component of the force vector, $P_j(t)$, is the resultant applied force on a mass, acting in the direction of the displacement x_j of that mass. (If x_j represents an angular displacement, $P_j(t)$ is the resultant applied moment on the relevant mass, corresponding to the rotation x_j.) The stiffness matrix $\mathbf{K}$ is square and symmetric, being defined as

$$\mathbf{K} = \begin{bmatrix} k_{11} & k_{12} & k_{13} & \cdots & k_{1n} \\ k_{21} & k_{22} & k_{23} & \cdots & k_{2n} \\ \cdot & \cdot & \cdot & \cdot & \cdot \\ k_{n1} & k_{n2} & k_{n3} & \cdots & k_{nn} \end{bmatrix} \tag{2.43}$$

with $k_{js} = k_{sj}$ by the reciprocal theorem. (The elements k_{js} of matrix $\mathbf{K}$ can be defined in terms of stiffnesses of a prescribed physical system by examination of the equations of motion; e.g. for the frame of Fig. 2.5 the jth row of $\mathbf{K}$ is $[0, 0, \ldots, 0, -k_j, (k_j + k_{j+1}), -k_{j+1}, 0, \ldots, 0]$ from equation (2.39).) The damping matrix $\mathbf{C}$ is also square and symmetric and can be defined similarly in terms of the damping coefficients c_{js}. The mass matrix $\mathbf{M}$ is square and symmetric, being defined as

$$\mathbf{M} \begin{bmatrix} m_{11} & m_{12} & m_{13} & \cdots & m_{1n} \\ m_{21} & m_{22} & m_{23} & \cdots & m_{2n} \\ \cdot & \cdot & \cdot & \cdot & \cdot \\ m_{n1} & m_{n2} & m_{n3} & \cdots & m_{nn} \end{bmatrix} \tag{2.44}$$

with $m_{js} = m_{sj}$. For the systems illustrated in Figs. 2.1, 2.2, 2.3 and 2.5, with the coordinates defined in those figures, the mass matrix is diagonal (i.e. $m_{js} = 0$ for $j \neq s$ and there is a single inertia term, typically $m_{jj}\ddot{x}_j$, in each equation). However, for a system with n degrees of freedom there must be n independent coordinates, but the latter can be chosen in many ways. In the above figures the absolute displacements of the centres of gravity of the masses have been adopted as coordinates. Alternative sets of coordinates can be defined in terms of relative displacements; for example, (a) relative displacements $y_1, y_2, y_3, \ldots, y_n$ for the frame of Fig. 2.5, where $y_1 = x_1$, $y_2 = x_2 - x_1$, $y_3 = x_3 - x_2$, $\ldots, y_j = x_j - x_{j-1}, \ldots$ (i.e. y_j is the deformation of the jth stanchion) and (b) the displacements of points A and B in Fig. 2.3. If, using these alternative sets of coordinates, the equations of motion are derived from Newton's second law of motion, the resulting mass and stiffness matrices in equation (2.40) are not symmetric. If the equations of motion are derived from energy expressions, using the Lagrange equation (A4.17), the resulting matrices are symmetric. Corresponding sets of equations can be shown to be equivalent by simple algebraic manipulation. In general, in this book absolute displacements will be used as coordinates, but if other sets of coordinates are used, the equations of motion will be derived from energy considerations so that symmetric matrices are obtained. The analysis of Sections 2.5, 2.6, 2.7 and 2.9 requires the matrices $\mathbf{M}$, $\mathbf{C}$ and $\mathbf{K}$ to be symmetric, but this imposes no restrictions on the types of linear multi-degree-of-freedom systems that can be represented by equation (2.40). The properties of matrices which will be used in this and subsequent sections are defined briefly in Appendix 2. Matrices are denoted by capital letters in bold type; vectors or column matrices by lower-case letters in bold type.

The significance of equation (2.40) must be stressed. Practical engineering structures are usually complex and their response to specified inputs can be determined only approximately. Three general approximate methods of analysis are Rayleigh–Ritz, finite elements and finite differences. The first two methods are described in Chapters 3 and 5 for beam, plate and shell problems. In these two methods the actual structures are replaced by approximate mathematical models, which are represented by equation (2.40) with symmetric mass, stiffness and damping matrices. Thus solutions of equation (2.40) are applicable to structures to be considered in later sections of this book, as well as to general linear multi-degree-of-freedom systems. These solutions are considered in subsequent sections. In Section 2.5 free undamped vibrations, obtained from equation (2.40) by putting $\mathbf{C}$ and $\mathbf{p}$ equal to zero, are studied; the concepts of orthogonality and normalization and the method of obtaining uncoupled equations through a transformation to principal coordinates are described. The latter are used in Section 2.6, where the response of the system, with damping included, to a general excitation vector $\mathbf{p}(t)$ is derived. Extensions to include excitation by imposed motion at the supports and by random forces are given in Sections 2.6 and 2.9, respectively. The uncoupling procedure places restrictions upon the form of the damping matrix $\mathbf{C}$; alternative methods of obtaining the response from equation (2.40) without imposing these restrictions are given for harmonic excitation and for a general excitation vector $\mathbf{p}(t)$ in Sections 2.7 and 2.8, respectively.

2.5. Free Undamped Vibration

The equations of motion for free undamped vibration of a multi-degree-of-freedom system can be established from first principles. The general, matrix form is obtained from equation (2.40) by making the vector of applied forces $\mathbf{p}$ and the damping matrix $\mathbf{C}$ zero, i.e.

$$\mathbf{M\ddot{x}} + \mathbf{Kx} = 0. \tag{2.45}$$

From earlier discussion (Section 2.4) the only restriction on the mass matrix $\mathbf{M}$ and the stiffness matrix $\mathbf{K}$ is that both are square and symmetric. For free vibrations we seek a solution of equation (2.45) in the form

$$\mathbf{x} = \mathbf{e} \sin{(\omega t + \alpha)}$$

or

$$x_j = e_j \sin{(\omega t + \alpha)}, \qquad j = 1, 2, 3, \ldots, n \tag{2.46}$$

where ω is a natural frequency, as yet unknown, and e_j is the amplitude of vibration at the point of the system where the displacement is measured. Substituting equation (2.46) into equation (2.45) and noting that $\sin{(\omega t + \alpha)}$ is common to all terms and cannot be zero for all values of time for a nontrivial solution,

$$(\mathbf{K} - \mathbf{M}\omega^2)\mathbf{e} = 0. \tag{2.47}$$

For equation (2.47) to have solutions, other than $\mathbf{e} = 0$, it is necessary that the determinant of the coefficients of $\mathbf{e}$ vanishes (see Appendix 2), i.e.

$$\det |\mathbf{K} - \mathbf{M}\omega^2|$$

$$\equiv \det \begin{vmatrix} k_{11} - m_{11}\omega^2 & k_{12} - m_{12}\omega^2 & k_{13} - m_{13}\omega^2 & \cdots \\ k_{21} - m_{21}\omega^2 & k_{22} - m_{22}\omega^2 & k_{23} - m_{23}\omega^2 & \cdots \\ k_{31} - m_{31}\omega^2 & k_{32} - m_{32}\omega^2 & k_{33} - m_{33}\omega^2 & \cdots \\ \cdot \quad \cdot \quad \cdot \quad \cdot \quad \cdot \quad \cdot \quad \cdot \quad \cdot \quad \cdot \quad \cdot \quad \cdot \quad \cdot \end{vmatrix} = 0. \tag{2.48}$$

In general, equation (2.48) gives n positive real roots for ω^2, say ω_1^2, ω_2^2, ω_3^2, $\ldots$, ω_n^2 with $\omega_1^2 < \omega_2^2 < \omega_3^2$, $\ldots$, $< \omega_n^2$. Then ω_1, ω_2, ω_3, $\ldots$, ω_n are the natural frequencies of the system. Corresponding to any frequency ω_r there is a set of values of the amplitudes e, obtained from equation (2.47). If $_r e_j$ is the amplitude at the jth coordinate in the rth mode, substitution of the known value of ω_r^2 into equation (2.47) yields only relative, rather than absolute, values of $_r e_j$; e.g. we can find $_r e_2/_r e_1, _r e_3/_r e_1, \ldots, _r e_n/_r e_1$. This indeterminancy has been met before when considering two-degree-of-freedom systems in Section 2.1.

For a large system, i.e. n large, the determination of some or all of the eigenvalues, i.e. values of ω_r^2, from equation (2.48) and their corresponding eigenvectors, $_r e \equiv [_r e_1, _r e_2, _r e_3, \ldots, _r e_n]^T$, from equation (2.47) requires appreciable computational effort. However, standard methods and computer programs for the determination of eigenvalues and eigenvectors exist. Detailed discussion is outside the scope of this book, but the appropriate method depends upon whether a few or all of the eigenvalues are required, the size n of the matrices and the number and distribution of zeros in the determinant. Full matrices have been assumed here, but in practice many zeros will occur in large matrices. Wilkinson[87] gives a general discussion of the problem; Bathe and Wilson[6] survey the important computational methods and comment on their suitability and efficiency for different types of problem. If only the lower eigenvalues are required, the eigenvalue economizer (or mass condensation) method of reducing the size of the matrices before determining any eigenvalues is useful (see Appendix 5).

Assuming that the eigenvalues and eigenvectors have been determined, the solution for any two different modes, denoted by the subscripts r and s, can be written

$$\mathbf{K}_r\mathbf{e} = \mathbf{M}\omega_r^2 \,_r\mathbf{e} \tag{2.49}$$

and

$$\mathbf{K}_s\mathbf{e} = \mathbf{M}\omega_s^2 \,_s\mathbf{e}. \tag{2.50}$$

The transpose of equation (2.49) is

$$(\mathbf{K}_r\mathbf{e})^T = \omega_r^2(\mathbf{M}_r\mathbf{e})^T$$

or

$$_r\mathbf{e}^T\mathbf{K} = \omega_r^2 \,_r\mathbf{e}^T\mathbf{M}, \tag{2.51}$$

noting that ω_r^2 is a scalar and the symmetric matrices $\mathbf{M} = \mathbf{M}^T$ and

$\mathbf{K} = \mathbf{K}^T$. If equation (2.51) is post-multiplied by $_s\mathbf{e}$ and equation (2.50) is pre-multiplied by $_r\mathbf{e}^T$,

$$_r\mathbf{e}^T\mathbf{K}\,_s\mathbf{c} = \omega_r^2\,_r\mathbf{e}^T\mathbf{M}\,_s\mathbf{e}$$

and

$$_r\mathbf{e}^T\mathbf{K}\,_s\mathbf{e} = \omega_s^2\,_r\mathbf{e}^T\mathbf{M}\,_s\mathbf{e}.$$

Subtracting,

$$(\omega_s^2 - \omega_r^2)_r\mathbf{e}^T\mathbf{M}\,_s\mathbf{e} = 0.$$

For two different modes $\omega_s \neq \omega_r$ and hence the orthogonality condition is established

$$_r\mathbf{e}^T\mathbf{M}\,_s\mathbf{e} = 0. \tag{2.52}$$

Expanding equation (2.52),

$$[_r e_1, \,_r e_2, \,\ldots, \,_r e_n]\begin{bmatrix} m_{11} & m_{12} & \cdots & m_{1n} \\ m_{21} & m_{22} & \cdots & m_{2n} \\ \cdot & \cdot & \cdots & \cdot \\ m_{n1} & m_{n2} & \cdots & m_{nn} \end{bmatrix}\begin{bmatrix} _se_1 \\ _se_2 \\ \cdot \\ _se_n \end{bmatrix} - 0.$$

For a diagonal mass matrix, i.e. $m_{ij} = 0$ for $i \neq j$, this reduces to

$$_r e_1 m_{11}\,_s e_1 + \,_r e_2 m_{22}\,_s e_2 + \cdots + \,_r e_n m_{nn}\,_s e_n = 0.$$

Introducing the vector

$$_r\mathbf{z} = (1/a_r)_r\mathbf{e} \tag{2.53}$$

where a_r is a scalar, such that

$$_r\mathbf{z}^T\mathbf{M}\,_r\mathbf{z} = 1, \tag{2.54}$$

the vectors $_r\mathbf{z}$ are normalized. Next we define a matrix $\mathbf{Z}$, in which the rth column is the vector $_r\mathbf{z}$, i.e.

$$\mathbf{Z} = \begin{bmatrix} _1z_1 & _2z_1 & \cdots & _nz_1 \\ _1z_2 & _2z_2 & \cdots & _nz_2 \\ \cdot & \cdot & \cdots & \cdots \\ \cdot & \cdot & \cdots & \cdots \\ _1z_n & _2z_n & \cdots & _nz_n \end{bmatrix}. \tag{2.55}$$

Equations (2.47) are rewritten in terms of the normalized vectors, expressing the first equation in terms of the first mode, the second equation in terms of the second mode, etc. Thus the rth equation becomes

$$k_{r1}\,{}_rz_1 + k_{r2}\,{}_rz_2 + \cdots + k_{rn}\,{}_rz_n = m_{r1}\,{}_rz_1\,\omega_r^2 + m_{r2}\,{}_rz_2\,\omega_r^2 + \cdots$$
$$+ m_{rn}\,{}_rz_n\,\omega_r^2. \tag{2.56}$$

The set of equations, corresponding to equation (2.56) with $r = 1, 2, \ldots, n$, can be written in matrix form as

$$\mathbf{KZ} = \mathbf{MZ\Omega} \tag{2.57}$$

where $\mathbf{\Omega}$ is the diagonal matrix:

$$\mathbf{\Omega} = \begin{bmatrix} \omega_1^2 & 0 & \cdots & 0 \\ 0 & \omega_2^2 & \cdots & 0 \\ \cdot & \cdot & \cdot & \cdot \\ 0 & 0 & \cdots & \omega_n^2 \end{bmatrix}. \tag{2.58}$$

From equations (2.52), (2.54) and (2.55)

$$\mathbf{Z}^T\mathbf{MZ} = \mathbf{I} \tag{2.59}$$

where $\mathbf{I}$ is the identity matrix [equation (A2.6)]. Pre-multiplying equation (2.57) by $\mathbf{Z}^T$ and using equation (2.59),

$$\mathbf{Z}^T\mathbf{KZ} = \mathbf{\Omega}. \tag{2.60}$$

Returning to the original equations for free undamped vibration in the matrix form

$$\mathbf{M\ddot{x}} + \mathbf{Kx} = 0 \tag{2.61}$$

and introducing a change of coordinates from $\mathbf{x}$ to the principal coordinates $\mathbf{q}$ by the transformation

$$\mathbf{x} = \mathbf{Zq}, \tag{2.62}$$

$$\mathbf{MZ\ddot{q}} + \mathbf{KZq} = 0. \tag{2.63}$$

Pre-multiplying equation (2.63) by $\mathbf{Z}^T$ and using equations (2.59) and (2.60),

$$\ddot{\mathbf{q}} + \mathbf{\Omega q} = 0. \tag{2.64}$$

Equation (2.64) gives n uncoupled equations of the form:

$$\ddot{q}_r + \omega_r^2 q_r = 0, \qquad r = 1, 2, 3, \ldots, n \tag{2.65}$$

with solutions

$$q_r = A_r \sin \omega_r t + B_r \cos \omega_r t \qquad (2.66)$$

where the constants A_r and B_r depend upon the initial conditions. From equation (2.62) the complete solution for one of the original coordinates x_j is

$$x_j = \sum_{r=1}^{n} {}_r z_j (A_r \sin \omega_r t + B_r \cos \omega_r t), \qquad j = 1, 2, 3, \ldots, n. \quad (2.67)$$

Equation (2.67) represents the free undamped vibrations of the multi-degree-of-freedom system for specified initial conditions. In general, the latter will be the values of the displacements and velocities, in terms of the original coordinates x, at time $t = 0$. Let these initial conditions be represented by the vectors $\mathbf{x}_0$ and $\dot{\mathbf{x}}_0$. From equation (2.66), we form vectors, $\mathbf{q}_0$ and $\dot{\mathbf{q}}_0$, of the initial values of q_r and $\dot{q}_r$, i.e.

$$\mathbf{q}_0 = [B_1, B_2, B_3, \ldots, B_n]^T$$

and

$$\dot{\mathbf{q}}_0 = [A_1 \omega_1, A_2 \omega_2, A_3 \omega_3, \ldots, A_n \omega_n]^T. \qquad (2.68)$$

Pre-multiplying equation (2.62) by $\mathbf{Z}^T \mathbf{M}$

$$\mathbf{Z}^T \mathbf{M} \mathbf{x} = \mathbf{Z}^T \mathbf{M} \mathbf{Z} \mathbf{q}$$

and using equation (2.59)

$$\mathbf{q} = \mathbf{Z}^T \mathbf{M} \mathbf{x}. \qquad (2.69)$$

Thus the constants A_r and B_r can be found in terms of the prescribed initial displacement and velocity vectors from the relations

$$\mathbf{q}_0 = \mathbf{Z}^T \mathbf{M} \mathbf{x}_0$$

and

$$\dot{\mathbf{q}}_0 = \mathbf{Z}^T \mathbf{M} \dot{\mathbf{x}}_0 . \qquad (2.70)$$

2.6. Response by the Normal Mode Method

In this method the response of the general system to prescribed time-dependent forces is obtained as a sum of contributions from individual modes, using the concepts of Section 2.5 and in particular the coordinate transformation (2.62), which has been shown to uncouple the equations of free vibration. If the transformation (2.62), i.e.

$$\mathbf{x} = \mathbf{Zq},$$

has to uncouple the general equation (2.40)

$$\mathbf{M\ddot{x}} + \mathbf{C\dot{x}} + \mathbf{Kx} = \mathbf{p}(t) \qquad (2.71)$$

after pre-multiplication by $\mathbf{Z}^T$, it is apparent that $\mathbf{Z}^T\mathbf{CZ}$ must be a diagonal matrix; i.e. $_s\mathbf{z}^T\mathbf{C}_r\mathbf{z} = 0$ for $r \neq s$. (From the previous section $_s\mathbf{z}^T\mathbf{M}_r\mathbf{z} = 0$ and $_s\mathbf{z}^T\mathbf{K}_r\mathbf{z} = 0$.) In general, this condition will not be satisfied and it is necessary to place a restriction on the form of the damping matrix. If it is assumed that

$$\mathbf{C} = \lambda_m \mathbf{M} + \lambda_k \mathbf{K} \qquad (2.72)$$

where λ_m and λ_k are constants, $_s\mathbf{z}^T\mathbf{C}_r\mathbf{z} = 0$ for $r \neq s$. More complex forms of equation (2.72) exist. If equation (2.49) is pre-multiplied by (a) $_s\mathbf{z}^T\mathbf{KM}^{-1}$ and (b) $_s\mathbf{z}^T\mathbf{MK}^{-1}$, additional orthogonal relations (a) $_s\mathbf{z}^T\mathbf{KM}^{-1}\mathbf{K}_r\mathbf{z} = 0$ and (b) $_s\mathbf{z}^T\mathbf{MK}^{-1}\mathbf{M}_r\mathbf{z} = 0$ are obtained. Thus additional terms $\lambda_1 \mathbf{KM}^{-1}\mathbf{K}$ and $\lambda_2 \mathbf{MK}^{-1}\mathbf{M}$ can be added to relation (2.72). This process can be extended. In equation (2.72) the damping matrix is assumed to be a linear combination of the mass and stiffness matrices. For practical structures, where light internal damping in the elastic members must be represented approximately, the assumption that $\mathbf{C}$ is proportional to $\mathbf{K}$ is common and reasonable. Thus equation (2.72) is not so restrictive as it first appears. Making assumption (2.72), substituting the transformation (2.62) in equation (2.71), pre-multiplying the resulting equation by $\mathbf{Z}^T$ and using equations (2.59) and (2.60), we obtain

$$\ddot{\mathbf{q}} + \lambda_m \dot{\mathbf{q}} + \lambda_k \mathbf{\Omega}\dot{\mathbf{q}} + \mathbf{\Omega q} = \mathbf{Z}^T\mathbf{p}(t) \qquad (2.73)$$

and this gives n uncoupled equations of the form

$$\ddot{q}_r + (\lambda_m + \lambda_k\omega_r^2)\dot{q}_r + \omega_r^2 q_r = \sum_{j=1}^{n} {}_r z_j P_j(t), \qquad r = 1, 2, 3, \ldots, n. \qquad (2.74)$$

From the Duhamel integral, equation (1.36), the solution of equation

(2.74) is

$$q_r = \frac{1}{\omega_r^1} \int_0^t f_r(\tau) \exp\left[-\gamma_r\omega_r(t-\tau)\right] \sin \omega_r^1(t-\tau)\, d\tau \qquad (2.75)$$

where

$$\omega_r^1 = \omega_r(1-\gamma_r^2)^{1/2}, \qquad 2\gamma_r\omega_r = \lambda_m + \lambda_k\omega_r^2 \quad \text{and} \quad f_r(t) = \sum_{j=1}^{n} {}_r z_j P_j(t)$$

provided that the initial conditions are zero. γ_r is the modal damping parameter; it is non-dimensional and critical damping for mode r occurs when $\gamma_r = 1$. Solution (2.75), but not the differential equation (2.74), assumes that $\gamma_r < 1$. Equation (2.75) and subsequent equations can be modified if $\gamma_r > 1$, but it is unlikely that a mode for which $\gamma_r > 1$ will make a significant contribution to the response.

For multi-degree-of-freedom systems of the type shown in Fig. 2.1, study of the actual damping coefficients will show whether the uncoupling condition (2.72) is satisfied. However, as shown in later chapters, the general equation (2.71) is obtained when either the finite element or Rayleigh–Ritz method is used for an approximate dynamic analysis of an elastic body. In these cases internal damping in the members of the structure may be the main contributor to the damping matrix and may not be known precisely. If the modal damping parameters γ_r are prescribed for two modes, values of λ_k and λ_m can be found to satisfy the uncoupling condition (2.72), but then the modal damping parameters for the other modes are fixed in terms of the natural frequencies and the two prescribed values of γ_r. If the parameters γ_r are prescribed for more than two modes, additional terms must be included in the uncoupling condition (2.72). For example, if γ_r is prescribed for $r = 1, 2, 3$ and 4, the additional terms containing λ_1 and λ_2, mentioned above, should be included. Equation (2.75) is still applicable, provided that γ_r is redefined as

$$2\gamma_r\omega_r = \lambda_m + \lambda_k\omega_r^2 + \lambda_1\omega_r^4 + \lambda_2\omega_r^{-2}.$$

The four constants, $\lambda_m, \lambda_k, \lambda_1$ and λ_2, can be found by solving the four simultaneous equations corresponding to $r = 1, 2, 3,$ and 4.

The conventional assumption for elastic bodies, namely C is proportional to K and thus $\lambda_m = 0$ in equation (2.72), means that λ_k can be chosen to satisfy one prescribed modal damping parameter. If this is γ_1, then $\lambda_k = 2\gamma_1/\omega_1$ and for the higher modes the uncoupling condition demands that

$$\gamma_r = \gamma_1 \omega_r/\omega_1, \quad r > 1,$$

i.e. the modal damping parameter increases as the mode number increases.

Approximations that allow the normal mode method to be used when the matrix $Z^T C Z$ is not diagonal are important in practice. Thomson et al.[72] have considered three approximations. In the simplest method, which was shown to be satisfactory by numerical studies on systems with three degrees of freedom, the non-diagonal matrix $Z^T C Z$ is replaced by a diagonal matrix with the same diagonal terms as the original matrix (i.e. the off-diagonal terms of $Z^T C Z$ are replaced by zeros); then the standard normal mode procedure is followed.

Returning to equation (2.75), evaluation of the principal coordinates from that equation and use of the transformation (2.62) yield the response x_s at any coordinate.

In practice numerical integration of equation (2.75) will usually be necessary. In general, the response from the higher modes (larger values of r) is insignificant and it is necessary to compute q_r only for a limited range of values of r, $r = 1, 2, 3, \ldots, n_1$ say with $n_1 < n$. Unfortunately no general rules can be given for deciding upon n_1; engineering judgement and trial computer runs are required to establish n_1 for a specific problem.

If the initial conditions are not zero, the solution for q_r, (2.75), will contain additional terms, $\exp(-\gamma_r \omega_r t)(A_r \sin \omega_r^1 t + B_r \cos \omega_r^1 t)$. The constants A_r and B_r can be found from the prescribed values of displacement and velocity at $t = 0$, i.e. from the vectors $\mathbf{x}_0$ and $\dot{\mathbf{x}}_0$, using equations (2.68) and (2.70), except that $\dot{\mathbf{q}}_0$ is redefined as

$$\dot{\mathbf{q}}_0 = [A_1 \omega_1^1, A_2 \omega_2^1, A_3 \omega_3^1, \ldots, A_n \omega_n^1]^T.$$

As the response x_j depends upon a sum of contributions from the normal modes, the Williams or modal acceleration method of improving the convergence is useful. This method was developed originally for the response of elastic structures[88] and will be illustrated in Section 3.6 when considering the transient response of beams in flexure. Here it will be applied to multi-degree-of-freedom systems. The method uses integration by parts of equation (2.75) and depends upon (a) the availability of a closed form sum for the resulting expressions outside the integrals and (b) the sum of the resulting integrals being more rapidly convergent than the original sum. For simplicity we consider only a single force $P_j(t)$ applied at coordinate j; for multiple force excitation summation with respect to j of the final expression (2.82) is possible. Subscripts s and r refer to the coordinate at which the response is determined and the mode number respectively. Noting that

$$\frac{1}{\omega_r^1} \int \exp\left[-\gamma_r \omega_r(t-\tau)\right] \sin \omega_r^1(t-\tau) \, d\tau$$

$$= \frac{1}{\omega_r^2} \exp\left[-\gamma_r \omega_r(t-\tau)\right]\left[\cos \omega_r^1(t-\tau) + \frac{\gamma_r \omega_r}{\omega_r^1} \sin \omega_r^1(t-\tau)\right]$$

$$(2.76)$$

integration by parts of equation (2.75) gives

$$q_r = \frac{{}_r z_j P_j(t)}{\omega_r^2} - \frac{{}_r z_j}{\omega_r^2} \int_0^t \dot{P}_j(\tau) \exp\left[-\gamma_r \omega_r(t-\tau)\right]$$

$$\times \left[\cos \omega_r^1(t-\tau) + \frac{\gamma_r \omega_r}{\omega_r^1} \sin \omega_r^1(t-\tau)\right] d\tau \qquad (2.77)$$

provided that $P_j(0) = 0$. $[\dot{P}_j(\tau) \equiv dP_j/d\tau.]$ From equation (2.62) the response at coordinate s is given by

$$x_s = \sum_{r=1}^{n} {}_r z_s q_r . \qquad (2.78)$$

Thus the contribution to x_s from terms outside the integral in equation (2.77) is

$$\sum_{r=1}^{n} \frac{{}_r z_s \, {}_r z_j P_j(t)}{\omega_r^2} . \qquad (2.79)$$

This summation equals the static deflection at coordinate s for a static force of magnitude $P_j(t)$ acting as coordinate j. This can be seen by considering the response in terms of normal modes to a force $P_0 \cos \omega t$ applied to the structure at j. Solving equation (2.74) for

$P_j(t) = P_0 \cos \omega t$ and neglecting damping,

$$q_r = \frac{{}_rz_j P_0 \cos \omega t}{\omega_r^2 - \omega^2}. \tag{2.80}$$

Using equations (2.78) and (2.80) and letting $\omega \to 0$, the static deflection at s for a static force P_0 at j is

$$\sum_{r=1}^{n} \frac{{}_rz_s\, {}_rz_j P_0}{\omega_r^2},$$

agreeing with expression (2.79). The closed form solution for the static deflection is obtained from equation (2.71) with inertia and damping terms omitted as

$$\mathbf{x} = \mathbf{K}^{-1}\mathbf{p}$$

$$= \frac{\mathbf{A}\mathbf{p}}{\det |\mathbf{K}|}$$

where $\mathbf{A}$ is the adjoint matrix of $\mathbf{K}$ (see Appendix 2). For the deflection x_s due to a single force P_j

$$x_s = \frac{a_{sj} P_j}{\det |\mathbf{K}|} \tag{2.81}$$

where a_{sj} is the element of $\mathbf{A}$ in the sth row and jth column. Using equation (2.81) for the sum of terms outside the integral in equation (2.77), the complete expression for the response x_s due to a single force $P_j(t)$ is

$$x_s = \frac{a_{sj} P_j(t)}{\det |\mathbf{K}|} - \sum_{r=1}^{n} \frac{{}_rz_s\, {}_rz_j}{\omega_r^2} \int_0^t \dot{P}_j(\tau) \exp\left[-\gamma_r \omega_r(t - \tau)\right]$$

$$\times \left[\cos \omega_r^1(t - \tau) + \frac{\gamma_r \omega_r}{\omega_r^1} \sin \omega_r^1(t - \tau)\right] d\tau. \tag{2.82}$$

The convergence of equation (2.82) is usually faster than the original solution based on equation (2.75).

In this section so far, the response has been determined when excitation consists of one or more forces applied at the coordinates x_j. Excitation is often caused by prescribed displacements or accelerations at the supports or boundaries of the structure, e.g. the vibrations of a structure due to earthquakes or passing traffic. This type of excitation can be represented by prescribed vertical displacements at support point A in Fig. 2.1 or by horizontal displacements at base EF of the frame of Fig. 2.2 or 2.5. The modifications to the normal mode method, required to accommodate excitation through the supports, will be outlined. It is assumed that there are no applied forces, other than those associated with support reactions, and that vibrations of the constrained or anchored structure are represented by a mathematical model with matrices of order $n \times n$, i.e. by equation (2.71) with $\mathbf{p} = 0$. (If applied forces and imposed displacements exist together, the response can be found by superposition of the two effects.) If inputs are prescribed at N coordinates, which have zero displacements in the constrained structure, a displacement vector $\mathbf{x}'$ is formed, consisting of the vector of displacements of the constrained structure $\mathbf{x}$ and the N additional displacements, i.e.

$$\mathbf{x}'^T = [\mathbf{x}^T \mid x_{n+1}, x_{n+2}, \ldots, x_{n+N}]. \tag{2.83}$$

For example, if the base EF of the frame of Fig. 2.5 is subjected to a horizontal displacement, $N = 1$ and x_{n+1} is this prescribed displacement; $x_1, x_2, \ldots, x_n$ are the displacements of the masses $m_1, m_2, \ldots, m_n$, respectively. A total of N additional displacements is used so that the theory is applicable to a foundation block, subjected to simultaneous vertical and horizontal translations and rocking. The equations of motion of the complete system, including support motion, are given by

$$\mathbf{M}'\ddot{\mathbf{x}}' + \mathbf{C}'\dot{\mathbf{x}}' + \mathbf{K}'\mathbf{x}' = \mathbf{p}' \tag{2.84}$$

where the primed quantities represent matrices and vectors of order $(n + N)$. The force vector $\mathbf{p}'$ has n zero elements and N non-zero elements, corresponding to the support forces where motion is prescribed. The stiffness matrix $\mathbf{K}'$ is defined as

$$\mathbf{K}' = \left[\begin{array}{ccc|ccc} & & & k_{1,n+1} & \cdots & k_{1,n+N} \\ & \mathbf{K} & & k_{2,n+1} & \cdots & k_{2,n+N} \\ & & & \cdot & \cdot & \cdot \\ \hline k_{n+1,1} & \cdots & k_{n+1,n} & k_{n+1,n+1} & \cdots & k_{n+1,n+N} \\ \cdot & \cdot & \cdot & \cdot & \cdot & \cdot \\ k_{n+N,1} & \cdots & k_{n+N,n} & k_{n+N,n+1} & \cdots & k_{n+N,n+N} \end{array} \right] \tag{2.85}$$

with similar definitions for $\mathbf{C}'$ and $\mathbf{M}'$; $\mathbf{K}$ is the stiffness matrix of the constrained system. If, for example, equation (2.84) is used for the motion of the frame of Fig. 2.5 when the base undergoes a horizontal displacement, $N = 1$; $k_{n+1, n+1} = -k_{1, n+1} = -k_{n+1, 1} =$ the combined stiffness of the lowest pair of stanchions and $k_{j, n+1} = 0$ for $j = 2, 3, 4, \ldots, n$; the $(n + 1)$th or additional equation represents dynamic equilibrium of the base or foundation.

In the normal mode method the damping matrix is assumed to be a linear combination of the mass and stiffness matrices, as in equation (2.72), i.e.

$$\mathbf{C}' = \lambda_k \mathbf{K}' + \lambda_m \mathbf{M}'. \tag{2.86}$$

The transformation

$$\mathbf{x}' = \mathbf{Z}'\mathbf{q}' \tag{2.87}$$

is used to uncouple equation (2.84). In equation (2.87)

$$\mathbf{Z}' = \begin{bmatrix} \mathbf{Z} & \mathbf{0} \\ \mathbf{0} & \mathbf{I} \end{bmatrix} \tag{2.88}$$

where $\mathbf{Z}$ is the matrix of the normalized eigenvectors of the *constrained* structure, defined by equation (2.55), $\mathbf{I}$ is the identity matrix and $\mathbf{0}$ the null matrix. Like $\mathbf{x}'$, $\mathbf{q}'$ can be partitioned into $\mathbf{q}$ and the additional principal coordinates $q_{n+1}, \ldots, q_{n+N}$. From equations (2.87) and (2.88)

$$\mathbf{x} = \mathbf{Z}\mathbf{q}$$

and

$$x_{n+j} = q_{n+j}, \qquad j = 1, 2, \ldots, N. \tag{2.89}$$

Redefining the partitioned stiffness matrix $\mathbf{K}'$ as

$$\mathbf{K}' = \left[\begin{array}{c|c} \mathbf{K} & \mathbf{K}_{SF} \\ \hline \mathbf{K}_{FS} & \mathbf{K}_F \end{array} \right]$$

and the vector

$$\mathbf{q}' = \left[\begin{array}{c} \mathbf{q} \\ \hline \mathbf{q}_F \end{array} \right]$$

the last term on the left-hand side of equation (2.84), $\mathbf{K'x'}$, becomes, after substitution of transformation (2.88) and pre-multiplication by $(\mathbf{Z'})^T$, in partitioned form

$$
\begin{bmatrix} \mathbf{Z}^T & \vdots & \mathbf{0} \\ \hline \mathbf{0} & \vdots & \mathbf{I} \end{bmatrix}
\begin{bmatrix} \mathbf{K} & \vdots & \mathbf{K}_{SF} \\ \hline \mathbf{K}_{FS} & \vdots & \mathbf{K}_F \end{bmatrix}
\begin{bmatrix} \mathbf{Z} & \vdots & \mathbf{0} \\ \hline \mathbf{0} & \vdots & \mathbf{I} \end{bmatrix}
\begin{bmatrix} \mathbf{q} \\ \hline \mathbf{q}_F \end{bmatrix}
=
\begin{bmatrix} \mathbf{\Omega q} + \mathbf{Z}^T\mathbf{K}_{SF}\,\mathbf{q}_F \\ \hline \mathbf{K}_{FS}\,\mathbf{Zq} + \mathbf{K}_F\,\mathbf{q}_F \end{bmatrix}
$$

using $\mathbf{Z}^T\mathbf{KZ} = \mathbf{\Omega}$ [equation (2.60)].

Performing the same operations on the other terms on the left-hand side of equation (2.84), using $\mathbf{Z}^T\mathbf{MZ} = \mathbf{I}$ and noting that the elements of the excitation vector $\mathbf{p'}$ above the partition are all zero, the first n equations, i.e. those above the partition, are in matrix notation (the remaining N equations give the support forces corresponding to the imposed motions and will not be considered further)

$$
\ddot{\mathbf{q}} + \mathbf{Z}^T\mathbf{M}_{SF}\,\ddot{\mathbf{q}}_F + \lambda_m\,\dot{\mathbf{q}} + \lambda_m\,\mathbf{Z}^T\mathbf{M}_{SF}\,\dot{\mathbf{q}}_F + \lambda_k\,\mathbf{\Omega}\dot{\mathbf{q}} + \lambda_k\,\mathbf{Z}^T\mathbf{K}_{SF}\,\dot{\mathbf{q}}_F + \mathbf{\Omega q}
$$
$$
+ \mathbf{Z}^T\mathbf{K}_{SF}\,\mathbf{q}_F = 0 \tag{2.90}
$$

where $\mathbf{M}_{SF}$ is the upper right-hand portion of the partitioned form of $\mathbf{M'}$. Equation (2.90) represents n uncoupled equations in the coordinates $q_1, q_2, \ldots, q_n$. Performing the matrix multiplication, rearranging the terms and using equation (2.85), a typical equation is

$$
\ddot{q}_r + (\lambda_m + \lambda_k\omega_r^2)\dot{q}_r + \omega_r^2 q_r
$$
$$
+ \sum_{i=1}^{n} {}_r z_i \sum_{j=1}^{N} [m_{i,\,n+j}(\ddot{x}_{n+j} + \lambda_m \dot{x}_{n+j}) + k_{i,\,n+j}(x_{n+j} + \lambda_k \dot{x}_{n+j})] = 0,
$$
$$
r = 1, 2, \ldots, n. \tag{2.91}
$$

The displacements x_{n+j}, velocities $\dot{x}_{n+j}$ and accelerations $\ddot{x}_{n+j}$ are the prescribed inputs at the supports of the structure. Comparing equations (2.74) and (2.91), the solution for support excitation is given by equation (2.75) [the solution for force excitation], if $f_r(t)$ is redefined as

$$
f_r(t) = -\sum_{i=1}^{n} {}_r z_i \sum_{j=1}^{N} [m_{i,\,n+j}(\ddot{x}_{n+j} + \lambda_m \dot{x}_{n+j}) + k_{i,\,n+j}(x_{n+j} + \lambda_k \dot{x}_{n+j})]. \tag{2.92}
$$

Thus the response of the structure to support excitation has been determined in terms of the eigenvalues ω_r^2 and normal modes of the *constrained* structure. In general, equation (2.92) includes prescribed displacements, velocities and accelerations. If the mass matrix is diagonal (as in the multi-degree-of-freedom systems of Figs. 2.1, 2.2, 2.3 and 2.5), $f_r(t)$ is defined in terms of prescribed displacements and velocities; if, further, λ_k is zero, $f_r(t)$ is defined in terms of prescribed displacements only.

Example. The base, *EF*, of the two-storey frame shown in Fig. 2.6 is subjected to a horizontal displacement,

$$x_0 = 0.01 \sin \pi t/t_0 \text{ (m)} \qquad 0 \le t \le t_0,$$
$$x_0 = 0 \qquad\qquad\qquad t \ge t_0.$$

The resulting displacements of the rigid horizontal members, *AB* and *CD*, are required. The combined stiffness in flexure of the light stanchions *AE* and *BF* and also of *CA* and *DB* are 10 MN/m; $m_1 = m_2 = 10^4$ kg.

Assume   (i) $t_0 = 0.2$ s,
        (ii) $t_0 = 0.075$ s.

Neglect damping.

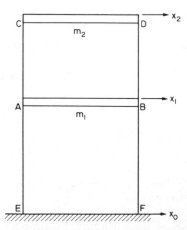

FIG. 2.6. Frame with two degrees of freedom subjected to a ground displacement x_0.

The equations of motion of AB and CD, respectively, are

$$10^4\ddot{x}_1 + 10^7(x_1 - x_0) + 10^7(x_1 - x_2) = 0$$

and

$$10^4\ddot{x}_2 - 10^7(x_1 - x_2) = 0,$$

i.e.

$$\ddot{x}_1 + 2 \times 10^3 x_1 - 10^3 x_2 = 10^3 x_0$$

and

$$\ddot{x}_2 + 10^3 x_2 - 10^3 x_1 = 0. \tag{a}$$

Considering free vibrations, in order to determine the natural frequencies, we have equations (a) with $x_0 = 0$. Substituting $x_1 = e_1 \sin(\omega t + \alpha)$, $x_2 = e_2 \sin(\omega t + \alpha)$, a solution exists, provided that

$$(2000 - \omega^2)e_1 - 1000 e_2 = 0,$$

$$-1000 e_1 + (1000 - \omega^2)e_2 = 0. \tag{b}$$

Eliminating the ratio e_2/e_1, or setting the determinant of equations (b) equal to zero, a quadratic equation in ω^2 is obtained with roots: $\omega_1^2 = 382 \cdot 0$, $\omega_2^2 = 2618 \cdot 0$. The corresponding natural frequencies $(= \omega_r/2\pi)$ are

$$f_1 = 3 \cdot 11 \quad \text{and} \quad f_2 = 8 \cdot 14 \text{ Hz.}$$

Substituting $\omega^2 = \omega_1^2$ in either of equations (b), we have for the first mode:

$$_1e_2 = 1 \cdot 6180 \, _1e_1. \tag{c}$$

Similarly, substituting $\omega^2 = \omega_2^2$,

$$_2e_2 = -0 \cdot 6180 \, _2e_1. \tag{d}$$

The orthogonality condition, (2.52),

$$_1e_1 m_1 \, _2e_1 + _1e_2 m_2 \, _2e_2 = 0$$

is satisfied by the above values.

From (c) and (d), respectively,

$$m_1 \,_1e_1^2 + m_2 \,_1e_2^2 = 10^4 \,_1e_1^2(1 + 1.6180^2)$$

$$= 3.6180 \times 10^4 \,_1e_1^2$$

and

$$m_1 \,_2e_1^2 + m_2 \,_2e_2^2 = 10^4 \,_2e_1^2(1 + 0.6180^2)$$

$$= 1.3820 \times 10^4 \,_2e_1^2.$$

Introducing the normalized vector $_r z$, defined in equations (2.53) and (2.54), it is apparent that

$$a_1 = (3.6180)^{1/2} \times 10^2 \,_1e_1$$

and

$$a_2 = (1.3820)^{1/2} \times 10^2 \,_2e_1.$$

Noting that $_r z_s = {}_r e_s / a_r$ and that in the matrix $\mathbf{Z}$, equation (2.55), the rth column consists of the components of the vector $_r z$

$$\mathbf{Z} \equiv \begin{bmatrix} {}_1e_1/a_1 & {}_2e_1/a_2 \\ {}_1e_2/a_1 & {}_2e_2/a_2 \end{bmatrix} = \begin{bmatrix} 0.5257 & 0.8506 \\ 0.8506 & -0.5257 \end{bmatrix} \times 10^{-2}. \qquad (e)$$

Introducing the change of coordinates $\mathbf{x} = \mathbf{Zq}$ [equation (2.62)], two uncoupled equations

$$\ddot{q}_r + \omega_r^2 q_r = {}_r z_1 P_1(t) + {}_r z_2 P_2(t), \qquad r = 1, 2 \qquad (f)$$

are obtained from equation (2.74). Comparing equations (2.71) with equations (a), or using equation (2.92),

$$P_1(t) = 10^7 x_0 = 10^5 \sin \pi t/t_0, \qquad 0 \le t \le t_0$$

and

$$P_2(t) = 0.$$

Substituting numerical values in equations (f)

$$\ddot{q}_1 + 382.0 q_1 = 525.7 \sin \pi t/t_0,$$

$$\ddot{q}_2 + 2618 q_2 = 850.6 \sin \pi t/t_0. \qquad (g)$$

Writing equations (g) in the form

$$\ddot{q}_r + \omega_r^2 q_r = A_r \sin \omega t$$

the solution could be obtained from Duhamel's integral, but for a sinusoidal excitation the complementary function and particular integral are known and the complete solution is:

$$q_r = \frac{A_r \sin \omega t}{\omega_r^2 - \omega^2} + B_r \cos \omega_r t + C_r \sin \omega_r t.$$

The constants B_r and C_r are chosen so that $q_r = 0$ and $\dot{q}_r = 0$ at $t = 0$. Thus

$$q_r = \frac{A_r}{\omega_r^2 - \omega^2}\left(\sin \omega t - \frac{\omega}{\omega_r} \sin \omega_r t\right), \qquad 0 \le t \le t_0. \tag{h}$$

For $t > t_0$, the excitation is zero, i.e. $x_0 = 0$. The response is free vibration, which with damping neglected can be represented by

$$q_r = D_r \sin \omega_r(t - t_0) + F_r \cos \omega_r(t - t_0), \qquad t \ge t_0, \quad r = 1, 2. \tag{j}$$

The constants D_r and F_r must give continuity of q_r and $\dot{q}_r$ at $t = t_0$. Thus equating equations (h) and (j) for $t = t_0$ and noting that $\omega t_0 = \pi$,

$$F_r = -\frac{A_r}{(\omega_r^2 - \omega^2)\,\omega_r} \frac{\omega}{} \sin \omega_r t_0. \tag{k}$$

From the continuity of $\dot{q}_r$

$$D_r = -\frac{A_r}{(\omega_r^2 - \omega^2)\,\omega_r} \frac{\omega}{} (1 + \cos \omega_r t_0). \tag{m}$$

Numerical values of q_r are found from equations (g), (h), (j), (k) and (m). The relations between the original coordinates, x_1 and x_2, and the principal coordinates, q_1 and q_2, are found by expanding $\mathbf{x} = \mathbf{Z}\mathbf{q}$ and using equation (e) as:

$$x_1 = [0.5257q_1 + 0.8506q_2] \times 10^{-2} \text{ m},$$

$$x_2 = [0.8506q_1 - 0.5257q_2] \times 10^{-2} \text{ m}. \tag{n}$$

Numerical values of q_1 and q_2 for the two assumptions regarding t_0 are given below.

Assumption (i) $t_0 = 0.2$ s.

For $0 \le t \le 0.2$ s, $q_1 = 3.8877[\sin 5\pi t - 0.8037 \sin 19.54t]$,

$$q_2 = 0.3249[\sin 5\pi t - 0.3070 \sin 51.17t]. \tag{p}$$

For $t > 0.2$ s, $q_1 = -0.8755 \sin 19.54(t - 0.2)$

$\qquad\qquad\qquad + 2.1691 \cos 19.54(t - 0.2)$,

$\qquad q_2 = -0.0341 \sin 51.17(t - 0.2)$

$\qquad\qquad\qquad + 0.0796 \cos 51.17(t - 0.2)$.

Assumption (ii) $t_0 = 0.075$ s.

For $0 \le t \le 0.075$ s, $q_1 = 0.3830 \left[-\sin \dfrac{40\pi t}{3} + 2.1433 \sin 19.54t \right]$,

$$q_2 = 0.9852 \left[\sin \frac{40\pi t}{3} - 0.8186 \sin 51.17t \right]. \quad \text{(r)}$$

For $t > 0.075$ s, $q_1 = 0.9069 \sin 19.54(t - 0.075)$

$\qquad\qquad\qquad + 0.8164 \cos 19.54(t - 0.075)$,

$\qquad q_2 = -0.1877 \sin 51.17(t - 0.075)$

$\qquad\qquad\qquad + 0.5172 \cos 51.17(t - 0.075)$.

Equations (n), together with either equations (p) or (r), give the complete response of the frame for the two assumed values of the pulse duration t_0. The response is plotted in Figs. 2.7 a and b for $t_0 = 0.2$ and 0.075 s, respectively. As can be seen from the figures or by comparison of equations (p) and (r), the response is dominated by the contribution from the first mode for $t_0 = 0.2$ s, but there are significant contributions from both modes when $t_0 = 0.075$ s. The maximum values of the displacements x_1 and x_2 are lower for the lower value of t_0. This example will be solved approximately by the methods of numerical integration in Section 2.8, and some further comments are given there.

2.7. Response to Harmonic Excitation

Harmonic applied forces occur frequently. As small changes in damping values cause large changes in response at or near resonant conditions, the restriction on damping with the normal mode method, imposed by condition (2.72), is a serious disadvantage when the excitation is harmonic. In this section, after a brief consideration of the normal mode method, two other methods, the frequency response and

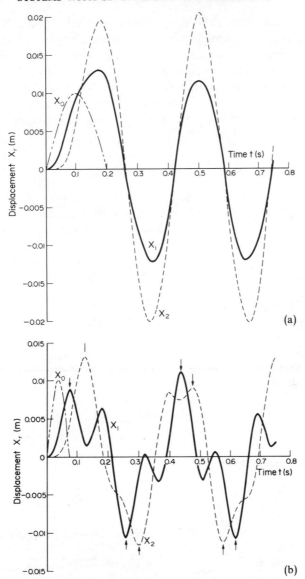

FIG. 2.7. Response of system of Fig. 2.6. (a) Duration of excitation $t_0 = 0{\cdot}2$ s, (b) $t_0 = 0{\cdot}075$ s.

complex eigenvalue methods, of determining the response to harmonic excitation are described. Both methods impose no restriction on damping, are limited to harmonic excitation and use complex arithmetic. The former is simpler in concept, but requires the inversion of a complex matrix at each excitation frequency for which the response is needed. The latter is similar to the normal mode method, but is formulated in terms of complex eigenvalues and eigenvectors.

Normal Mode Method

A typical term in the force vector $\mathbf{p}(t)$ of equation (2.71) is $P_j(t) = P_j \sin \omega t$, where P_j is a constant and ω is the frequency of the applied force. However, we consider a force vector $\mathbf{p} \exp(i\omega t)$, where $\mathbf{p}$ is a vector of constants, obtain the response in complex form and then take the real or imaginary part of the complex solution for excitation proportional to $\cos \omega t$ or $\sin \omega t$, respectively. Thus equation (2.74) becomes

$$\ddot{q}_r + 2\gamma_r \omega_r \dot{q}_r + \omega_r^2 q_r = \exp(i\omega t) \sum_{j=1}^{n} {}_r z_j P_j, \qquad r = 1, 2, \ldots, n.$$

$$(2.93)$$

Noting that $\dot{q}_r = i\omega q_r$ and $\ddot{q}_r = -\omega^2 q_r$ and we require only the steady-state solution,

$$q_r = \frac{\exp(i\omega t) \sum_{j=1}^{n} {}_r z_j P_j}{\omega_r^2 - \omega^2 + 2i\gamma_r \omega_r \omega}.$$

A typical displacement x_s is given by

$$x_s = \sum_{r=1}^{n} {}_r z_s q_r$$

$$= \sum_{r=1}^{n} \left[\frac{{}_r z_s \left(\sum_{j=1}^{n} {}_r z_j P_j \right) (\omega_r^2 - \omega^2 - 2i\gamma_r \omega_r \omega) \exp(i\omega t)}{(\omega_r^2 - \omega^2)^2 + 4\gamma_r^2 \omega_r^2 \omega^2} \right]. \quad (2.94)$$

Finally, for an excitation vector proportional to $\cos \omega t$ or $\sin \omega t$ the real or imaginary part of the complete complex summation is taken.

Alternatively, if the damping is hysteretic, the equation of motion for the system becomes

$$\mathbf{M}\ddot{x} + \mathbf{H}\dot{x}/\omega + \mathbf{K}x = \mathbf{p}\exp(i\omega t) \qquad (2.95)$$

where $\mathbf{H}$ is a symmetric matrix of hysteretic damping constants. For the transformation $x = \mathbf{Z}q$ to yield uncoupled equations, matrix $\mathbf{H}$ has to satisfy the condition

$$\mathbf{H} = a_m\,\mathbf{M} + a_k\,\mathbf{K} \qquad (2.96)$$

where a_m and a_k are constants. Following the standard procedure uncoupled equations in q_r are obtained of the form

$$\ddot{q}_r + \omega_r^2(1 + i\mu_r)q_r = \sum_{j=1}^{n} {}_r z_j P_j \exp(i\omega t) \qquad (2.97)$$

where

$$\mu_r = a_k + a_m/\omega_r^2. \qquad (2.98)$$

The modal damping parameters μ_r must satisfy condition (2.98) in order to obtain uncoupled equations. Thus for a structure with specified distributions of stiffness and mass, which fix the values of the natural frequencies ω_r, the constants a_k and a_m can be chosen to give desired values to any two modal damping parameters, but the other values of μ_r are then prescribed by equation (2.98). (As discussed in Section 2.6, additional terms can be included in equations (2.96) and (2.98) if more than two modal damping parameters are prescribed.) The steady-state solution of equation (2.97) is

$$q_r = \frac{\exp(i\omega t)\sum_{j=1}^{n} {}_r z_j P_j}{\omega_r^2 - \omega^2 + i\mu_r\,\omega_r^2}$$

and a typical displacement

$$x_s = \sum_{r=1}^{n}\left[\frac{{}_r z_s\left(\sum_{j=1}^{n} {}_r z_j P_j\right)(\omega_r^2 - \omega^2 - i\mu_r\omega_r^2)\exp(i\omega t)}{(\omega_r^2 - \omega^2)^2 + \mu_r^2\omega_r^4}\right]. \qquad (2.99)$$

Again, the real or imaginary part of equation (2.99) is taken, depending upon whether the excitation is proportional to $\cos \omega t$ or $\sin \omega t$.

Equations (2.94) and (2.99) will be used to generate solutions to random excitation in Section 2.9. Comparing these equations, a change from viscous to hysteretic damping is effected by replacing $2\gamma_r\omega$ by $\mu_r\omega_r$. This conforms to the differences in the equations of motion, where $c_{ij}\omega$ in equation (2.71) is replaced by h_{ij} in equation (2.95). In Section 2.9, where the normal mode method is used, the interchangeability of $2\gamma_r\omega$ and $\mu_r\omega_r$ will be used to avoid the presentation of parallel analysis for viscous and hysteretic damping.

Often resonant amplitudes are of greater interest than complete response curves. For resonance in the rth mode, i.e. $\omega = \omega_r$, where r is a specified integer, the response can be approximated by considering only the term corresponding to r in the summations (2.94) and (2.99). For viscous damping when $\omega = \omega_r$,

$$x_s \simeq {}_rz_s \frac{\left(\sum_{j=1}^{n} {}_rz_jP_j\right) \sin\left(\omega_r t - \pi/2\right)}{2\gamma_r\omega_r^2} \tag{2.100}$$

for excitation proportional to $\sin \omega t$. For this approximation to be reasonably accurate damping must be small, the natural frequency ω_r should not be close to another natural frequency and r should be small, i.e. one of the lower modes. (Usually for large systems only the lower resonances are of interest.) For hysteretic damping the denominator is $\mu_r\omega_r^2$. However, when the damping matrix is proportional to the stiffness matrix (i.e., $\lambda_m = 0$ or $a_m = 0$), resonant amplitudes are proportional to $1/\omega_r^3$ and $1/\omega_r^2$ for viscous and hysteretic damping respectively. For a particular system and excitation resonant amplitudes decrease more rapidly as r increases for viscous damping than for hysteretic damping.

Frequency Response Method

As in the previous method the actual force vector is replaced by $\mathbf{p} \exp(i\omega t)$. Thus for viscous damping the matrix equation (2.71) is rewritten as

$$\mathbf{M\ddot{x}} + \mathbf{C\dot{x}} + \mathbf{Kx} = \mathbf{p} \exp(i\omega t). \tag{2.101}$$

As the steady-state solution of equation (2.101) is proportional to $\exp(i\omega t)$, $\dot{x} = i\omega x$ and $\ddot{x} = -\omega^2 x$; thus equation (2.101) becomes

$$Jx = p \exp(i\omega t)$$

where

$$J = K - \omega^2 M + i\omega C.$$

Thus the response is obtained from

$$x = J^{-1} p \exp(i\omega t). \tag{2.102}$$

For excitation proportional to $\sin \omega t$ the imaginary part of equation (2.102) is taken, i.e.

$$x = \text{Im}\,[J^{-1} p \exp(i\omega t)]. \tag{2.103}$$

In general, the elements of the matrix J are complex and frequency dependent. Thus a complex matrix must be inverted at each excitation frequency for which the response is required.

If the damping is hysteretic, the equation of motion is equation (2.95), and the response is given by equation (2.103) with J redefined as

$$J = K - \omega^2 M + iH.$$

Complex Eigenvalue Method

Hurty and Rubinstein[40] describe a method of determining response in terms of damped normal modes. In their method it is necessary to replace the second-order differential equations by first-order equations, treating all displacements and velocities as unknown variables. Thus for a system with n degrees of freedom matrix equations of order $2n$ are obtained. However, the method can be used with any form of linear damping and any excitation force. In the method to be given here only hysteretic damping is considered, because with this limitation the complete analysis is in terms of matrices of order n; as hysteretic damping is considered, harmonic excitation is assumed.

Equation (2.95) gives the general equation for systems with hysteretic damping. This can be rewritten:

$$M\ddot{x} + K'x = p \exp(i\omega t) \tag{2.104}$$

where the complex matrix $K' = K + iH$. Assuming that $x = e \exp(\lambda t)$ and putting the right-hand side of equation (2.104) equal to zero, the complex eigenvalues λ^2 are found from

$$[\lambda^2 M + K']e = 0. \qquad (2.105)$$

If these eigenvalues are $\lambda_1^2, \lambda_2^2, \lambda_3^2, \ldots, \lambda_n^2$, there will be corresponding complex eigenvectors ${}_1z, {}_2z, {}_3z, \ldots, {}_nz$. Vector ${}_rz$ is the value of e from equation (2.105) when $\lambda^2 = \lambda_r^2$ and the vector has been scaled to satisfy the normalizing condition

$$_rz^T M\,_r z = 1. \qquad (2.106)$$

The eigenvectors satisfy the orthogonality condition

$$_rz^T M\,_s z = 0, \qquad r \neq s. \qquad (2.107)$$

This can be proved by analysis similar to equations (2.49) to (2.52), noting that the complex matrix K' is symmetric. Rewriting the equations with p zero in terms of successive eigenvalues,

$$k'_{11}\,_1z_1 + k'_{12}\,_1z_2 + \cdots + k'_{1n}\,_1z_n + m_{11}\,_1z_1\lambda_1^2 + m_{12}\,_1z_2\lambda_1^2 + \cdots$$
$$+ m_{1n}\,_1z_n\lambda_1^2 = 0,$$
$$k'_{21}\,_2z_1 + k'_{22}\,_2z_2 + \cdots + k'_{2n}\,_2z_n + m_{21}\,_2z_1\lambda_2^2 + m_{22}\,_2z_2\lambda_2^2 + \cdots$$
$$+ m_{2n}\,_2z_n\lambda_2^2 = 0$$

$$\cdot \quad \cdot \quad \cdot \quad \cdot \quad \cdot \quad \cdot \quad \cdot \quad \cdot \quad \cdot \quad \cdot \quad \cdot \quad \cdot \quad \cdot \quad \cdot \quad \cdot \quad \cdot \quad \cdot \quad \cdot \quad (2.108)$$

or in matrix form

$$K'Z + MZ\Omega = 0 \qquad (2.109)$$

where Z consists of the normalized complex eigenvectors, i.e.

$$Z = \begin{bmatrix} {}_1z_1 & {}_2z_1 & \cdots & {}_nz_1 \\ {}_1z_2 & {}_2z_2 & \cdots & {}_nz_2 \\ \cdot & \cdot & \cdot & \cdot \\ {}_1z_n & {}_2z_n & \cdots & {}_nz_n \end{bmatrix}$$

and

$$\Omega = \begin{bmatrix} \lambda_1^2 & 0 & \cdots & 0 \\ 0 & \lambda_2^2 & \cdots & 0 \\ \cdot & \cdot & \cdots & \cdot \\ 0 & 0 & \cdots & \lambda_n^2 \end{bmatrix}.$$

Using the normalizing and orthogonality conditions

$$\mathbf{Z}^T \mathbf{M} \mathbf{Z} = \mathbf{I}. \tag{2.110}$$

Pre-multiplying equation (2.109) by $\mathbf{Z}^T$

$$\mathbf{Z}^T \mathbf{K}' \mathbf{Z} + \Omega = 0. \tag{2.111}$$

Returning to equation (2.104), introducing the change of coordinates

$$\mathbf{x} = \mathbf{Z} \mathbf{q}, \tag{2.112}$$

pre-multiplying by $\mathbf{Z}^T$ and using equations (2.110) and (2.111), we obtain the set of uncoupled equations

$$\ddot{\mathbf{q}} - \Omega \mathbf{q} = \mathbf{Z}^T \mathbf{p} \exp(i\omega t) \tag{2.113}$$

with typical solution

$$q_r = -\sum_j \frac{(_r z_j P_j) \exp(i\omega t)}{\omega^2 + \lambda_r^2}, \qquad r = 1, 2, \ldots, n. \tag{2.114}$$

The steady-state response is determined from equations (2.112) and (2.114) by taking the real or imaginary part of the complete (complex) solution for excitation proportional to $\cos \omega t$ or $\sin \omega t$ respectively.

Figure 2.4 shows typical response curves for a two-degree-of-freedom system. For systems with several degrees of freedom and light damping there will be maximum amplitudes when the excitation frequency is approximately equal to any of the natural frequencies ω_1, ω_2, ω_3, $\ldots$.

Example. Figure 2.8 shows a two-degree-of-freedom system with the masses subjected to harmonic forces $P_1 \sin \omega t$ and $P_2 \sin \omega t$. Using the frequency response method, determine the conditions for which the displacements x_1 and x_2 have the same phase angle. Investigate this in detail for ω equal to each natural frequency.

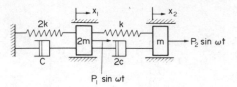

FIG. 2.8. Two-degree-of-freedom system.

If the viscous dampers c and $2c$ are replaced by hysteretic dampers with constants h and $2h$, where $h/k = 0.1$, use the complex eigenvalue method to determine the response of each mass when ω is equal to the lower natural frequency and $P_1/P_2 = -0.5$.

The equations of motion are

$$2m\ddot{x}_1 + c\dot{x}_1 + 2kx_1 + 2c(\dot{x}_1 - \dot{x}_2) + k(x_1 - x_2) = P_1 \sin \omega t$$

and

$$m\ddot{x}_2 + 2c(\dot{x}_2 - \dot{x}_1) + k(x_2 - x_1) = P_2 \sin \omega t,$$

i.e.

$$2m\ddot{x}_1 + 3c\dot{x}_1 - 2c\dot{x}_2 + 3kx_1 - kx_2 = P_1 \sin \omega t$$

and

$$m\ddot{x}_2 + 2c\dot{x}_2 - 2c\dot{x}_1 + kx_2 - kx_1 = P_2 \sin \omega t. \quad \text{(a)}$$

Natural Frequencies

The equations for free undamped vibrations are

$$2m\ddot{x}_1 + 3kx_1 - kx_2 = 0$$

and

$$m\ddot{x}_2 + kx_2 - kx_1 = 0.$$

Putting $x_j = X_j \sin (\omega t + \alpha)$, $j = 1, 2$, we obtain the frequency determinant

$$\det \begin{vmatrix} 3k - 2m\omega^2 & -k \\ -k & k - m\omega^2 \end{vmatrix} = 0.$$

The natural frequencies are given by

$$\omega_1^2/\omega_0^2 = \tfrac{1}{2} \quad \text{and} \quad \omega_2^2/\omega_0^2 = 2$$

where $\omega_0^2 = k/m$. For mode 1, $X_2/X_1 = 2$; for mode 2, $X_2/X_1 = -1$. (It is noted that these values of X_2/X_1 satisfy the orthogonality condition.)

Frequency Response Method

Putting $\omega/\omega_0 = r$, $c/m = \gamma_0 \omega_0$, equations (a) are rewritten in the form

$$\mathbf{J}\mathbf{x} = \mathbf{p} \exp (i\omega t)$$

where

$$\mathbf{J} = \begin{bmatrix} 3 - 2r^2 + 3i\gamma_0 r & -1 - 2i\gamma_0 r \\ -1 - 2i\gamma_0 r & 1 - r^2 + 2i\gamma_0 r \end{bmatrix},$$

$$\mathbf{p} = \begin{bmatrix} P_1/k \\ P_2/k \end{bmatrix}.$$

(In this example γ_0 is simply a convenient non-dimensional damping parameter and should not be confused with the modal damping parameter γ_r, used in the general analysis.) The solution is

$$\mathbf{x} = \mathbf{J}^{-1}\mathbf{p} \exp (i\omega t)$$

$$= \frac{\mathbf{A}\mathbf{p} \exp (i\omega t)}{\det |\mathbf{J}|}$$

where the adjoint matrix $\mathbf{A}$, formed from the co-factors of $\mathbf{J}$, is given by

$$\mathbf{A} = \begin{bmatrix} 1 - r^2 + 2i\gamma_0 r & 1 + 2i\gamma_0 r \\ 1 + 2i\gamma_0 r & 3 - 2r^2 + 3i\gamma_0 r \end{bmatrix}.$$

Thus

$$x_1 = [(1 - r^2 + 2i\gamma_0 r)P_1 + (1 + 2i\gamma_0 r)P_2] \exp (i\omega t)/k \det |\mathbf{J}|$$

and

$$x_2 = [(1 + 2i\gamma_0 r)P_1 + (3 - 2r^2 + 3i\gamma_0 r)P_2] \exp (i\omega t)/k \det |\mathbf{J}|. \quad \text{(b)}$$

The phase of x_1 and x_2, relative to that of $\mathbf{p}$, will be the same if

$$\frac{2\gamma_0 rP_1 + 2\gamma_0 rP_2}{(1 - r^2)P_1 + P_2} = \frac{2\gamma_0 rP_1 + 3\gamma_0 rP_2}{P_1 + (3 - 2r^2)P_2}.$$

This reduces to

$$2r^2(P_1/P_2)^2 + (3 - r^2)(P_1/P_2) + (3 - 4r^2) = 0. \quad \text{(c)}$$

This equation is independent of γ_0; for a specified excitation frequency there are two values of P_1/P_2 for which the phase angles of the two masses are equal.

Equations (b) can be written

$$x_j = \frac{(C_j + iD_j)}{A + iB} \exp (i\omega t), \quad j = 1, 2.$$

Rationalizing and taking the imaginary part

$$x_j = \left[\frac{C_j^2 + D_j^2}{A^2 + B^2} \right]^{1/2} \sin (\omega t - \alpha_j)$$

$$= X_j \sin (\omega t - \alpha_j)$$

where

$$\tan \alpha_j = \frac{BC_j - AD_j}{AC_j + BD_j}.$$

Equation (c) gives the condition for $\alpha_1 = \alpha_2$.

(i) If $\omega = \omega_1$, $r^2 = \frac{1}{2}$ and from equation (c) $P_1/P_2 = -0.5$ or -2.
Also $A = -\gamma_0^2 k$ and $B = 1.5\gamma_0 k/2^{1/2}$.
For $P_1/P_2 = -0.5$, $\quad C_1 = \frac{1}{2}P_1 + P_2 = \frac{3}{4}P_2$,

$$D_1 = 2^{1/2}\gamma_0(P_1 + P_2) = \gamma_0 P_2/2^{1/2},$$

$$C_2 = P_1 + 2P_2 = 1.5P_2,$$

$$D_2 = \gamma_0(2P_1 + 3P_2)/2^{1/2} = 2^{1/2}\gamma_0 P_2.$$

Thus $X_2/X_1 = 2$ for all values of γ_0, $\alpha_j = \pi/2$ and

$$\frac{kX_1}{P_2} = \left[\frac{0.75^2 + 0.5\gamma_0^2}{\gamma_0^4 + 1.125\gamma_0^2}\right]^{1/2}$$

$$\simeq 0.707/\gamma_0 \quad \text{for} \quad \gamma_0 \text{ small.}$$

Similarly, for $P_1/P_2 = -2$, $X_2/X_1 = 0.5$ for all values of γ_0, tan $\alpha_j = 0.943\gamma_0$ and

$$\frac{kX_1}{P_2} = \left[\frac{2\gamma_0^2}{\gamma_0^4 + 1.125\gamma_0^2}\right]^{1/2}$$

$$\simeq 1.33 \quad \text{for} \quad \gamma_0 \text{ small.}$$

(ii) If $\omega = \omega_2$, $r^2 = 2$ and from equation (c) $P_1/P_2 = -1.25$ or 1.

For $P_1/P_2 = -1.25$, $X_2/X_1 = -1$, $\alpha_j = \pi/2$ and

$kX_1/P_2 \simeq 0.177/\gamma_0$ for γ_0 small.

For $P_1/P_2 = 1$, $X_2/X_1 = 1.25$, tan $\alpha_j = -0.314\gamma_0$ and

$kX_1/P_2 \simeq 0.444$ for γ_0 small.

This example illustrates a simple application of the theory of characteristic phase lags[33]. For a system with n degrees of freedom with the masses m_j subjected to forces $P_j \sin \omega t$, $j = 1, 2, \ldots, n$, the steady-state response is of the form $x_j = X_j \sin (\omega t - \alpha_j)$. For any excitation frequency ω there are n values of the ratio $P_1 : P_2 : P_3 : \cdots : P_n$ for which $\alpha_1 = \alpha_2 = \alpha_3 = \cdots = \alpha_n$; these values of α are the characteristic phase lags. If $\omega = \omega_r$, where ω_r is a natural frequency, *one* set of the values $P_1 : P_2 : P_3 : \cdots : P_n$ for which $\alpha_1 = \alpha_2 = \alpha_3 = \cdots = \alpha_n$ makes the amplitude ratio $X_1 : X_2 : X_3 : \cdots : X_n$ exactly equal to the normal mode shape for mode r. (In the example $P_1/P_2 = -0.5$ with $\omega = \omega_1$ and $P_1/P_2 = -1.25$ with $\omega = \omega_2$ make the response identical with the first and second mode shapes, respectively.) This behaviour has important practical applications. If it is required to determine accurately the natural frequencies and normal modes of complex structures, e.g. aircraft structures, large dams, etc., from experimental tests on the structures or models, standard resonance tests, in which a single harmonic force is applied to the structure and the resulting structural amplitude recorded, are unsatisfactory if two or more natural frequencies are close together and there is significant damping[11]. However, if multi-point excitation $P_j \sin \omega t$, $j = 1, 2, 3 \ldots$, is applied to the structure and the force amplitude ratio $P_1 : P_2 : P_3 \ldots$ and the excitation frequency ω adjusted until $\alpha_j = \pi/2$ at each excitation point, ω is equal to a natural frequency and the measured amplitude ratio $X_1 : X_2 : X_3 \ldots$ gives the corresponding mode shape[8].

The standard normal mode method of Section 2.6 cannot be applied directly to this example, because the viscous damping matrix for the system of Fig. 2.8 does not satisfy the uncoupling condition (2.72). However, if Thomson's approximation[72] is used, i.e. the non-zero off-diagonal terms in the matrix $\mathbf{Z}^T \mathbf{C} \mathbf{Z}$ are replaced by zeros, the response can be obtained approximately by this method. For γ_0 small the resulting dynamic magnification factors agree with those previously determined for $\omega = \omega_1$ and ω_2 when $\alpha_j = \pi/2$; the relatively small magnification factors, occurring when $\omega = \omega_1$ or ω_2 and $\alpha_j \neq \pi/2$, are not obtained accurately.

Complex Eigenvalue Method

Replacing the viscous dampers by hysteretic dampers equations (a) are replaced by

$$2m\ddot{x}_1 + k(3 + 0\cdot3i)x_1 - k(1 + 0\cdot2i)x_2 = P_1 \exp(i\omega t),$$
$$m\ddot{x}_2 + k(1 + 0\cdot2i)x_2 - k(1 + 0\cdot2i)x_1 = P_2 \exp(i\omega t). \quad \text{(d)}$$

Substituting $x_j = X_j \exp(\lambda t)$ in equations (d) with the right-hand sides zero, the complex eigenvalues are the roots of the determinant

$$\det \begin{vmatrix} k(3 + 0\cdot3i) + 2m\lambda^2 & -k(1 + 0\cdot2i) \\ -k(1 + 0\cdot2i) & k(1 + 0\cdot2i) + m\lambda^2 \end{vmatrix} = 0.$$

Putting $\omega_0^2 = k/m$, the roots are

$$\lambda_1^2/\omega_0^2 = -0\cdot5032 - 0\cdot0494i, \qquad \lambda_2^2/\omega_0^2 = -1\cdot9967 - 0\cdot3005i.$$

For mode j,

$$\left(\frac{X_2}{X_1}\right)_j = \frac{jz_2}{jz_1} = \frac{3 + 0\cdot3i + (2\lambda_j^2/\omega_0^2)}{1 + 0\cdot2i}.$$

Substituting for λ_j^2,

$$_1z_2/_1z_1 = 1\!\cdot\!9555 - 0\!\cdot\!1900i$$

and

$$_2z_2/_2z_1 = -1\!\cdot\!0132 - 0\!\cdot\!0984i.$$

(At this stage it is advisable to check that the orthogonality condition

$$_1z_1 2m \,_2z_1 + \,_1z_2 m \,_2z_2 = 0$$

is satisfied.) The normalizing conditions are

$$m(_jz_1)^2[2 + (_jz_2/_jz_1)^2] = 1.$$

Using the above ratios $_jz_1$ can be determined and the matrix $\mathbf{Z}$ formed as

$$\mathbf{Z} = \begin{bmatrix} 0\!\cdot\!4131 + 0\!\cdot\!0264i & 0\!\cdot\!5748 - 0\!\cdot\!0190i \\ 0\!\cdot\!8129 - 0\!\cdot\!0268i & -0\!\cdot\!5842 - 0\!\cdot\!0373i \end{bmatrix} \times m^{-1/2}.$$

For the response when $\omega = \omega_1$ (i.e. $\omega^2 = 0\!\cdot\!5\omega_0^2$) and $P_1/P_2 = -0\!\cdot\!5$

$$\mathbf{x} = \mathbf{Zq}$$

and

$$q_j = \frac{-(-0\!\cdot\!5 \,_jz_1 + \,_jz_2)P_2 \exp{(i\omega t)}}{0\!\cdot\!5\omega_0^2 + \lambda_j^2}.$$

Substituting numerical values,

$$q_1 = (-0\!\cdot\!004 - 12\!\cdot\!26i)P_2 \exp{(i\omega t)}/m^{1/2}\omega_0^2,$$

$$q_2 = (-0\!\cdot\!5634 + 0\!\cdot\!0945i)P_2 \exp{(i\omega t)}/m^{1/2}\omega_0^2,$$

$$x_1 = -5\!\cdot\!000iP_2 \exp{(i\omega t)}/k,$$

$$x_2 = -10\!\cdot\!000iP_2 \exp{(i\omega t)}/k.$$

Taking the imaginary parts of x_j,

$$x_1 = 5\!\cdot\!000(P_2/k) \sin{(\omega t - \pi/2)},$$

$$x_2 = 10\!\cdot\!000(P_2/k) \sin{(\omega t - \pi/2)}.$$

Again the phase angles of x_1 and x_2, relative to that of the applied forces, are $\pi/2$ and the amplitude ratio X_2/X_1 is identical to the normal mode shape of mode 1.

2.8. Numerical Integration Methods

In these methods of integrating approximately the general dynamic equation (2.40) assumptions are made about the variation of either the displacements or accelerations during small time intervals; e.g. it may be assumed that during a small interval the displacement is a cubic function of time or the acceleration varies linearly. With the aid of these assumptions the set of n second-order differential equations (2.40) is replaced by n simultaneous equations, in general; the solution of the latter gives the displacements at the end of the short time interval for known conditions at the beginning of the interval. Successive application of this procedure leads to a complete solution.

Many possible methods exist. From those commonly used to determine the response of structures the Newmark β method and the central difference method will be described and used to illustrate some of the general points relating to these methods. Unfortunately insufficient numerical evidence exists to recommend an optimum method for all structural dynamic response problems.

The following notation is used for both methods:

$\mathbf{x}_s$ is the value of vector $\mathbf{x}$ at time t_s,

$\mathbf{x}_{s+1}$ is its value at time t_{s+1}, $\Delta t = t_{s+1} - t_s$,

Δt is a small time interval.

Newmark β Method

It is assumed that the displacement and velocity at the end of a time interval can be expressed in terms of the displacement, velocity and acceleration at the beginning of the interval and the acceleration at the end by the relations:

$$\dot{x}_{s+1} = \dot{x}_s + \tfrac{1}{2}(\Delta t)[\ddot{x}_s + \ddot{x}_{s+1}],$$

$$x_{s+1} = x_s + (\Delta t)\dot{x}_s + (\tfrac{1}{2} - \beta)(\Delta t)^2\ddot{x}_s + \beta(\Delta t)^2\ddot{x}_{s+1}. \qquad (2.115)$$

Some values of the parameter β can be given a physical explanation. For example, if the acceleration during Δt is assumed to be constant and equal to the mean of $\ddot{x}_s$ and $\ddot{x}_{s+1}$, then $\beta = 1/4$. In this case,

$$\ddot{x}(\tau) = \tfrac{1}{2}[\ddot{x}_s + \ddot{x}_{s+1}], \qquad 0 \le \tau \le \Delta t.$$

Integration twice with respect to the auxiliary time variable τ and substitution of $\tau = \Delta t$ gives expressions (2.115) with $\beta = 1/4$. If the acceleration is assumed to vary linearly from $\ddot{x}_s$ to $\ddot{x}_{s+1}$ over the time interval, then $\beta = 1/6$. In this case

$$\ddot{x}(\tau) = \ddot{x}_s + (\ddot{x}_{s+1} - \ddot{x}_s)\tau/\Delta t.$$

Integrating

$$\dot{x}(\tau) = \dot{x}_s + \ddot{x}_s\tau + (\ddot{x}_{s+1} - \ddot{x}_s)\tau^2/(2\Delta t)$$

and

$$x(\tau) = x_s + \dot{x}_s\tau + \tfrac{1}{2}\ddot{x}_s\tau^2 + (\ddot{x}_{s+1} - \ddot{x}_s)\tau^3/(6\Delta t).$$

Putting $\tau = \Delta t$, we obtain equations (2.115) with $\beta = 1/6$.

We write the matrix equation (2.40) at three successive time intervals in the form:

$$M\ddot{x}_{s+1} + C\dot{x}_{s+1} + Kx_{s+1} = p_{s+1},$$

$$M\ddot{x}_s + C\dot{x}_s + Kx_s = p_s,$$

and

$$M\ddot{x}_{s-1} + C\dot{x}_{s-1} + Kx_{s-1} = p_{s-1}, \qquad (2.116)$$

where p_s is the value of the vector of applied forces at time t_s. Multiplying the first and last of equations (2.116) by $(\Delta t)^2\beta$, the middle equation by $(\Delta t)^2(1 - 2\beta)$, adding and rearranging terms

$$(\Delta t)^2 \mathbf{M}[\{\beta \ddot{\mathbf{x}}_{s+1} + (\tfrac{1}{2} - \beta)\ddot{\mathbf{x}}_s\} - \{\beta \ddot{\mathbf{x}}_s + (\tfrac{1}{2} - \beta)\ddot{\mathbf{x}}_{s-1}\} + \{\tfrac{1}{2}\ddot{\mathbf{x}}_s + \tfrac{1}{2}\ddot{\mathbf{x}}_{s-1}\}]$$
$$+ (\Delta t)^2 \mathbf{C}[\{\tfrac{1}{2}\dot{\mathbf{x}}_s\} + \{\tfrac{1}{2}\dot{\mathbf{x}}_{s-1}\} + \{\beta(\dot{\mathbf{x}}_{s+1} - \dot{\mathbf{x}}_s)\} + \{(\tfrac{1}{2} - \beta)(\dot{\mathbf{x}}_s - \dot{\mathbf{x}}_{s-1})\}]$$
$$+ (\Delta t)^2 \mathbf{K}[\beta \mathbf{x}_{s+1} + (1 - 2\beta)\mathbf{x}_s + \beta \mathbf{x}_{s-1}]$$
$$= (\Delta t)^2[\beta \mathbf{p}_{s+1} + (1 - 2\beta)\mathbf{p}_s + \beta \mathbf{p}_{s-1}].$$

Substituting from equations (2.115)

$$\mathbf{M}[\{\mathbf{x}_{s+1} - \mathbf{x}_s - (\Delta t)\dot{\mathbf{x}}_s\} - \{\mathbf{x}_s - \mathbf{x}_{s-1} - (\Delta t)\dot{\mathbf{x}}_{s-1}\} + \{\dot{\mathbf{x}}_s - \dot{\mathbf{x}}_{s-1}\}(\Delta t)]$$
$$+ (\Delta t)\mathbf{C}[\{\tfrac{1}{2}\mathbf{x}_{s+1} - \tfrac{1}{2}\mathbf{x}_s - \tfrac{1}{2}(\tfrac{1}{2} - \beta)(\Delta t)^2 \ddot{\mathbf{x}}_s - \tfrac{1}{2}\beta(\Delta t)^2 \ddot{\mathbf{x}}_{s+1}\}$$
$$+ \{\tfrac{1}{2}\mathbf{x}_s - \tfrac{1}{2}\mathbf{x}_{s-1} - \tfrac{1}{2}(\tfrac{1}{2} - \beta)(\Delta t)^2 \ddot{\mathbf{x}}_{s-1} - \tfrac{1}{2}\beta(\Delta t)^2 \ddot{\mathbf{x}}_s\}$$
$$+ \{\tfrac{1}{2}\beta(\ddot{\mathbf{x}}_s + \ddot{\mathbf{x}}_{s+1})(\Delta t)^2\} + \{\tfrac{1}{2}(\tfrac{1}{2} - \beta)(\ddot{\mathbf{x}}_{s-1} + \ddot{\mathbf{x}}_s)(\Delta t)^2\}]$$
$$+ (\Delta t)^2 \mathbf{K}[\beta \mathbf{x}_{s+1} + (1 - 2\beta)\mathbf{x}_s + \beta \mathbf{x}_{s-1}]$$
$$= (\Delta t)^2[\beta \mathbf{p}_{s+1} + (1 - 2\beta)\mathbf{p}_s + \beta \mathbf{p}_{s-1}].$$

Simplifying and rearranging terms

$$[\mathbf{M} + \tfrac{1}{2}(\Delta t)\mathbf{C} + \beta(\Delta t)^2 \mathbf{K}]\mathbf{x}_{s+1} = (\Delta t)^2[\beta \mathbf{p}_{s+1} + (1 - 2\beta)\mathbf{p}_s + \beta \mathbf{p}_{s-1}]$$
$$+ [2\mathbf{M} - (\Delta t)^2(1 - 2\beta)\mathbf{K}]\mathbf{x}_s$$
$$- [\mathbf{M} - \tfrac{1}{2}(\Delta t)\mathbf{C} + \beta(\Delta t)^2 \mathbf{K}]\mathbf{x}_{s-1}.$$
$$\text{(2.117)}$$

Using equation (2.117) the displacement vector at time t_{s+1} can be obtained from previously determined earlier values of the vector and known values of the force vector. This applies for $s = 1, 2, 3, \ldots$ and gives the response at times $2\Delta t, 3\Delta t, 4\Delta t, \ldots$. For the vector $\mathbf{x}_1$ at time Δt a modified form of equation (2.117) is obtained from the first and second of equations (2.116) and equations (2.115) and is

$$[\mathbf{M} + \tfrac{1}{2}(\Delta t)\mathbf{C} + \beta(\Delta t)^2 \mathbf{K}]\mathbf{x}_1$$
$$= (\Delta t)^2 \beta \mathbf{p}_1 + (\Delta t)^2[(\tfrac{1}{2} - \beta)\mathbf{I} + (\tfrac{1}{4} - \beta)\,\Delta t \mathbf{C} \mathbf{M}^{-1}]\mathbf{p}_0 \qquad \text{(2.117a)}$$

if it is assumed that at $t = 0$ $\mathbf{x}_0 = \dot{\mathbf{x}}_0 = 0$.

In general, the solution of n simultaneous equations is required at each time step. When $\mathbf{M}$ is diagonal, $\mathbf{C} = 0$ and $\beta = 0$, equation (2.117) reduces to a set of n uncoupled equations; the significance of this will be discussed later.

Central Difference Method

Using the Taylor series, if y is a function of time t, its value at $t + \Delta t$ can be expressed in terms of y and its derivatives with respect to t at time t by

$$y(t + \Delta t) = y(t) + (\Delta t) \cdot \dot{y} \cdot (t) + \frac{(\Delta t)^2}{1.2} \ddot{y}(t) + \cdots.$$

Also

$$y(t - \Delta t) = y(t) - (\Delta t) \dot{y}(t) + \frac{(\Delta t)^2}{1.2} \ddot{y}(t) - \cdots.$$

In our vector notation

$$\mathbf{x}_{s+1} = \mathbf{x}_s + (\Delta t)\dot{\mathbf{x}}_s + \tfrac{1}{2}(\Delta t)^2 \ddot{\mathbf{x}}_s + \cdots,$$
$$\mathbf{x}_{s-1} = \mathbf{x}_s - (\Delta t)\dot{\mathbf{x}}_s + \tfrac{1}{2}(\Delta t)^2 \ddot{\mathbf{x}}_s - \cdots. \tag{2.118}$$

Neglecting higher derivatives, subtracting and adding the two equations (2.118) and rearranging terms, expressions for the velocity and acceleration vectors in terms of displacement vectors are:

$$\dot{\mathbf{x}}_s = [\mathbf{x}_{s+1} - \mathbf{x}_{s-1}]/(2\Delta t),$$
$$\ddot{\mathbf{x}}_s = [\mathbf{x}_{s+1} - 2\mathbf{x}_s + \mathbf{x}_{s-1}]/(\Delta t)^2. \tag{2.119}$$

Direct substitution in the middle of equations (2.116) and rearrangement of terms gives

$$[\mathbf{M} + \tfrac{1}{2}(\Delta t)\mathbf{C}]\mathbf{x}_{s+1} = (\Delta t)^2 \mathbf{p}_s + [2\mathbf{M} - (\Delta t)^2 \mathbf{K}]\mathbf{x}_s$$
$$+ [\tfrac{1}{2}(\Delta t)\mathbf{C} - \mathbf{M}]\mathbf{x}_{s-1}, \qquad s = 1, 2, 3 \ldots. \tag{2.120}$$

Comparison of equations (2.117) and (2.120) shows that this is a special case of the Newmark method with $\beta = 0$. (The modified form of equation (2.120) for $s = 0$ is obtained from equation (2.117a) by putting $\beta = 0$.)

In the central difference method if the mass matrix is diagonal and the damping matrix is either diagonal or zero, we have a set of uncoupled equations, whereas in the Newmark method with $\beta > 0$ we have a set of simultaneous equations. In practical problems, where the number of equations may be large and the number of time steps to give response curves may also be large, the saving in computing time required to solve uncoupled equations, as opposed to simultaneous equations, will be significant. However, accuracy and stability of computation must be considered also before selecting a method. Some methods of numerical integration become unstable, if too large a time interval Δt is used, i.e. the solution oscillates with increasing amplitude. It should be noted that this is a computational instability, as the physical systems are stable. Stability analysis shows that the Newmark method is unconditionally stable for $\beta \geq \frac{1}{4}$, i.e. the solution is stable for any length of time step. For $\beta < \frac{1}{4}$, the method is conditionally stable. If T_n is the period of the nth or highest mode of the system, the stability condition is: $\Delta t/T_n$ must be less than 0·318, 0·450 and 0·551 for $\beta = 0, \frac{1}{8}$ and $\frac{1}{6}$, respectively.

For the Newmark method with $\beta = \frac{1}{4}$, which is the standard form of this method, computations are stable and a value of Δt must be chosen to give results of acceptable accuracy. For the central difference method $\Delta t/T_n$ must be less than 0·318 for stability. In general, considering the response to comprise contributions from normal modes the approximations in these methods cause artificial attenuation of the response and some error in the period of the mode predicted by the numerical solution; these two effects increase as $\Delta t/T_r$ increases, where T_r is the period of the mode whose contribution is under consideration. However, for $\beta = \frac{1}{4}$ the period error occurs without the artificial attenuation. Bathe and Wilson[7] survey several methods of numerical integration and comment on stability and accuracy. For some simple examples with the Newmark method ($\beta = \frac{1}{4}$) they show that the percentage error in period is 1 and 10 per cent approximately for $\Delta t/T_r$ equal to 0·06 and 0·18, respectively. Thus a possible criterion for reasonable accuracy is $\Delta t/T_j \leq 0·06$, where T_j is the period of the highest mode making a significant contribution to the response. As mentioned in Section 2.6, an estimate for T_j requires considerable experience and judgement. However, for large problems only a fraction of the total number of modes will make a significant contribution, i.e. j/n is small; thus T_j/T_n is large and the criterion $\Delta t/T_j \leq 0·06$ may lead to a *larger* value of Δt than that given by the stability criterion for the central difference method.

As in the normal mode method the analysis has been formulated initially for excitation by applied forces, but can be used with small modifications to determine the response of a structure to displacements or accelerations prescribed at the supports or boundaries. Using the notation of Section 2.6 the constrained system has n degrees of freedom and support displacements are prescribed at N additional coordinates. The equations of motion of the complete system can be written

$$\mathbf{M'\ddot{x}' + C'\dot{x}' + K'x' = p'} \tag{2.121}$$

where the primed quantities represent matrices and vectors of order $(n + N)$. Assuming that vibration is caused solely by support excitation and expanding equation (2.121), the first n equations can be written in the matrix form

$$\mathbf{M\ddot{x} + C\dot{x} + Kx = p''} \tag{2.122}$$

where a typical element of the vector $\mathbf{p''}$ is defined by

$$p_r'' = -\sum_{j=1}^{N} (k_{r,\,n+j}x_{n+j} + c_{r,\,n+j}\dot{x}_{n+j} + m_{r,\,n+j}\ddot{x}_{n+j}), \qquad r = 1, 2, \ldots, n. \tag{2.123}$$

Equation (2.122) is of the standard form so the methods of numerical integration described above can be applied directly, provided that equation (2.123) is used to give the elements of $\mathbf{p''}$ in terms of prescribed displacements, velocities and accelerations.

The analysis presented here is for linear systems, but the numerical integration methods, unlike the normal mode method, can be used to predict the response of non-linear systems. This will not be considered further here.

Example. The example on the response of a two-degree-of-freedom system of Section 2.6 (Fig. 2.6) is to be repeated, using the Newmark method with $\beta = \frac{1}{4}$ and the central difference method, and accuracy of these solutions assessed by comparison with that from the normal mode method.

From the data given in Section 2.6 the damping matrix is zero and the stiffness and mass matrices are

$$\mathbf{K} = \begin{bmatrix} 2 & -1 \\ -1 & 1 \end{bmatrix} \times 10^7, \qquad \mathbf{M} = \begin{bmatrix} 1 & 0 \\ 0 & 1 \end{bmatrix} \times 10^4.$$

The excitation vector

$$\mathbf{p} = \begin{bmatrix} 10^7 x_0 \\ 0 \end{bmatrix}.$$

Thus

$$\mathbf{p} = \begin{bmatrix} \sin \pi t/t_0 \\ 0 \end{bmatrix} \times 10^5 \quad \text{for} \quad 0 \le t \le t_0$$

and

$$\mathbf{p} = \begin{bmatrix} 0 \\ 0 \end{bmatrix} \quad \text{for} \quad t \ge t_0.$$

Substituting in equation (2.117), putting $\beta = \frac{1}{4}$, dividing by 10^4 and using $\xi = \frac{1}{4} \times 10^3 (\Delta t)^2$, the Newmark method gives the pair of simultaneous equations

$$\begin{bmatrix} 1 + 2\xi & -\xi \\ -\xi & 1 + \xi \end{bmatrix} \begin{bmatrix} x_1 \\ x_2 \end{bmatrix}_{s+1} = \begin{bmatrix} (a_{s+1} + 2a_s + a_{s-1}) \\ 0 \end{bmatrix} \times 10^{-2} \xi$$

$$+ \begin{bmatrix} 2 - 4\xi & 2\xi \\ 2\xi & 2 - 2\xi \end{bmatrix} \begin{bmatrix} x_1 \\ x_2 \end{bmatrix}_s$$

$$- \begin{bmatrix} 1 + 2\xi & -\xi \\ -\xi & 1 + \xi \end{bmatrix} \begin{bmatrix} x_1 \\ x_2 \end{bmatrix}_{s-1} \qquad \text{(a)}$$

where a_s is the value of $\sin \pi t/t_0$ at $t = t_s$ if $t \le t_0$ and is zero for $t \ge t_0$. For this example, where $\mathbf{C} = 0$ and $\mathbf{p}_0 = 0$, equations (a) are applicable for $s = 0$. Thus successive solutions of equations (a) give the vector $\mathbf{x}$ at times Δt, $2\Delta t$, $3\Delta t \ldots$. Numerical results have been obtained for $t_0 = 0.075$ s, using $\Delta t = 0.00625$ and 0.0125 s, and for $t_0 = 0.2$ s, using $\Delta t = 0.0125$ and 0.025 s.

Substituting the above matrix expressions in equation (2.120), dividing by 10^4, the central difference method gives the uncoupled equations

$$\begin{bmatrix} x_1 \\ x_2 \end{bmatrix}_{s+1} = \begin{bmatrix} a_s \\ 0 \end{bmatrix} \times 10^{-2}\xi + \begin{bmatrix} 2 - 2\xi & \xi \\ \xi & 2 - \xi \end{bmatrix} \begin{bmatrix} x_1 \\ x_2 \end{bmatrix}_s - \begin{bmatrix} x_1 \\ x_2 \end{bmatrix}_{s-1} \quad \text{(b)}$$

where $\xi = 10^3(\Delta t)^2$ and a_s is defined above. Successive solutions of equations (b) give the vector x at times $\Delta t, 2\Delta t, 3\Delta t \dots$. For numerical stability it is essential that $\Delta t/T_2 < 0.318$, where T_2, the period of the higher mode, is equal to $1/8.14$ from Section 2.6, i.e. $\Delta t < 0.0391$ s for stability. Thus, in order to compare with the Newmark method, the same values of Δt and t_0 have been used when obtaining numerical results with the central difference method.

Response curves, obtained by the two approximate methods, are substantially similar to those in Fig. 2.7. Two methods of assessing errors will be given. For engineering problems the maximum values of displacement are of major interest and the exact form of the response curves is less significant. Thus response curves for x_1 and x_2, obtained by the normal mode, Newmark and central difference methods, have been plotted and the magnitudes of successive peaks are given in Table 2.2. For $t_0 = 0.2$ s the first four peaks are shown in Fig. 2.7a and their values are given in Table 2.2. For $t_0 = 0.075$ s, due to the substantial contributions from the second mode, there are some double peaks and the four peaks selected for inclusion in Table 2.2 are denoted by arrows in Fig. 2.7b. In compiling the table no account is taken of the fact that the approximate methods may predict peak amplitudes at slightly different times compared with the normal mode method. In general, these time shifts are very small. However, for the double peak in x_2, occurring between $t = 0.38$ and 0.5 s (Fig. 2.7b), the right-hand peak is larger according to the normal mode and central difference methods, but the Newmark method leads to a larger left-hand peak. In each case the value of the larger peak has been tabulated.

A more stringent assessment of accuracy is to compare values of x_1 and x_2 from the three methods at small time intervals. In Table 2.3 maximum errors in the two displacements predicted by the two approximate methods are given. These errors tend to increase as the

TABLE 2.2. Peak amplitudes for the two-degree-of-freedom system of Fig. 2.6 from the normal mode, Newmark and central difference methods. For the normal mode method the peaks are shown in Figs. 2.7 a and b for $t_0 = 0.2$ and 0.075 s, respectively.

t_0 (s)	Δt (s)	Peak no.	x_1 (cm) from			x_2 (cm) from		
			Normal mode	Newmark	Central difference	Normal mode	Newmark	Central difference
0·075	0·00625	1	0·88	0·87	0·88	1·31	1·30	1·31
		2	−1·04	−1·04	−1·04	−1·17	−1·15	−1·18
		3	1·11	1·10	1·11	0·91	0·87	0·93
		4	−1·07	−1·04	−1·08	−1·12	−1·12	−1·10
0·075	0·0125	1	0·88	0·87	0·89	1·31	1·29	1·32
		2	−1·04	−1·02	−1·06	−1·17	−1·05	−1·21
		3	1·11	1·06	1·12	0·91	0·96	1·00
		4	−1·07	−0·92	−1·12	−1·12	−1·19	−1·03
0·2	0·0125	1	1·31	1·28	1·34	1·96	1·96	1·96
		2	−1·23	−1·20	−1·26	−2·00	−2·00	−2·01
		3	1·15	1·18	1·16	2·06	2·00	2·06
		4	−1·20	−1·24	−1·18	−2·02	−1·97	−2·05

response time t increases and are given for three ranges of response time. Errors were computed at times Δt, $2\Delta t$, $3\Delta t$, ... and the maximum error, irrespective of sign, in the appropriate time interval is recorded in the table.

Considering the results of Tables 2.2 and 2.3, the peak values are predicted with acceptable accuracy by the two methods; the deterioration in accuracy as response time increases (a characteristic of approximate methods) is apparent. If the contribution from mode r should be $A_r \sin \omega_r t$, the Newmark method with $\beta = \frac{1}{4}$ predicts a contribution $A_r \sin (\omega_r - \delta\omega_r)t$, where $\delta\omega_r$ increases with Δt. Thus the error for this modal contribution is

$$-2A_r \cos (\omega_r - \tfrac{1}{2} \delta\omega_r)t \sin (\tfrac{1}{2} \delta\omega_r t);$$

for $\delta\omega_r t$ small, this is approximately

$$A_r \, \delta\omega_r t \cos \omega_r t$$

and is oscillating with amplitude increasing with t. For this example the central difference method is generally more accurate than the Newmark method. The restriction upon Δt to ensure numerical stability

TABLE 2.3. Maximum errors in displacement, obtained by the Newmark and central difference methods, for the two-degree-of-freedom system of Fig. 2.6.

t_0 (s)	Δt (s)	Time interval considered (s)	Maximum error (cm) for			
			x_1 from		x_2 from	
			Newmark	Central difference	Newmark	Central difference
0·075	0·00625	0 to 0·2	0·028	0·013	0·019	0·012
		0 to 0·5	0·086	0·048	0·054	0·025
		0 to 0·75	0·129	0·068	0·102	0·049
0·075	0·0125	0 to 0·2	0·101	0·053	0·073	0·048
		0 to 0·5	0·321	0·196	0·197	0·101
		0 to 0·75	0·474	0·285	0·371	0·198
0·2	0·0125	0 to 0·2	0·038	0·024	0·027	0·013
		0 to 0·5	0·081	0·049	0·075	0·052
		0 to 0·75	0·111	0·073	0·156	0·081
0·2	0·025	0 to 0·2	0·122	0·104	0·099	0·051
		0 to 0·5	0·227	0·224	0·219	0·248
		0 to 0·75	0·253	0·413	0·502	0·309

Note. The above maximum errors in x_1 and x_2 should be compared with the maximum values of x_1 and x_2, which are 1·11 and 1·32 cm, respectively, for $t_0 = 0.075$ s and 1·31 and 2·06 cm, respectively, for $t_0 = 0.2$ s.

with the central difference method is not a disadvantage here, as lower values of Δt are required to obtain results of acceptable accuracy.

Changing the excitation period t_0 from 0·2 to 0·075 s causes a noticeable decrease in accuracy. When $t_0 = 0.075$ s, but not for $t_0 = 0.2$ s, the response includes a significant contribution from the second mode (Fig. 2.7). Thus if T_j is the period of the highest mode making a significant contribution to the response:

For $t_0 = 0.075$ s, $\Delta t/T_j = 8.14\Delta t = 0.051$ and 0·102 for $\Delta t = 0.00625$ and 0·0125 s, respectively.

For $t_0 = 0.2$ s, $\Delta t/T_j = 3.11\Delta t = 0.039$ and 0·078 for $\Delta t = 0.0125$ and 0·025 s, respectively.

Thus the results of Table 2.3 confirm that $\Delta t/T_j \leq 0.06$ is a necessary criterion to give accurate response curves, but the time interval for which the response is required is important and a further criterion, possibly in terms of this time interval/T_1, is necessary. However, this cannot be formulated from the evidence of a single simple example.

The response of the two-degree-of-freedom system of Problem 5, which is subjected to a step function input, is also accurate if $\Delta t/T_j \leq 0.06$. Considering maximum errors, as defined in Table 2.3, there is a significant improvement in accuracy for any value of Δt in the range 0.025 to 0.003125 s when the damping is included in the problem.

2.9. Random Vibrations

We consider the response of the general multi-degree-of-freedom system, represented by the matrix equation (2.40), to a random disturbance. This could be a random force applied at one of the degrees of freedom or a random acceleration applied at the base of the structure. It must be emphasized that only a *single* random disturbance is considered. In previous sections of this chapter it has been possible to determine the response to multi-point excitation by introducing an appropriate summation with respect to an integer. For random multi-point excitation the response can be obtained by summation with respect to an integer, *only if* all the cross-spectral densities, and hence all the cross-correlation functions, between pairs of exciting forces are zero. The general case, including cross-correlation effects, will not be considered, but is studied for beam vibrations by Robson[66].

The concepts of random processes, introduced in Section 1.8, and the general analysis of random vibrations, equations (1.46) to (1.52) of Section 1.9, are applicable to this section. It is recalled that we consider stationary and ergodic random processes and assume Gaussian distributions. It is assumed that a random force $P_j(t)$ is applied at degree of freedom j and it has a spectral density $S_p(\omega)$. The displacement $x_s(t)$ at degree of freedom s is of interest and this has a spectral density (as yet unknown) $S_x(\omega)$. From equations (1.51) and (1.52) the spectral densities are related by

$$S_x(\omega) = |H_{sj}(\omega)|^2 S_p(\omega) \qquad (2.124)$$

and the mean square value of the response is given by

$$\overline{x_s^2(t)} = \frac{1}{2\pi} \int_0^\infty |H_{sj}(\omega)|^2 S_p(\omega) \, d\omega \qquad (2.125)$$

where $H_{sj}(\omega)$ is the complex frequency response or receptance, i.e. $H_{sj}(\omega) \exp(i\omega t)$ is the response at s to a force $\exp(i\omega t)$ at j. Series expressions for $H_{sj}(\omega)$ are obtained from equations (2.94) and (2.99) as

$$H_{sj}(\omega) = \sum_{r=1}^{n} \left[\frac{{}_r z_s \, {}_r z_j (\omega_r^2 - \omega^2 - 2i\gamma_r \omega_r \omega)}{(\omega_r^2 - \omega^2)^2 + 4\gamma_r^2 \omega_r^2 \omega^2} \right] \qquad (2.126)$$

for viscous damping and

$$H_{sj}(\omega) = \sum_{r=1}^{n} \left[\frac{{}_r z_s \, {}_r z_j (\omega_r^2 - \omega^2 - i\mu_r \omega_r^2)}{(\omega_r^2 - \omega^2)^2 + \mu_r^2 \omega_r^4} \right] \qquad (2.127)$$

for hysteretic damping. It is recalled that for viscous and hysteretic damping the damping matrices must satisfy equations (2.72) and (2.96), respectively, or extended forms of these equations. (The frequency response method of Section 2.7 gives a closed form expression in terms of the system matrices for $H_{sj}(\omega)$, but this form is less suitable for use in equation (2.125).)

In the remainder of this section structures with hysteretic damping are considered; to obtain comparable results for viscous damping $\mu_r \omega_r$ must be replaced by $2\gamma_r \omega$. Writing equation (2.127) in the form

$$H_{sj}(\omega) = \sum_{r=1}^{n} (A_r + iB_r)$$

$$|H_{sj}(\omega)|^2 = \left[\sum_{r=1}^{n} A_r \right]^2 + \left[\sum_{r=1}^{n} B_r \right]^2. \qquad (2.128)$$

Equation (2.128) cannot be simplified unless approximations are introduced, because of the terms $A_r A_q$ and $B_r B_q$, $r \neq q$. The first approximation is to neglect the latter terms in comparison with A_r^2 and B_r^2, respectively; this is justified because for small damping the major contributions to $|H_{sj}(\omega)|^2$, as ω varies, come from the terms B_r^2 when ω is near to ω_r. Ultimately we want to evaluate the integral in equation (2.125), so a good approximation must give accurate values of $|H_{sj}(\omega)|^2$ when the latter is large. With this assumption

$$|H_{sj}(\omega)|^2 = \sum_{r=1}^{n} (A_r^2 + B_r^2)$$

$$= \sum_{r=1}^{n} \left[\frac{{}_r z_s^2 \, {}_r z_j^2}{(\omega_r^2 - \omega^2)^2 + \mu_r^2 \omega_r^4} \right]. \qquad (2.129)$$

There are alternative forms for a further approximation. Firstly, the series in equation (2.129) is replaced by the dominant term, i.e. for $\omega \simeq \omega_1$, $|H_{sj}(\omega)|^2$ is approximated by the term for $r = 1$; for $\omega \simeq \omega_2$ by the term for $r = 2$, etc. For structures with well-spaced natural frequencies and low damping this gives accurate values of $|H_{sj}(\omega)|^2$ for $\omega = \omega_r$, but leads to inaccurate values when ω is well removed from a resonant value. However, for the latter ranges of ω, $|H_{sj}(\omega)|^2$ is very small compared with its value at adjacent resonances.

The alternative approximation is used when evaluating $\overline{x^2(t)}$. If $S_p(\omega)$ varies slowly, it is replaced by the value $S_p(\omega_r)$ when the rth term is considered. Thus

$$\overline{x_s^2(t)} = \frac{1}{2\pi} \sum_{r=1}^{n} \int_0^\infty \frac{S_p(\omega_r)_r z_s^2 {}_r z_j^2 \, d\omega}{(\omega_r^2 - \omega^2)^2 + \mu_r^2 \omega_r^4}$$

$$= \sum_{r=1}^{n} \left[\frac{S_p(\omega_r)_r z_s^2 {}_r z_j^2}{4\mu_r \omega_r^3} \right] \tag{2.130}$$

using equation (1.59) and assuming that $\mu_r^2 \ll 1$. It is relatively simple to obtain approximate values for the mean square response from equation (2.130). The effects of the approximations involved in equation (2.130) are illustrated by the two examples for two-degree-of-freedom systems at the end of this section.

In the analysis of this section a single random force $P_j(t)$ of spectral density $S_p(\omega)$ has been applied at degree of freedom j; by definition this force acts in the direction of the coordinate x_j. In practice. random vibration is often caused by random motion at a support. In Section 2.6 the determination of response by the normal mode method for force excitation and for support excitation has been discussed in detail.

As the analysis of this section is restricted to a single disturbance, random motion can be prescribed at only one support coordinate; for this case the analysis of Section 2.6 is simplified as the integer j has a single value and the summation with respect to j in equation (2.92) vanishes. Thus the modifications required to equations (2.124) to (2.130), if random motion at a support replaces the random force $P_j(t)$, can be inferred. (This is illustrated for a simple system in the examples at the end of the section.) When support motion is prescribed, the response is often formulated in terms of relative displacements, i.e. the displacement of each mass relative to the moving support. This procedure will not be considered here, but is discussed in Section 6.1 for plane frames.

Example 1. Considering the example of Section 2.6 (Fig. 2.6), the base EF of the structure is subjected to a random displacement x_0 with spectral density

$$S_0(\omega) = 10^{-6}(1 - 0.0001\omega^2) \text{ m}^2 \text{ s}, \quad 0 \le \omega \le 100,$$

$$S_0(\omega) = 0, \quad \omega > 100.$$

If the non-dimensional hysteretic damping factors μ_1 and μ_2 are both equal to (a) 0·01, (b) 0·1, find the mean square values of the response of the two masses, $\overline{x_1^2(t)}$ and $\overline{x_2^2(t)}$.

The equations of motion of the frame of Fig. 2.6, when the base is subjected to a harmonic displacement $X_0 \exp(i\omega t)$ and hysteretic damping is included, are

$$m_1 \ddot{x}_1 + (k_1 + ih_1)\dot{x}_1 + (k_2 + ih_2)(x_1 - x_2) = (k_1 + ih_1)X_0 \exp(i\omega t)$$

and

$$m_2 \ddot{x}_2 - (k_2 + ih_2)(x_1 - x_2) = 0 \tag{a}$$

where $m_1 = m_2 = 10^4$ kg, $k_1 = k_2 = 10$ MN/m. The hysteretic damping parameters h_1 and h_2 are expressed in terms of the non-dimensional factors μ_r as follows. The damping matrix $\mathbf{H}$ has to satisfy the uncoupling condition (2.96)

$$\mathbf{H} = a_m \mathbf{M} + a_k \mathbf{K}$$

and then the modal damping factors μ_r are given by condition (2.98) as

$$\mu_r = a_k + a_m/\omega_r^2.$$

As

$$\mu_1 = \mu_2 = \mu \text{ (say)}, \quad a_k = \mu \quad \text{and} \quad a_m = 0,$$

so

$$h_j = \mu k_j, \quad j = 1, 2.$$

Thus equations (a) become

$$m_1 \ddot{x}_1 + (k_1 + k_2)(1 + i\mu)x_1 - k_2(1 + i\mu)x_2 = k'X_0 \exp[i(\omega t + \alpha)],$$

$$m_2 \ddot{x}_2 - k_2(1 + i\mu)(x_1 - x_2) = 0$$

where $k^1 = k_1(1 + \mu^2)^{1/2}$ and $\tan \alpha = \mu$.

From equation (2.127) the complex frequency response $H_{sj}(\omega)$, relating the response at s for a force 1 exp $(i\omega t)$ at degree of freedom 1, is

$$H_{s1}(\omega) = \sum_{r=1}^{2} \left[\frac{{}_r z_{sr} z_1(\omega_r^2 - \omega^2 - i\mu\omega_r^2)}{(\omega_r^2 - \omega^2)^2 + \mu^2\omega_r^4} \right]. \tag{b}$$

The effective applied force in the equation for m_1 is of magnitude $k_1(1 + \mu^2)^{1/2}X_0$. Thus for a random displacement x_0 with spectral density $S_0(\omega)$, the spectral density of the applied force

$$S_p(\omega) = k_1^2(1 + \mu^2)S_0(\omega). \tag{c}$$

From equations (2.125) and (c) the mean square value of the response at degree of freedom s is given by

$$\overline{x_s^2(t)} = \frac{1}{2\pi} \int_0^\infty |H_{s1}(\omega)|^2 k_1^2(1 + \mu^2)S_0(\omega) \, d\omega. \tag{d}$$

Substitution of equation (b) in (d) and numerical integration, effectively from $\omega = 0$ to $\omega = \omega_c$ where ω_c is the cut-off frequency above which $S_0(\omega)$ is zero, yields the mean square response.

For this example from Section 2.6, $\omega_1 = 19.54$ and $\omega_2 = 51.17$ rad/s. Also

$$\mathbf{Z} = \begin{bmatrix} {}_1 z_1 & {}_2 z_1 \\ {}_1 z_2 & {}_2 z_2 \end{bmatrix} = \begin{bmatrix} 0.5257 & 0.8506 \\ 0.8506 & -0.5257 \end{bmatrix} \times 10^{-2}.$$

For $\mu = 0.01$ numerical integration gives

$$\overline{x_1^2(t)} = 318.7 \text{ mm}^2$$

and $\overline{x_2^2(t)} = 672.7$ mm².

For $\mu = 0.1$ numerical integration gives

$$\overline{x_1^2(t)} = 32.45 \text{ mm}^2$$

and $\overline{x_2^2(t)} = 67.32$ mm².

For comparison the mean square value of the input displacement

$$\overline{x_0^2(t)} = \frac{1}{2\pi} \int_0^\infty S_0(\omega) \, d\omega$$

$$= \frac{1}{2\pi} \int_0^{100} (1 - 0\cdot0001\omega^2) \, d\omega \ \text{mm}^2$$

$$= 10\cdot61 \ \text{mm}^2.$$

Using equations (2.130) and (c) the approximate method gives

$$\overline{x_s^2(t)} = \sum_{r=1}^{2} \frac{k_1^2(1 + \mu^2)S_0(\omega_r)_{,r}z_s^2 {}_{,r}z_1^2}{4\mu\omega_r^3} \tag{e}$$

yielding

$$\overline{x_1^2(t)} = 318\cdot3 \ \text{mm}^2 \quad \text{and} \quad \overline{x_2^2(t)} = 672\cdot1 \ \text{mm}^2 \quad \text{when} \ \mu = 0\cdot01;$$

and

$$\overline{x_1^2(t)} = 32\cdot14 \ \text{mm}^2 \quad \text{and} \quad \overline{x_2^2(t)} = 67\cdot87 \ \text{mm}^2 \quad \text{when} \ \mu = 0\cdot1.$$

For this example the mean square values for response from the approximate expression agree closely with values from the complete expression. This is to be expected as the two natural frequencies are well separated. In practice complex structures may have two or more natural frequencies that lie close together (e.g. the cylindrical shell whose response is plotted in Fig. 5.18). In the next example a system with two close natural frequencies is considered.

Example 2. Repeat Example 1 with the revised data: $m_1 = 1000$ kg, $m_2 = 2\cdot271$ kg, $k_1 = 104\cdot88$ kN/m, $k_2 = 238\cdot2$ N/m;

$$S_0(\omega) = 10^{-6}(1 - 0\cdot001\omega^2) \ \text{m}^2 \ \text{s}, \qquad 0 \le \omega \le 30;$$

$$S_0(\omega) = 0, \qquad \omega > 30.$$

Equations (a) to (d) apply to this problem. Putting $h_j = 0$ and $X_0 = 0$ in equations (a) and solving the free vibration problem,

$$\omega_1 = 10\cdot0 \quad \text{and} \quad \omega_2 = 10\cdot488 \ \text{rad/s}.$$

After normalizing the modal matrix

$$\mathbf{Z} = \begin{bmatrix} 0\cdot02209 & 0\cdot02263 \\ 0\cdot4748 & -0\cdot4636 \end{bmatrix}.$$

The mean square values of the response $\overline{x_1^2(t)}$ and $\overline{x_2^2(t)}$ are found by numerical integration from the complete expressions (b) and (d) and approximately from equation (e) for the two values of the modal damping factor 0·01 and 0·1; they are given in Table 2.4 with the mean square value of the input $\overline{x_0^2(t)}$.

TABLE 2.4. Mean square values of response and input in mm^2

μ	$\overline{x_0^2(t)}$	$\overline{x_1^2(t)}$ From equation (d)	(e)	$\overline{x_2^2(t)}$ From equation (d)	(e)
0·01	3·34	119·5	114·6	48470	50580
0·1	3·34	20·9	11·6	958	5108

Even for this example with relatively close natural frequencies (approximately 5 per cent apart) the approximate expression (e) gives response values of acceptable accuracy for $\mu = 0.01$, but it gives erroneous values when $\mu = 0.1$.

Problems

1. A frame consists of four light, elastic vertical members and two rigid horizontal members (Fig. 2.2a) and can execute swaying motion in its own plane. If $m_1 = 777$ kg, $m_2 = 518$ kg and the relevant flexural stiffness of each of the four stanchions is 740 kN/m, determine the natural frequencies and mode shapes. Check the orthogonality condition.

2. If the mass m_1 of the frame of Problem 1 is given an initial displacement, $x_1 = 0.01$ m, and the frame released from rest, obtain the equation for the resulting motion of m_1. Neglect damping.
   If m_1 is given the above initial displacement and mass m_2 is displaced initially a distance x_2, find x_2 if the resulting motion is restricted to (a) the fundamental mode, (b) the higher mode.

3. The frame of Problem 1 is subjected to a horizontal transient force:

$$P(t) = 20 \text{ kN}, \qquad 0 \le t \le 0.2 \text{ s},$$
$$P(t) = 0, \qquad\qquad t \quad > 0.2 \text{ s}.$$

   Obtain expressions for the displacements of the masses m_1 and m_2 for $t \le 0.2$ s, if the force is applied (a) to mass m_1, (b) simultaneously to m_1 and m_2 and (i) damping is neglected, (ii) there is a viscous damping force in each stanchion of 3·7 kN for a relative velocity between the ends of the stanchion of 1 m/s.

4. The mass m_1 of the frame of Problem 1 is subjected to a horizontal harmonic force

2 sin ωt kN. Obtain an expression for the steady-state amplitude of m_1 for any excitation frequency ω and determine its value when ω is equal to zero and to each of the natural frequencies, ω_1 and ω_2: (a) using the normal mode and frequency response methods with the damping specified in (ii) of Problem 3; (b) using the frequency response and complex eigenvalue methods with hysteretic damping represented by a complex stiffness $k(1 + i\mu_j)$ for each stanchion, where $\mu_j = 0.05$ for AC and BD and $\mu_j = 0.1$ for AE and BF (Fig. 2.2a).

5. Use the central difference and Newmark methods to determine the values of x_1 and x_2 at $t = 0.1$ s, if the force of Problem 3 is applied to mass m_1 for the frame of Problem 1. Consider damping conditions (i) and (ii) of Problem 3. For both methods use $\Delta t = 0.025$ s and also $\Delta t = 0.0125$ s. By comparison with the solutions to Problem 3, what are the errors in x_1 and x_2 for each case for the two methods?

6. The base EF of the frame of Problem 1 is subjected to a horizontal acceleration:

$$\ddot{x}_0(t) = 1.0 \sin (10\pi t) \text{ m/s}^2, \qquad 0 \le t \le 0.1 \text{ s},$$

$$\ddot{x}_0(t) = 0, \qquad\qquad\qquad t > 0.1 \text{ s}.$$

Derive expressions for the displacements of the masses m_1 and m_2 relative to that of the base during the time interval $0 \le t \le 0.1$ s; neglect damping.

7. Considering the frame of Fig. 2.3, in which only extensional deformation of the light vertical members AC and BD is permitted, G is the centre of mass of AB, $AG = 8$ m, $BG = 4$ m, the radius of gyration of AB about an axis through G is 3 m and the mass of AB is 100 kg. The longitudinal stiffness of each of the members AC and BD is 15 MN/m. Determine the natural frequencies and mode shapes. Check the orthogonality condition.

8. The frame of Problem 7 is subjected to a vertical harmonic force $1.0 \sin 100\pi t$ kN acting through G. Find the steady-state amplitudes of the points A and B, using (i) the frequency response method, and (ii) the normal mode method. Neglect damping.

9. If a hysteretic damping moment of $30\theta/\omega$ MN m is applied to the member AB of Fig. 2.3, where θ is the rotation of AB and ω is the excitation frequency, use the data of Problems 7 and 8 and the complex eigenvalue method to determine the steady-state amplitudes and the phase angles for the response at points A and B, if the applied force (i) is as given in Problem 8, and (ii) has an excitation frequency equal to the lower natural frequency of the system.

10. If a vertical displacement

$$x_0(t) = 0.02 \sin (100\pi t) \text{ m}, \qquad 0 \le t \le 0.01 \text{ s},$$

$$x_0(t) = 0, \qquad\qquad\qquad t > 0.01 \text{ s}$$

is imposed on the base C of the frame of Fig. 2.3, use the data of Problem 7 to determine the displacements of the points A and B when $t = 0.005, 0.0075$ and 0.01 s by (i) the normal mode method, (ii) the Newmark method, and (iii) the central difference method. Neglect damping.

11. The base EF of the multi-degree-of-freedom system of Fig. 2.5 is subjected to a random acceleration $\ddot{x}_0(t)$, which has a spectral density $S_0(\omega)$. Hysteretic damping exists in the vertical members of the structure and can be allowed for by replacing a typical stiffness k_j in Fig. 2.5 by $k_j(1 + i\mu)$, where $\mu^2 \ll 1$. Considering the displacement of mass m_s relative to the displacement of the base, i.e. $y_s = x_s - x_0$, and making the standard assumptions and approximations, show that

$$\overline{y_s^2(t)} \simeq \sum_{r=1}^{n} \frac{S_0(\omega_r)\,_r z_s^2 \left[\sum_{j=1}^{n} {}_r z_j m_j\right]^2}{4\mu\omega_r^3}$$

where the symbols have their usual meanings.

12. The member AB of the frame of Fig. 2.3 is subjected to a vertical random force at G. The spectral density of this force has a uniform value S_0 up to a cut-off frequency ω_c, which is above the higher natural frequency ω_2 (as shown in Fig. 1.13b). The stiffness of each of the vertical members, AC and BD, is k. There are viscous damping forces (not shown in Fig. 2.3) between A and C and between B and D with coefficients c. Making the assumptions used to establish equation (2.130), but allowing for the finite cut-off frequency, show that the mean square value of the vertical displacement of G is given by

$$\overline{x_1^2(t)} \simeq \sum_{r=1}^{2} \left\{ \frac{S_0 k_r z_1^4}{4c\omega_r^4} \left[1, - \frac{2}{3\pi} \frac{c\omega_r}{k(\omega_c/\omega_r)^3} \cdots \right] \right\}.$$

13. The mass m_1 in the frame of Problem 1 (Fig. 2.2a) is subjected to a horizontal, ramp function, force,

$$P(t) = 10t \text{ kN}, \qquad t \geq 0.$$

Determine the displacements of the two masses, x_1 and x_2, at time $t = 0.1$ s, using (a) the normal mode method, (b) the central difference method, and (c) the Newmark method. Use $\Delta t = 0.00625$ s in the numerical integration methods. Neglect damping.

CHAPTER 3

Vibrations of Beams—I

INITIALLY in this chapter the extensional (or longitudinal) and torsional vibrations of bars are considered but the main topic of this, and the following, chapter is the flexural vibrations of beams. Extensional and torsional vibrations are treated first, because the governing equations are simpler than that for flexural vibrations. However, flexural vibrations are of greater importance in practice, because the natural frequencies in flexure of a particular beam tend to be considerably lower than those in extension and torsion. For extension and torsion only free undamped vibrations are considered. For flexure natural frequencies and the response to applied forces for uniform and non-uniform beams are obtained. The three types of deformation are considered separately in this chapter, but beam systems, where coupled motion is possible, are analysed in Section 4.5. Other specialized problems, relating to flexural vibrations, are discussed in Chapter 4. In general, the analysis is applicable only to slender beams, for which the effects of rotatory inertia and transverse shear deformation are negligible, but these effects are considered in Section 4.6.

For non-uniform beams approximate methods of analysis are usually necessary. From the many available methods the Rayleigh–Ritz and finite element methods have been chosen, because, in addition to their suitability for beam problems, they can be used for more complex elastic bodies, such as plates and shells in Chapter 5. It will be shown that with these methods elastic bodies, which possess an infinite number of degrees of freedom, are replaced effectively by approximate multi-degree-of-freedom systems, governed by the equations which have been studied in Chapter 2. (Although the finite difference method can also be used to provide approximate mathematical models of complex structures, it has not been included as in its conventional form it leads to matrix equation (2.40) with an asymmetric stiffness matrix **K** and the analysis of most of Chapter 2 depends upon **K** being symmetric.)

3.1. Extensional Vibrations

Throughout this chapter it is assumed that the material is elastic, homogeneous and isotropic. For extensional vibrations it is assumed that cross-sections, which are initially plane and perpendicular to the axis of the bar, remain plane and perpendicular to that axis and that the normal stress in the axial direction is the only component of stress. The axis of the bar in Fig. 3.1 coincides with the X-axis; the displacement at any section x at time t is denoted by u. (In order to maintain a consistent notation in this and subsequent chapters, where the vibrations of plates and shells are considered, u, v and w are assumed to be the components of displacement at any point in the X-, Y-, and Z-directions, respectively. In general, u, v and w are functions of the three space coordinates x, y and z and of the time t. For extensional vibrations of bars only the component of displacement u is of interest and it is a function of x and t only.)

Considering the element of the bar between sections x and $(x + dx)$, if P is the axial force at section x (positive when tensile), the axial force at section $(x + dx)$ is $P + (\partial P/\partial x)\, dx$. Also the displacement at section $x + dx$ is $u + (\partial u/\partial x)\, dx$. Thus the axial strain

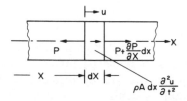

FIG. 3.1. Element of bar for extensional vibrations.

ε_x = increase of length/original length of the element

$$= \frac{\partial u}{\partial x} \, dx / dx$$

$$= \frac{\partial u}{\partial x}.$$

From Hooke's law, the axial stress $\sigma_x = E\varepsilon_x$, where E is Young's modulus. Also $\sigma_x = P/A$ where A is the cross-sectional area of the bar. Combining these relations,

$$P = AE \frac{\partial u}{\partial x}. \tag{3.1}$$

Considering the dynamic equilibrium in the X-direction of the element dx,

$$P + \frac{\partial P}{\partial x} \, dx - P = \rho A \, dx \frac{\partial^2 u}{\partial t^2}$$

where ρ = density, i.e.

$$\frac{\partial P}{\partial x} = \rho A \frac{\partial^2 u}{\partial t^2}. \tag{3.2}$$

Eliminating P between equations (3.1) and (3.2),

$$AE \frac{\partial^2 u}{\partial x^2} = \rho A \frac{\partial^2 u}{\partial t^2},$$

i.e.

$$C_1^2 \frac{\partial^2 u}{\partial x^2} = \frac{\partial^2 u}{\partial t^2} \tag{3.3}$$

where $C_1 = (E/\rho)^{1/2}$ and is the velocity of propagation of extensional waves in the bar. For free vibrations it is assumed that the displacement is a harmonic function of time, i.e.

$$u(x, t) = U(x) \sin (\omega t + \alpha).$$

Substituting in equation (3.3), the auxiliary equation for U is

$$C_1^2 \frac{d^2 U}{dx^2} + \omega^2 U = 0$$

with solution

$$U = B_1 \sin \frac{\omega x}{C_1} + B_2 \cos \frac{\omega x}{C_1}. \qquad (3.4)$$

It is noted that equation (3.3) and its solution (3.4) are independent of the cross-sectional dimensions of the bar.

End conditions. The standard simple end conditions are:
(a) Free, for which $P = 0$, i.e. $dU/dx = 0$ from equation (3.1),
(b) Clamped, for which $U = 0$.

The frequency equations and natural frequencies will be determined now for some particular end conditions.

Both ends clamped. For this bar the end conditions are $U = 0$ at $x = 0$ and at $x = l$. Substituting in equation (3.4),

$$0 = B_2$$

and

$$0 = B_1 \sin \frac{\omega l}{C_1}.$$

The only non-trivial solution is $\sin \omega l/C_1 = 0$, which leads to

$$\frac{\omega l}{C_1} = n\pi, \qquad n = 1, 2, 3, \ldots$$

or

$$\omega_n = \frac{n\pi}{l}\left(\frac{E}{\rho}\right)^{1/2}, \qquad n = 1, 2, 3, \ldots. \qquad (3.5)$$

The complete solution is:

$$u(x, t) = \sum_n B_n \sin \frac{n\pi x}{l} \sin \left(\frac{n\pi C_1 t}{l} + \alpha_n\right). \qquad (3.6)$$

The constants B_n and α_n are obtained from the initial conditions, i.e. the values of u and $\partial u/\partial t$ at $t = 0$.

Bar clamped at $x = 0$, free at $x = l$.

From $U = 0$ at $x = 0$, $\qquad 0 = B_2$.

From $P = 0$ at $x = l$, $\qquad \dfrac{\omega}{C_1} B_1 \cos \dfrac{\omega l}{C_1} = 0$.

Thus

$$\frac{\omega l}{C_1} = (2n - 1)\pi/2, \qquad n = 1, 2, 3, \ldots$$

and the natural frequencies are given by

$$\omega_n = \frac{(2n - 1)\pi}{2l} \left(\frac{E}{\rho}\right)^{1/2}, \qquad n = 1, 2, 3, \ldots. \tag{3.7}$$

The complete solution is:

$$u(x, t) = \sum_n B_n \sin \frac{(2n - 1)\pi x}{2l} \sin \left[\frac{(2n - 1)\pi C_1 t}{2l} + \alpha_n\right]. \tag{3.8}$$

As an illustration of the determination of B_n and α_n for specified initial conditions, we assume that the above bar is subjected to a displacement u_0 at the free end and released from this position at $t = 0$. Thus the initial conditions are:

$$u(x, 0) = u_0 x/l,$$

$$\frac{\partial u}{\partial t}(x, 0) = 0.$$

From the latter condition

$$\sum_n \frac{(2n - 1)\pi C_1}{2l} B_n \sin \frac{(2n - 1)\pi x}{2l} \cos \alpha_n = 0$$

for all values of x; thus $\cos \alpha_n = 0$ and $\alpha_n = \pi/2$. From the former condition

$$\frac{u_0 x}{l} = \sum_n B_n \sin \frac{(2n - 1)\pi x}{2l}.$$

Multiplying by $\sin [(2j - 1)\pi x/2l]$ and integrating with respect to x from 0 to l,

$$\int_0^l \frac{u_0 x}{l} \sin \frac{(2j - 1)\pi x}{2l} dx = \frac{l}{2} B_j$$

as

$$\int_0^l \sin \frac{(2j-1)\pi x}{2l} \sin \frac{(2n-1)\pi x}{2l} \, dx = 0 \text{ for } j \neq n,$$

$$= \frac{l}{2} \text{ for } j = n.$$

Integrating by parts,

$$\frac{l}{2} B_n = \left| -\frac{u_0 x}{l} \frac{2l}{(2n-1)\pi} \cos \frac{(2n-1)\pi x}{2l} \right.$$

$$\left. + \frac{u_0}{l} \frac{4l^2}{(2n-1)^2\pi^2} \sin \frac{(2n-1)\pi x}{2l} \right|_0^l$$

$$= \frac{4u_0 l(-1)^{n+1}}{(2n-1)^2\pi^2}.$$

Thus for this set of initial conditions the complete solution is

$$u(x, t) = \frac{8u_0}{\pi^2} \sum_n \frac{(-1)^{n+1}}{(2n-1)^2} \sin \frac{(2n-1)\pi x}{2l} \cos \frac{(2n-1)\pi C_1 t}{2l}.$$

3.2. Torsional Vibrations

Initially it is assumed that the cross-section of the bar is circular, cross-sections of the bar initially perpendicular to the axis remain plane and radii remain straight during deformation. Then from elementary elasticity theory the torque–twist relation is

$$T = GJ \frac{\partial \theta}{\partial x} \tag{3.9}$$

where T is the torque at section x, θ is the twist, or angular rotation about the X-axis, at section x, G is the shear modulus and J is the polar second moment of area. Considering the dynamic equilibrium of an element of the bar between sections x and $x + dx$,

$$\left(T + \frac{\partial T}{\partial x} \, dx \right) - T = \rho J \, dx \frac{\partial^2 \theta}{\partial t^2},$$

i.e.

$$\frac{\partial T}{\partial x} = \rho J \frac{\partial^2 \theta}{\partial t^2}.$$

Thus

$$\frac{\partial^2 \theta}{\partial t^2} = C_2^2 \frac{\partial^2 \theta}{\partial x^2} \tag{3.10}$$

where $C_2 = (G/\rho)^{1/2}$ and is the velocity of propagation of shear waves in the material. The similarity of equations (3.3) and (3.10) is apparent. Thus the solution is

$$\theta(x,\,t) = \left(B_1 \sin \frac{\omega x}{C_2} + B_2 \cos \frac{\omega x}{C_2}\right) \sin{(\omega t + \alpha)}. \tag{3.11}$$

The standard simple end conditions are free, for which $T = 0$ and hence $\partial \theta/\partial x = 0$, and clamped, for which $\theta = 0$. For specified end conditions the frequency equations and natural frequencies are similar to those in Section 3.1 with G replacing E.

For non-circular cross-sections the torque–twist relation can be expressed as

$$T = \kappa G J \frac{\partial \theta}{\partial x}$$

where κ is a numerical factor depending upon the cross-sectional dimensions. The only change in the analysis is that C_2 is redefined as $(\kappa G/\rho)^{1/2}$. Thus for non-circular cross-sections the solution depends upon the cross-sectional dimensions. Timoshenko and Goodier[73] give an extensive treatment of the torsion of non-circular bars and include data from which the factor κ can be determined.

3.3. Flexural Vibrations: Natural Frequencies of Uniform Beams

Equation of motion and its solution. In deriving the equation governing free undamped vibrations in flexure of beams it is assumed that vibration occurs in one of the principal planes of the beam. The effects of rotatory inertia and of transverse shear deformation are

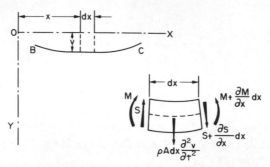

FIG. 3.2. Element of beam in flexure and coordinate axes.

neglected. In Fig. 3.2 BC represents the centre line of the beam during vibration; the displacement at any section x at time t is denoted by v. Gravity forces will be neglected by measuring the displacement from the position of static equilibrium of the beam. The forces and moments on an element of length dx are shown also in Fig. 3.2; S and M are the shear force and bending moment at section x; the inertia force on the element is $\rho A \, dx \, \partial^2 v / \partial t^2$, where ρ is the density of the material of the beam and A is its cross-sectional area. Taking moments about the centre line of the element (neglecting products of small quantities), and resolving for forces in the Y-direction,

$$S \, dx + M - \left(M + \frac{\partial M}{\partial x} \, dx\right) = 0$$

or

$$S = \frac{\partial M}{\partial x}$$

and

$$\frac{\partial S}{\partial x} = \rho A \frac{\partial^2 v}{\partial t^2}. \tag{3.12}$$

From the relation between bending moment and curvature and the approximate curvature–displacement relation, used in determining static deflections of beams,

$$M = -EI \frac{\partial^2 v}{\partial x^2} \qquad (3.13)$$

where E is Young's modulus and I is the relevant second moment of area of the cross-section. Combining these three equations,

$$\frac{\partial^2}{\partial x^2}\left(-EI \frac{\partial^2 v}{\partial x^2}\right) = \rho A \frac{\partial^2 v}{\partial t^2}. \qquad (3.14)$$

Equation (3.14) can be used for uniform and non-uniform beams; for the latter the flexural rigidity EI and the mass per unit length ρA are functions of the coordinate x. For a beam of uniform cross-section, equation (3.14) reduces to

$$EI \frac{\partial^4 v}{\partial x^4} + \rho A \frac{\partial^2 v}{\partial t^2} = 0. \qquad (3.15)$$

For free vibrations, $v(x, t)$ must be a harmonic function of time, i.e.

$$v(x, t) = V(x) \sin (\omega t + \alpha). \qquad (3.16)$$

Substituting equation (3.16) in (3.15)

$$\frac{d^4 V}{dx^4} - \frac{\rho A \omega^2}{EI} V = 0. \qquad (3.17)$$

A solution of equation (3.17) of the form

$$V = B \exp (\lambda_0 x)$$

is satisfactory, if $\lambda_0^4 = \rho A \omega^2 / EI$. This has the four roots $\lambda_0 = \pm \lambda$, $\lambda_0 = \pm i\lambda$, where $\lambda = (\rho A \omega^2 / EI)^{1/4}$, so the general solution is

$$V = B_1 \sin \lambda x + B_2 \cos \lambda x + B_3 \sinh \lambda x + B_4 \cosh \lambda x. \qquad (3.18)$$

End conditions. The four constants are determined from the end conditions; the standard end conditions are (a) simply supported or pinned, for which the displacement is zero and the bending moment is zero as there is no rotational constraint; (b) fixed or clamped, for which the displacement and slope are zero; (c) free, for which the bending moment and shear force are zero. In terms of the function $V(x)$ these conditions for a uniform beam are:

D.B.S.—E

(a) Simply supported:

$$V = 0 \quad \text{and} \quad d^2V/dx^2 = 0. \tag{3.19}$$

(b) Clamped:

$$V = 0 \quad \text{and} \quad dV/dx = 0. \tag{3.20}$$

(c) Free:

$$d^2V/dx^2 = 0 \quad \text{and} \quad d^3V/dx^3 = 0. \tag{3.21}$$

The natural frequencies will be determined now for some particular end conditions.

Both ends simply supported. For this beam, shown in Fig. 3.3, the

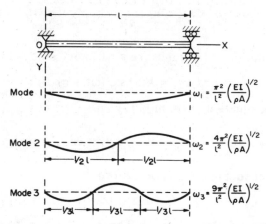

FIG. 3.3. Natural frequencies and mode shapes of a uniform simply supported beam.

end conditions are $V = 0$ and $d^2V/dx^2 = 0$ at $x = 0$ and at $x = l$. Substituting the conditions at $x = 0$ in equation (3.18)

$$0 = B_2 + B_4$$

and

$$0 = -\lambda^2 B_2 + \lambda^2 B_4 \,.$$

Thus $B_2 = B_4 = 0$.

Applying the conditions at $x = l$,

$$0 = B_1 \sin \lambda l + B_3 \sinh \lambda l$$

and

$$0 = -\lambda^2 B_1 \sin \lambda l + \lambda^2 B_3 \sinh \lambda l.$$

The only non-trivial solution is $B_3 = 0$ and $\sin \lambda l = 0$ leading to

$$\lambda l = n\pi, \qquad n = 1, 2, 3, \ldots$$

or

$$\omega_n^2 = \left(\frac{n\pi}{l}\right)^4 \frac{EI}{\rho A}, \qquad n = 1, 2, 3, \ldots. \tag{3.22}$$

Thus for a uniform beam with simply supported ends the modes of vibration are given by:

$$v(x, t) = B_n \sin \frac{n\pi x}{l} \sin (\omega_n t + \alpha_n) \tag{3.23}$$

with the natural frequencies, $\omega_n/2\pi$, obtained from equation (3.22). The first three modes of vibration, corresponding to $n = 1, 2$ and 3, and the associated natural frequencies, are given in Fig. 3.3.

Cantilever. With the origin at the fixed end, as in Fig. 3.4, and using equations (3.20) and (3.21), the end conditions are:

At $x = 0$, $V = 0$, i.e. $0 = B_2 + B_4$.

At $x = 0$, $\dfrac{dV}{dx} = 0$, i.e. $0 = \lambda B_1 + \lambda B_3$.

At $x = l$, $\dfrac{d^2 V}{dx^2} = 0$, i.e. $0 = -\lambda^2 B_1 \sin \lambda l - \lambda^2 B_2 \cos \lambda l + \lambda^2 B_3 \sinh \lambda l$

$$+ \lambda^2 B_4 \cosh \lambda l.$$

At $x = l$, $\dfrac{d^3 V}{dx^3} = 0$, i.e. $0 = -\lambda^3 B_1 \cos \lambda l + \lambda^3 B_2 \sin \lambda l + \lambda^3 B_3 \cosh \lambda l$

$$+ \lambda^3 B_4 \sinh \lambda l.$$

Hence

$$B_1(-\sin \lambda l - \sinh \lambda l) + B_2(-\cos \lambda l - \cosh \lambda l) = 0$$

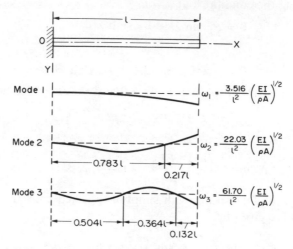

FIG. 3.4. Natural frequencies and mode shapes of a uniform cantilever.

and

$$B_1(-\cos \lambda l - \cosh \lambda l) + B_2(\sin \lambda l - \sinh \lambda l) = 0.$$

Eliminating B_1/B_2, the frequency equation is

$$(\sin \lambda l + \sinh \lambda l)(\sinh \lambda l - \sin \lambda l) - (\cos \lambda l + \cosh \lambda l)^2 = 0,$$

i.e.

$$\cos \lambda l \cosh \lambda l + 1 = 0 \tag{3.24}$$

(noting that $\cos^2 \lambda l + \sin^2 \lambda l = 1$ and $\cosh^2 \lambda l - \sinh^2 \lambda l = 1$). The successive roots $\lambda_1, \lambda_2, \lambda_3, \ldots$ of equation (3.24), from which the natural frequencies can be obtained, are given by $\lambda_1 l = 1 \cdot 875$, $\lambda_2 l = 4 \cdot 694$, $\lambda_3 l = 7 \cdot 855$, $\lambda_r l \simeq (r - \frac{1}{2})\pi$ for $r \geq 4$; and $\omega_n = \lambda_n^2 (EI/\rho A)^{1/2}$. The shape of the rth mode, in terms of a single arbitrary constant, is

$$V_r(x) = B_r[\cosh \lambda_r x - \cos \lambda_r x - \eta_r(\sinh \lambda_r x - \sin \lambda_r x)]$$

where

$$\eta_r = \frac{\cos \lambda_r l + \cosh \lambda_r l}{\sin \lambda_r l + \sinh \lambda_r l}, \qquad r = 1, 2, 3, \ldots. \tag{3.25}$$

The first three modes of vibration, and the associated natural frequencies, are given in Fig. 3.4.

Both ends clamped. Following the above procedure and using the appropriate end conditions, it can be shown that the frequency equation is

$$\cos \lambda l \cosh \lambda l - 1 = 0 \qquad (3.26)$$

with roots: $\lambda_1 l = 4\cdot730$, $\lambda_2 l = 7\cdot853$, $\lambda_r l \simeq (r + \frac{1}{2})\pi$ for $r \geq 3$.

For several end conditions, including those given above, the natural frequencies and the mode shapes for the first few modes have been tabulated[9, 10].

General End Conditions

In addition to the simple end conditions previously defined more complex conditions are possible, where such effects as lumped masses, and springs giving linear and rotational restraint are included. We consider a general end condition at the left-hand end of a beam, $x = 0$, where there exists (Fig. 3.5):

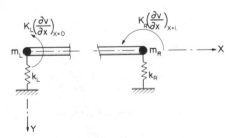

FIG. 3.5. General end conditions.

   (i) a concentrated mass m_L,
   (ii) a linear spring of stiffness k_L, opposing transverse displacement,
   (iii) a rotational spring of stiffness K_L, opposing rotation.
Quantities m_R, k_R and K_R are defined similarly for the right-hand end of the beam, $x = l$.

Recalling that $M = -EI(\partial^2 v/\partial x^2)$ and $S = -EI(\partial^3 v/\partial x^3)$ and the sign convention for bending moments and shear forces (Fig. 3.2), the end conditions are:

At $x = 0$,

$$-EI\left(\frac{\partial^2 v}{\partial x^2}\right)_{x=0} = -K_L\left(\frac{\partial v}{\partial x}\right)_{x=0} \tag{3.27}$$

and

$$-EI\left(\frac{\partial^3 v}{\partial x^3}\right)_{x=0} = k_L v_{x=0} + m_L\left(\frac{\partial^2 v}{\partial t^2}\right)_{x=0}. \tag{3.28}$$

At $x = l$,

$$-EI\left(\frac{\partial^2 v}{\partial x^2}\right)_{x=l} = K_R\left(\frac{\partial v}{\partial x}\right)_{x=l} \tag{3.29}$$

and

$$-EI\left(\frac{\partial^3 v}{\partial x^3}\right)_{x=l} = -k_R v_{x=l} - m_R\left(\frac{\partial^2 v}{\partial t^2}\right)_{x=l}. \tag{3.30}$$

Example. Determine the frequency equation for the uniform propped cantilever beam shown in Fig. 3.6. If $kl^3/EI = 5$, find the value of λl corresponding to the fundamental natural frequency.

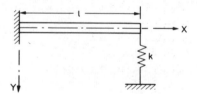

FIG. 3.6. Propped cantilever.

From equation (3.18)

$$V = B_1 \sin \lambda x + B_2 \cos \lambda x + B_3 \sinh \lambda x + B_4 \cosh \lambda x.$$

At the fixed end, $x = 0$, $V = 0$; thus

$$0 = B_2 + B_4.$$

Also from

$$dV/dx = 0, \qquad 0 = \lambda B_1 + \lambda B_3.$$

At $x = l$, the bending moment is zero; thus

$$0 = -EI\lambda^2(-B_1 \sin \lambda l - B_2 \cos \lambda l + B_3 \sinh \lambda l + B_4 \cosh \lambda l).$$

From equation (3.30), putting $k_R = k$ and $m_R = 0$,

$$-EI\lambda^3(-B_1 \cos \lambda l + B_2 \sin \lambda l + B_3 \cosh \lambda l + B_4 \sinh \lambda l)$$
$$= -k(B_1 \sin \lambda l + B_2 \cos \lambda l + B_3 \sinh \lambda l + B_4 \cosh \lambda l).$$

Eliminating the constants B_i from the above four equations yields the frequency equation

$$\frac{EI(\lambda l)^3}{kl^3}(1 + \cos \lambda l \cosh \lambda l) + \sin \lambda l \cosh \lambda l - \cos \lambda l \sinh \lambda l = 0.$$

$$(3.31)$$

For a specified value of EI/kl^3 successive roots λl of equation (3.31) give the natural frequencies, recalling that $(\lambda l)^4 = \rho A l^4 \omega^2/EI$. (It is noted that putting $k = 0$ and ∞ gives the frequency equations for the cantilever beam and the clamped-simply supported beam respectively.)

The first root of equation (3.31) will lie between 1·875 and 3·926, the values fo $\lambda_1 l$ for $k = 0$ and ∞, respectively. For $kl^3/EI = 5$, successive trial values of λl lead to the root

$$\lambda_1 l = 2·3668.$$

Orthogonality

It can be shown that mode shapes are orthogonal. For a uniform beam

$$\int_0^l V_r(x)V_s(x)\,dx = 0, \qquad r \neq s \qquad (3.32)$$

provided that there is not a concentrated mass at either end of the beam. Equation (3.32) is valid for all other end conditions. The analogy between orthogonality condition (3.32) and that for multi-degree-of-freedom systems [equation (2.52)] is seen. The orthogonality conditions for uniform and nonuniform beams with simple and general end conditions are derived in Appendix 3.

The natural frequencies and mode shapes, obtained in this section, will be used later to determine the dynamic response of beams. It should be noted that, as rotatory inertia and deformation due to shear have been neglected, this analysis is restricted to slender beams. The effects of rotatory inertia and shear deformation upon natural frequencies are considered in Section 4.6, where it is shown that natural frequencies, obtained by the method of this section, are reasonably accurate, provided that $l/nK_r > 20$, where K_r is the radius of gyration of the cross-section and n is the mode number.

3.4. Response of a Beam to an Applied Force: General Theory

If an applied force, which is a function of time $f(t)$ and acts in the Y-direction, is distributed along the length of the beam, the equation of motion for the element of the beam (Fig. 3.2) will include an additional term $p(x)f(t)$, equal to the applied force per unit length, added to equation (3.14), i.e.

$$\rho A \frac{\partial^2 v}{\partial t^2} + \frac{\partial^2}{\partial x^2}\left[EI\frac{\partial^2 v}{\partial x^2}\right] = p(x)f(t). \tag{3.33}$$

This is valid for uniform and non-uniform beams. In equation (3.33) it is assumed that the applied force varies with time in the same way for all points on the beam. Before solving equation (3.33) it will be modified to include damping.

For the elastic beam the stress σ_x and the strain ε_x in the longitudinal direction at a point distant y from the centre line are related by Hooke's law, i.e.

$$\sigma_x = E\varepsilon_x.$$

In order to include damping we consider a viscoelastic material with the stress–strain relation:

$$\sigma_x = E\left(\varepsilon_x + c\frac{\partial \varepsilon_x}{\partial t}\right). \tag{3.34}$$

The strain in a fibre distance y from the neutral axis is

$$\varepsilon_x = y/R$$

$$= -y \, \partial^2 v/\partial x^2$$

where R is the radius of curvature.

The relation between bending moment M and stress σ_x is

$$M = \int \sigma_x b(y) y \, dy$$

where $b(y)$ is the breadth of the beam at y and the integration is performed over the depth. Thus the effect of including damping is to modify the bending moment–curvature relation from equation (3.13) to

$$M = -EI \left[\frac{\partial^2 v}{\partial x^2} + c \frac{\partial^3 v}{\partial x^2 \, \partial t} \right] \tag{3.35}$$

noting that the second moment of area

$$I = \int b(y) y^2 \, dy.$$

Using expression (3.35) in the equilibrium equation for an element of the beam, the equation of motion becomes

$$\rho A \frac{\partial^2 v}{\partial t^2} + \frac{\partial^2}{\partial x^2} \left[EI \left(\frac{\partial^2 v}{\partial x^2} + c \frac{\partial^3 v}{\partial x^2 \, \partial t} \right) \right] = p(x) f(t). \tag{3.36}$$

A solution of equation (3.36) will be sought in terms of the normal modes, $\phi_r(x)$; each mode shape is associated with a natural frequency ω_r and

$$v = \sum_r B_r \phi_r(x) \sin(\omega_r t + \alpha_r)$$

is the general solution for *free vibrations* of the beam, i.e. the solution of equation (3.14), or of equation (3.36) with the right-hand side equal to zero and $c = 0$. The general solution of equation (3.36) will be of the form

$$v = \sum_r \phi_r(x) q_r(t) \tag{3.37}$$

where $q_r(t)$, a principal coordinate, is a function of time. Substituting for v from equation (3.37) in equation (3.36), multiplying by ϕ_s and integrating with respect to x over the length of the beam,

$$\int_0^l \rho A \phi_s \sum_r (\phi_r \ddot{q}_r) \, dx + \int_0^l \phi_s \frac{d^2}{dx^2} \left[EI \sum_r \left(\frac{d^2\phi_r}{dx^2} q_r + c \frac{d^2\phi_r}{dx^2} \dot{q}_r \right) \right] dx$$

$$= \int_0^l p(x) \phi_s f(t) \, dx. \tag{3.38}$$

From the orthogonal relations, equations (A3.7) and (A3.9),

$$\int_0^l \rho A \phi_r \phi_s \, dx = 0,$$

$$\int_0^l \phi_s \frac{d^2}{dx^2} \left[EI \frac{d^2\phi_r}{dx^2} \right] dx = 0, \qquad r \neq s.$$

Also, from equation (A3.14),

$$\int_0^l \phi_s \frac{d^2}{dx^2} \left[EI \frac{d^2\phi_s}{dx^2} \right] dx = \omega_s^2 \int_0^l \rho A \phi_s^2 \, dx$$

so that equation (3.38) becomes

$$\ddot{q}_s + c\omega_s^2 \dot{q}_s + \omega_s^2 q_s = \frac{f(t) \int_0^l p(x) \phi_s(x) \, dx}{\int_0^l \rho A [\phi_s(x)]^2 \, dx} \tag{3.39}$$

For a known distribution of applied forces $p(x)$ and known mode shapes the integrals in equation (3.39) can be evaluated; if their ratio is K_s, then

$$\ddot{q}_s + c\omega_s^2 \dot{q}_s + \omega_s^2 q_s = K_s f(t). \tag{3.40}$$

Equation (3.40) is of the same form as equation (1.32) so that if the beam is at rest in its equilibrium position at $t = 0$ the solution is given by the Duhamel integral [equation (1.36)] as

$$q_s = \frac{K_s}{\omega_s'} \int_0^t f(\tau) \exp \left[-\tfrac{1}{2} c\omega_s^2 (t - \tau) \right]$$

$$\times \sin \omega_s'(t - \tau) \, d\tau, \qquad s = 1, 2, 3 \dots \tag{3.41}$$

where $\omega_s' = \omega_s [1 - (\tfrac{1}{2} c\omega_s)^2]^{1/2}$. If the variation of the applied force with time is given, the principal coordinates q_s may be determined from equation (3.41), using numerical integration techniques if necessary;

then the complete dynamic response is found by substituting in equation (3.37).

The damping constant c has the dimensions of time, as shown by the definition (3.34). A non-dimensional modal damping parameter γ_s can be defined by the relation

$$\gamma_s = \tfrac{1}{2}c\omega_s .$$

Critical damping occurs when $\gamma_s = 1$. This damping parameter γ_s is analogous to that used in Chapter 2 with $\lambda_m = 0$ and increases with mode number s.

If the applied force is located at one point along the beam as in

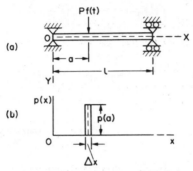

FIG. 3.7. (a) Beam with an applied force at $x = a$. (b) Representation of this force required in deriving equation (3.42).

Fig. 3.7a, where a force $Pf(t)$ acts at $x = a$, the distribution of $p(x)$ is shown in Fig. 3.7b and $p(a) . \Delta x \to P$ as $\Delta x \to 0$. Thus the upper integral in equation (3.39) becomes $P\phi_s(a)$, and

$$\ddot{q}_s + c\omega_s^2\dot{q}_s + \omega_s^2 q_s = \frac{P\phi_s(a)f(t)}{\displaystyle\int_0^l \rho A[\phi_s(x)]^2 \, dx} . \tag{3.42}$$

If the mode shapes are normalized so that

$$\int_0^l \rho A[\phi_s(x)]^2 \, dx = m, \qquad s = 1, 2, 3 \dots \tag{3.43}$$

where m is the mass of the beam, then for an applied force $Pf(t)$ at $x = a$,

$$\ddot{q}_s + c\omega_s^2\dot{q}_s + \omega_s^2 q_s = \frac{P\phi_s(a)f(t)}{m}. \tag{3.44}$$

(It will be noted that for the special case of a uniform beam the normalizing condition (3.43) reduces to

$$\int_0^l [\phi_s(x)]^2 \, dx = l$$

where l is the length of the beam, as in this case $m = \rho Al$.) The response is given by

$$q_s = \frac{P\phi_s(a)}{m\omega_s'} \int_0^t f(t) \exp\left[-\tfrac{1}{2}c\omega_s^2(t - \tau)\right] \sin \omega_s'(t - \tau) \, d\tau \tag{3.45}$$

and

$$v(x, t) = \sum_s \left[\frac{P\phi_s(a)\phi_s(x)}{m\omega_s'} \int_0^t f(\tau) \exp\left\{-\tfrac{1}{2}c\omega_s^2(t - \tau)\right\} \sin \omega_s'(t - \tau) \, d\tau\right]. \tag{3.46}$$

The theory given above will be applied to some specific problems in the next two sections.

3.5. Response of a Uniform Beam to a Harmonic Applied Force

If a uniform beam is subjected to a harmonic applied force $P \sin \omega t$ at $x = a$ (Fig. 3.7), equation (3.44) becomes

$$\ddot{q}_s + c\omega_s^2\dot{q}_s + \omega_s^2 q_s = P\phi_s(a)(\sin \omega t)/m. \tag{3.47}$$

Although the complete solution can be found from equation (3.45), it is simpler to take advantage of the fact that the particular integral is known in this case. The complete solution consists of the complementary function and particular integral; by analogy with solution (1.21) for equation (1.14) it is

$$q_s = \exp\left(-\tfrac{1}{2}c\omega_s^2 t\right)(B_s \sin \omega_s' t + D_s \cos \omega_s' t)$$

$$+ \frac{P\phi_s(a) \sin (\omega t - \alpha_s)}{m[(\omega_s^2 - \omega^2)^2 + (c\omega_s^2\omega)^2]^{1/2}} \tag{3.48}$$

where $\tan \alpha_s = c\omega_s^2\omega/(\omega_s^2 - \omega^2)$. The constants B_s and D_s depend upon the initial conditions. The free damped vibration terms decrease gradually and are neglected when determining the steady-state response; from equations (3.37) and (3.48) the latter is

$$v(x, t) = \sum_s \left\{ \frac{P\phi_s(x)\phi_s(a) \sin (\omega t - \alpha_s)}{m[(\omega_s^2 - \omega^2)^2 + (c\omega_s^2\omega)^2]^{1/2}} \right\}. \tag{3.49}$$

For light damping the response at resonance, i.e. when the excitation frequency ω is equal to ω_r, one of the natural frequencies ω_s, is dominated by the contribution to $v(x, t)$ from the resonant mode r. This assumes that $(\omega_s^2 - \omega_r^2)$ for $s \neq r$ is large compared with $c\omega_r^3$ and that $\phi_r(x)$ and $\phi_r(a)$ are non-zero. With these assumptions the amplitude $V(x)$ at the resonant frequency $\omega = \omega_r$ is:

$$V(x) \simeq \frac{P\phi_r(x)\phi_r(a)}{mc\omega_r^3}. \tag{3.50}$$

(Equations (3.49) and (3.50) are of similar form to equations (2.94) and (2.100), respectively, for the harmonic response of multi-degree-of-freedom systems.) Equation (3.50) predicts that the resonant peaks, associated with $r = 1, 2, 3 \ldots$, decrease rapidly in magnitude because the term ω_r^3 occurs in the denominator. (For example, for a simply supported beam $\omega_r^3 \propto r^6$ from equation (3.22).) Although increased information concerning damping in practical structures is required, it is known that this damping mechanism underestimates the resonant amplitudes associated with higher modes, because the modal damping parameter, γ_r $(= \frac{1}{2}c\omega_r)$, increases with the mode number.

We consider now hysteretic damping (discussed for single- and multi-degree-of-freedom systems in Sections 1.6 and 2.7), which leads to a better representation of the dynamic behaviour of elastic bodies. For hysteretic damping equation (3.34) for a viscoelastic material is replaced by

$$\sigma_x = E\left(\varepsilon_x + \frac{\mu}{\omega}\frac{\partial \varepsilon_x}{\partial t}\right) \tag{3.51}$$

where μ is the hysteretic damping constant and ω is the excitation frequency. (The introduction of ω in equation (3.51) is essential to obtain the improved dynamic representation, but cannot be made easily in the general analysis of Section 3.4, where no excitation frequency is prescribed.) As we are concerned here with the steady-state response to harmonic excitation, all physical quantities can be represented as functions of time by exp $(i\omega t)$. Thus $\partial \varepsilon_x / \partial t = i\omega \varepsilon_x$ and from equation (3.51)

$$\sigma_x = E(1 + i\mu)\varepsilon_x . \tag{3.52}$$

Equation (3.52) is the basis of the "complex modulus" approach to the inclusion of structural damping; Young's modulus E in equations for the undamped elastic beam is replaced by the complex quantity $E(1 + i\mu)$. The excitation is assumed to be proportional to exp $(i\omega t)$; the response is determined as a complex quantity by standard analysis and the real or imaginary part taken if the actual excitation is proportional to cos ωt or sin ωt, respectively. This approach can be applied to the analysis of this section; alternatively results can be obtained by substituting μ for $c\omega$ in equation (3.49). In either case the steady-state response for the beam with hysteretic damping is

$$v(x, t) = \sum_s \frac{P\phi_s(x)\phi_s(a) \sin (\omega t - \alpha_s)}{m[(\omega_s^2 - \omega^2)^2 + (\mu\omega_s^2)^2]^{1/2}} \tag{3.53}$$

where tan $\alpha_s = \mu\omega_s^2/(\omega_s^2 - \omega^2)$. With the above assumptions related to light damping the amplitude at the resonant frequency $\omega = \omega_r$ (replacing equation (3.50)) is

$$V(x) \simeq \frac{P\phi_r(x)\phi_r(a)}{m\mu\omega_r^2} .$$

Thus for hysteretic damping successive resonant peaks decrease less rapidly in magnitude than for a viscoelastic beam.

The response of uniform beams to harmonic forces can be evaluated in closed form using dynamic stiffness matrices or receptances. The series solution has been presented here because of its applicability to other problems, including the response to transient forces and the response of non-uniform beams. However, the dynamic stiffness matrix will be introduced when considering the vibrations of beam systems (i.e. continuous beams and frames) in Section 4.5.

The use of equation (3.53) will be illustrated by an example.

Example. Determine the steady-state response at mid-span and quarter-span $(x = l/2$ and $x = l/4)$ for a uniform beam with simply supported ends, if an applied force $P \sin \omega t$ acts at the quarter-span point. The hysteretic damping constant μ is $0 \cdot 01$. Consider the range of excitation frequencies from $\omega = 0$ to $\omega = 12\omega_1$.

From equation (3.22) the natural frequencies are given by

$$\omega_s^2 = \frac{s^4 \pi^4 EI}{\rho A l^4}, \qquad s = 1, 2, 3 \ldots$$

$$= s^4 \pi^4 EI/ml^3.$$

Putting $\omega/\omega_1 = r$, equation (3.53) can be written in the form

$$v(x, t) = \sum_s \frac{Pl^3 \phi_s(x)\phi_s(l/4) \sin (\omega t - \alpha_s)}{\pi^4 EI[(s^4 - r^2)^2 + (\mu s^4)^2]^{1/2}}.$$

For the mode shapes

$$\phi_s(x) = 2^{1/2} \sin \frac{s\pi x}{l}.$$

The factor $2^{1/2}$ is required to satisfy the normalizing condition (3.43). Thus

$$\phi_s(l/4) = 2^{1/2} \sin s\pi/4$$

$$= 1, 2^{1/2} \text{ and } 1 \text{ for } s = 1, 2 \text{ and } 3, \text{ respectively.}$$

$$\phi_s(l/2) = 2^{1/2} \sin s\pi/2$$

$$= 2^{1/2}, 0 \text{ and } -2^{1/2} \text{ for } s = 1, 2 \text{ and } 3, \text{ respectively.}$$

We consider the approximate resonant response associated with the first three modes by putting: $\omega = \omega_1$ and taking only the first term in the series; $\omega = \omega_2$ and taking only the second term, etc. For the second mode $\phi_2(l/2) = 0$, and there is no contribution to $v(l/2, t)$ from this mode.

TABLE 3.1. Approximate values of $V(x)\pi^4 EI/Pl^3$ at resonance no.

	1	2	3
$x = l/4$	100	12·5	1·235
$x = l/2$	141·4	—	1·746

Next we consider the accuracy of the above values by investigating $\psi(s) = [(s^4 - r^2)^2 + (\mu s^4)^2]^{-1/2}$ for the off-resonant terms. For $\omega = \omega_1$ the largest off-resonant value of $\psi(s)$ occurs for $s = 2$; when $\omega = \omega_2$ and ω_3 the corresponding values are $\psi(1)$ and $\psi(2)$, respectively.

    1st resonance: $r = 1$, $\psi(2) = [(16 - 1)^2 + (0.16)^2]^{-1/2} = 0.067$,
    2nd resonance: $r = 4$, $\psi(1) = [(1 - 16)^2 + (0.01)^2]^{-1/2} = 0.067$,
    3rd resonance: $r = 9$, $\psi(2) = [(16 - 81)^2 + (0.16)^2]^{-1/2} = 0.015$.

Noting that the off-resonant contributions are approximately proportional to $\sin \omega t$ for the small damping of this problem and the resonant contributions are proportional to $\cos \omega t$, the above off-resonant terms will have negligible effects upon the resonant amplitudes.

To obtain response curves values of v at excitation frequencies away from resonance are required and several terms in the series will be required. Consider $\omega = 2\omega_1$, i.e. $r = 2$,

$$\frac{\pi^4 E I v(l/4, t)}{Pl^3} = \left[\frac{\sin (\omega t - \alpha_1)}{\{(1 - 4)^2 + 0.01^2\}^{1/2}} + \frac{2 \sin (\omega t - \alpha_2)}{\{(16 - 4)^2 + 0.16^2\}^{1/2}} \right.$$

$$\left. + \frac{\sin (\omega t - \alpha_3)}{\{(81 - 4)^2 + 0.81^2\}^{1/2}} + \cdots \right]$$

and

$$\frac{\pi^4 E I v(l/2, t)}{Pl^3} = \left[\frac{2^{1/2} \sin (\omega t - \alpha_1)}{\{(1 - 4)^2 + 0.01^2\}^{1/2}} \right.$$

$$\left. + 0 - \frac{2^{1/2} \sin (\omega t - \alpha_3)}{\{(81 - 4)^2 + 0.81^2\}^{1/2}} + \cdots \right].$$

The phase angles α_s are approximately equal to π for $s = 1$ and 0 for other values of s. For low values of damping, as in this example, this approximation has a small effect upon numerical values and saves considerable time if hand computation is undertaken. Hence

$$\frac{\pi^4 E I V(l/4)}{Pl^3} = \left| -0.3333 + 0.1667 + 0.0130 + \cdots \right| \simeq 0.1537$$

and

$$\frac{\pi^4 E I V(l/2)}{Pl^3} = \left| -0.3333 + 0 - 0.0130 \cdots \right| . 2^{1/2} \simeq 0.4898.$$

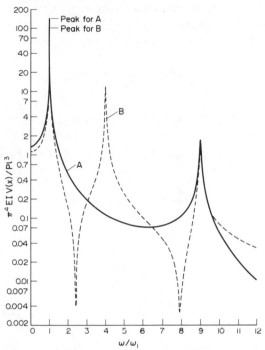

FIG. 3.8. Response of uniform simply supported beam to a harmonic force at $x = l/4$; $\mu = 0.01$. Curves A and B give response at $x = l/2$ and $l/4$, respectively.

Amplitudes $V(l/4)$ and $V(l/2)$ have been computed for $0 \le \omega/\omega_1 \le 12$ using eight modes and are plotted non-dimensionally in Fig. 3.8. For parts of the range of excitation frequencies a smaller number of modes would give adequate accuracy, but convergence is slow in the vicinity of the anti-resonant points; two of these are seen in Fig. 3.8 for $x = l/4$, but none for $x = l/2$.

3.6. Response of a Beam to a Transient Force

Equations (3.45) and (3.46) give the response of a beam to a force $Pf(t)$ applied at $x = a$. Solutions can be obtained for particular forms of transient force from these equations, but more rapid convergence of the series solution is obtained by the method of Williams[88], which has been described in Section 2.6 for multi-degree-of-freedom systems. If, using equation (2.76), equation (3.45) is integrated by parts,

$$
\begin{aligned}
q_s = \frac{P\phi_s(a)}{m\omega_s^2} & \left[\left| f(\tau) \exp\left\{ -\frac{c}{2}\omega_s^2(t-\tau) \right\} \right. \right. \\
& \left. \times \left\{ \cos \omega_s'(t-\tau) + \frac{c\omega_s^2}{2\omega_s'} \sin \omega_s'(t-\tau) \right\} \right|_0^t \\
& - \int_0^t \dot{f}(\tau) \exp\left\{ -\frac{c}{2}\omega_s^2(t-\tau) \right\} \\
& \left. \times \left\{ \cos \omega_s'(t-\tau) + \frac{c\omega_s^2}{2\omega_s'} \sin \omega_s'(t-\tau) \right\} d\tau \right] \\
= \frac{P\phi_s(a)}{m\omega_s^2} & \left[f(t) - \int_0^t \dot{f}(\tau) \exp\left\{ -\frac{c}{2}\omega_s^2(t-\tau) \right\} \right. \\
& \left. \times \left\{ \cos \omega_s'(t-\tau) + \frac{c\omega_s^2}{2\omega_s'} \sin \omega_s'(t-\tau) \right\} d\tau \right]
\end{aligned}
\tag{3.54}
$$

provided that $f(t) = 0$ when $t = 0$; $\dot{f}(\tau) = d[f(\tau)]/d\tau$ and

$$
(\omega_s')^2 = \omega_s^2\left(1 - \frac{c^2\omega_s^2}{4} \right).
$$

Using equation (3.37),

$$
v(x, t) = \sum_{s=1}^{\infty} q_s(t)\phi_s(x),
$$

the summation of the terms, which contain $f(t)$ and are outside the integral in equation (3.54), gives the static deflection of the beam at position x for a static force of magnitude $Pf(t)$ acting at $x = a$.

This can be seen by considering the response in terms of normal modes when a force $P_0 \cos \omega t$ is applied to the beam at $x = a$; from equation (3.49)

$$
v = \sum_s \frac{P_0\,\phi_s(x)\phi_s(a) \cos\,(\omega t - \alpha_s)}{m[(\omega_s^2 - \omega^2)^2 + (c\omega_s^2\omega)^2]^{1/2}}.
$$

The static deflection for a force P_0 is obtained by making $\omega \to 0$, and putting the damping coefficient $c = 0$.

$$v = \sum_s \frac{P_0 \phi_s(x) \phi_s(a)}{m \omega_s^2} .$$

Thus, if the flexibility function d_{xa} is defined as the static displacement at any point x for a unit force applied at point a, this function can be expressed in terms of normal modes as

$$d_{xa} = \sum_s \frac{\phi_s(x) \phi_s(a)}{m \omega_s^2} . \tag{3.55}$$

Returning to the transient vibration analysis and using equation (3.55), the response $v(x, t)$, obtained from equations (3.37) and (3.54), can be written

$$v(x, t) = d_{xa} Pf(t) - \sum_s \left[\frac{P \phi_s(a) \phi_s(x)}{m \omega_s^2} \int_0^t \dot{f}(\tau) \exp\left\{ -\frac{c \omega_s^2}{2}(t - \tau) \right\} \right.$$

$$\left. \times \left\{ \cos \omega_s'(t - \tau) + \frac{c \omega_s^2}{2 \omega_s'} \sin \omega_s'(t - \tau) \right\} d\tau \right]. \tag{3.56}$$

In general, the series of integrals in equation (3.56) converges more rapidly than that in equation (3.46); the flexibility function d_{xa} can be obtained without difficulty for specified end conditions. Thus there are advantages in using equation (3.56), rather than equation (3.46). Equation (3.56) is applicable to non-uniform beams, as well as to uniform beams, provided that the mode shapes $\phi_s(x)$ are normalized to satisfy equation (3.43); however, although the flexibility function can be obtained for non-uniform beams, analytical determination of $\phi_s(x)$ is difficult except for some special types of non-uniformity.

The integral in equation (3.54) could be integrated by parts again; then the resulting expression for q_s contains an additional term $\dot{f}(0) \sin \omega_s t$ and the integral includes $\ddot{f}(\tau)$ instead of $\dot{f}(\tau)$. In this form the expression for v is similar to that obtained when discussing time dependent boundary conditions in Section 4.2 and will not be considered here. The use of equations (3.46) and (3.56) will be illustrated by an example.

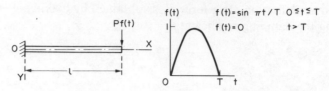

FIG. 3.9. Cantilever subjected to a transient force (half-sine pulse) at the tip.

Example. The tip deflection and maximum bending stress at the root section of a uniform cantilever beam subjected at the tip to the transient force of Fig. 3.9 are required. (T is the period of the fundamental mode of vibration of the cantilever.) Neglect damping.

From equations (3.24) and (3.25), or from tabulated data [9, 10], the natural frequencies of a uniform cantilever beam are given by

$$\omega_s^2 = \lambda_s^4 \frac{EI}{\rho A} = (\lambda_s l)^4 \frac{EI}{ml^3},$$

where

$$\lambda_1 l = 1 \cdot 875, \qquad \lambda_2 l = 4 \cdot 694, \qquad \lambda_3 l = 7 \cdot 855,$$

or

$$(\lambda_1 l)^4 = 12 \cdot 361, \qquad (\lambda_2 l)^4 = 485 \cdot 5, \qquad (\lambda_3 l)^4 = 3807.$$

Also

$$\phi_s(l) = 2 \quad \text{for} \quad s = 1, 3, 5, \ldots$$

and

$$\phi_s(l) = -2 \quad \text{for} \quad s = 2, 4, 6, \ldots$$

(i) *Response from equation* (3.54) *for* $0 \leq t \leq T$.
For

$$0 \leq t \leq T, \qquad f(t) = \sin \pi t/T.$$

As $f(0) = 0$, equation (3.54) may be used to obtain the response. As $\dot{f}(t) = \pi/T \cos \pi t/T$

$$q_s = \frac{P\phi_s(l)}{m\omega_s^2} \left[\sin \frac{\pi t}{T} - \int_0^t \frac{\pi}{T} \cos \frac{\pi \tau}{T} \cos \omega_s(t - \tau) \, . \, d\tau \right].$$

Now

$$\int_0^t \frac{\pi}{T} \cos \frac{\pi\tau}{T} \cos \omega_s(t - \tau) \, d\tau = \frac{\pi}{2T} \int_0^t \left[\cos \left\{ \omega_s(t - \tau) + \frac{\pi\tau}{T} \right\} \right.$$

$$\left. + \cos \left\{ \omega_s(t - \tau) - \frac{\pi\tau}{T} \right\} \right] d\tau$$

$$= \frac{\pi}{2T} \left[\frac{\sin (\pi t/T) - \sin \omega_s t}{\pi/T - \omega_s} + \frac{\sin (\pi t/T) + \sin \omega_s t}{\pi/T + \omega_s} \right]$$

$$= \frac{\pi}{T} \left[\frac{\pi/T \sin (\pi t/T) - \omega_s \sin \omega_s t}{\pi^2/T^2 - \omega_s^2} \right]$$

$$= \frac{\sin (\pi t/T) - (\omega_s T/\pi) \sin \omega_s t}{1 - (\omega_s T/\pi)^2}.$$

The sum of the terms outside the integral in q_s is

$$\sum_s \frac{P}{m\omega_s^2} [\phi_s(l)]^2 \sin \frac{\pi t}{T}$$

for the tip displacement and is equal to the deflection for a static force $P \sin \pi t/T$ at the tip; thus the sum is $Pl^3 \sin (\pi t/T)/3EI$. Hence the tip deflection

$$v(l, t) = \frac{Pl^3 \sin (\pi t/T)}{3EI} - \sum_s \left[\frac{P\{\phi_s(l)\}^2}{m\omega_s^2} \left\{ \frac{\sin (\pi t/T) - (\omega_s T/\pi) \sin \omega_s t}{1 - (\omega_s T/\pi)^2} \right\} \right]$$

$$= \frac{Pl^3}{EI} \left[\frac{1}{3} \sin \frac{\pi t}{T} - \sum_s \frac{\{\phi_s(l)\}^2}{(\lambda_s l)^4} \left\{ \frac{\sin (\pi t/T) - (\omega_s T/\pi) \sin \omega_s t}{1 - (\omega_s T/\pi)^2} \right\} \right].$$

Now

$$T = 2\pi/\omega_1,$$

so

$$\frac{\omega_s T}{\pi} = \frac{2\omega_s}{\omega_1} = 2 \left(\frac{\lambda_s l}{\lambda_1 l} \right)^2.$$

Substituting numerical values,

$$v(l, t) = \frac{Pl^3}{EI} \left[0.333 \sin \frac{\pi t}{T} + \left(0.108 \sin \frac{\pi t}{T} - 0.216 \sin \omega_1 t \right) \right.$$

$$+ \left(0\cdot000052 \sin \frac{\pi t}{T} - 0\cdot00066 \sin \omega_2 t \right) + \cdots \right]. \quad \text{(a)}$$

It is seen that the terms corresponding to $s = 2$ and to higher values of s may be neglected, so

$$v(l,\, t) = \frac{Pl^3}{EI} \left[0\cdot441 \sin \pi t/T - 0\cdot216 \sin 2\pi t/T \right], \quad 0 \leq t \leq T. \quad \text{(b)}$$

The maximum displacement at the tip during the time interval 0 to T is found by solving $dv/dt = 0$ to be

$$v_{max} = 0\cdot569 Pl^3/EI \qquad \qquad \text{(c)}$$

occurring when $t = 0\cdot666T$. In Fig. 3.10 the tip displacement from equation (b) is plotted against t/T; the curve for $0 \leq t/T \leq 1$ gives the response during the time of application of the transient force; the

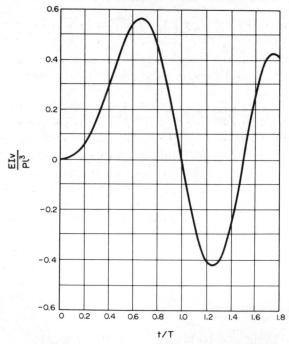

FIG. 3.10. Variation of tip displacement of cantilever of Fig. 3.9 with time.

curve for $t/T > 1$ represents the resulting free vibrations, which will be discussed later.

The maximum bending stress at any section is given by $\sigma = M/Z$, where Z is the section modulus and M is the bending moment at the section.

Now $M = -EI \, \partial^2 v/\partial x^2$, and assuming that the maximum bending moment occurs at $x = 0$ (a correct assumption for a response with small contributions from the second and higher modes), the stress at the root will be investigated.

$$\sigma(0, t) = -\frac{EI}{Z} \left(\frac{\partial^2 v}{\partial x^2}\right)_{x=0} = -P\frac{EI}{Z} \times$$

$$\sum_s \left[\left(\frac{d^2\phi_s}{dx^2}\right)_{x=0} \frac{\phi_s(l)}{m\omega_s^2} \left\{ f(t) - \int_0^t f(\tau) \cos \omega_s(t - \tau) \, . \, d\tau \right\} \right] \quad \text{(d)}$$

from equations (3.37) and (3.54). The sum from $s = 1$ to $s = \infty$ of the terms outside the integral is $[-Plf(t)]/Z$, i.e. the static stress at the root due to a tip force of magnitude $Pf(t)$. This can be seen by reasoning similar to that used when discussing equation (3.54). If a force $P_0 \cos \omega t$ is applied to a beam at $x = a$ and the response evaluated in terms of normal modes, then

$$\sigma(x, t) = -\frac{EI}{Z} \sum_s q_s \, d^2\phi_s/dx^2$$

and

$$q_s = \frac{P_0 \phi_s(a)}{m(\omega_s^2 - \omega^2)}.$$

As $\omega \to 0$, the summation gives the static stress at x and in terms of normal modes this is:

$$-\frac{EI}{Z} \sum_s \frac{P_0 \phi_s(a) \, d^2\phi_s/dx^2}{m\omega_s^2}.$$

Comparing this expression with equation (d), it is seen that the sum of the terms not containing an integral in equation (d) is the static stress at $x = 0$ due to a tip force $Pf(t)$, and this stress is: $-Plf(t)/Z$. Thus for $0 \le t \le T$,

$$\sigma(0, t) = -\frac{Pl}{Z} \left[\sin \frac{\pi t}{T} - \sum_s \frac{EI}{l} \left(\frac{d^2\phi_s}{dx^2} \right)_{x=0} \right.$$
$$\left. \times \frac{\phi_s(l)}{m\omega_s^2} \left\{ \frac{\sin(\pi t/T) - (\omega_s T/\pi)\sin \omega_s t}{1 - (\omega_s T/\pi)^2} \right\} \right].$$

Noting that

$$\left(\frac{d^2\phi_s}{dx^2} \right)_{x=0} = 2\lambda_s^2,$$

$$\sigma(0, t) = -\frac{Pl}{Z} \left[\sin \frac{\pi t}{T} - \sum_s \frac{2\phi_s(l)}{(\lambda_s l)^2} \left\{ \frac{\sin(\pi t/T) - (\omega_s T/\pi)\sin \omega_s t}{1 - (\omega_s T/\pi)^2} \right\} \right].$$

Substituting numerical values,

$$\sigma(0, t) = -\frac{Pl}{Z} \left[\sin \frac{\pi t}{T} + \left(0.379 \sin \frac{\pi t}{T} - 0.758 \sin \omega_1 t \right) \right.$$
$$\left. - \left(0.00116 \sin \frac{\pi t}{T} - 0.0144 \sin \omega_2 t \right) \cdots \right]. \qquad \text{(e)}$$

The accuracy will be reasonable (almost 1 per cent) if terms corresponding to $s \ge 2$ are neglected, i.e. the second bracket in equation (e) is neglected. The maximum stress occurring during the time interval 0 to T is given by:

$$\left| \frac{\sigma Z}{Pl} \right| = 1.851,$$

occurring when $t = 0.678T$.

(ii) *Response from equation* (3.46) *for* $0 \le t \le T$.
From equation (3.46) with $c = 0$

$$v(l, t) = \frac{P}{m} \sum_s \left[\frac{\{\phi_s(l)\}^2}{\omega_s} \int_0^t \sin \frac{\pi \tau}{T} \sin \omega_s(t - \tau) \, d\tau \right].$$

Integrating and substituting numerical values,

$$v(l, t) = \frac{Pl^3}{EI} \left[\left(0{\cdot}431 \sin \frac{\pi t}{T} - 0{\cdot}216 \sin \omega_1 t \right) \right.$$

$$+ \left(0{\cdot}0083 \sin \frac{\pi t}{T} - 0{\cdot}00066 \sin \omega_2 t \right)$$

$$\left. + \left(0{\cdot}0011 \sin \frac{\pi t}{T} - 0{\cdot}00003 \sin \omega_3 t \right) + \cdots \right]. \qquad (f)$$

It is necessary to consider the $\sin \pi t/T$ terms in q_1, q_2 and q_3 and the $\sin \omega_1 t$ in q_1 in equation (f) to give comparable accuracy with equation (b), where only the $\sin \pi t/T$ and $\sin \omega_1 t$ terms in q_1 and the "static" term in equation (a) were retained. This illustrates the better convergence of solutions based on equation (3.56) compared with those from equation (3.46).

The maximum stress at the root section is obtained from equation (3.46) as, after integrating and substituting numerical values,

$$\sigma(0, t) = -\frac{Pl}{Z} \left[\left(1{\cdot}514 \sin \frac{\pi t}{T} - 0{\cdot}757 \sin \omega_1 t \right) \right.$$

$$- \left(0{\cdot}182 \sin \frac{\pi t}{T} - 0{\cdot}0145 \sin \omega_2 t \right)$$

$$+ \left(0{\cdot}0649 \sin \frac{\pi t}{T} - 0{\cdot}0018 \sin \omega_3 t \right)$$

$$\left. - \left(0{\cdot}0331 \sin \frac{\pi t}{T} - 0{\cdot}0005 \sin \omega_4 t \right) + \cdots \right]. \qquad (g)$$

The series for stress converges more slowly than that for displacement. In equation (e) it was possible to neglect terms corresponding to $s \geq 2$, but for comparable accuracy from solution (g) it is necessary to consider the terms in $\sin \pi t/T$ in q_1, q_2, q_3, q_4, and q_5.

(iii) *Response for $t \geq T$.*

The free vibrations that occur for $t \geq T$ will be the sum of the normal modes and can be expressed in the form:

$$v(x, t) = \frac{Pl^3}{EI} \sum_s [\phi_s(x)\{B_s \sin \omega_s(t - T) + C_s \cos \omega_s(t - T)\}]. \quad \text{(h)}$$

The constants B_s and C_s are obtained from the conditions of continuity of displacement and velocity for all values of x at time T. During the application of the pulse the displacement is:

$$v(x, t) = \sum_s q_s(t)\phi_s(x). \quad \text{(i)}$$

For continuity equations (h) and (i) must give the same values for $v(x, T)$. Similarly, if $\partial v/\partial t$ is found from equations (h) and (i), it must have the same value at time T. This leads to the following set of equations:

$$\frac{Pl^3}{EI} C_s = q_s(T),$$

$$\frac{Pl^3}{EI} B_s\omega_s = \dot{q}_s(T).$$

From equation (3.46)

$$q_s(t) = \frac{Pl^3}{EI} \frac{\phi_s(l)}{(\lambda_s l)^4} \left[\frac{(\pi/\omega_s T) \sin \omega_s t - \sin (\pi t/T)}{(\pi/\omega_s T)^2 - 1} \right].$$

Thus

$$C_s = \frac{\phi_s(l)}{(\lambda_s l)^4} \left[\frac{(\pi/\omega_s T) \sin \omega_s T}{(\pi/\omega_s T)^2 - 1} \right]$$

and

$$B_s = \frac{\phi_s(l)}{(\lambda_s l)^4} \left[\frac{(\pi/\omega_s T)(\cos \omega_s T + 1)}{(\pi/\omega_s T)^2 - 1} \right].$$

Substituting numerical values, $C_1 = 0$; $B_1 = -0.216$; $C_2 = 0.00329$; $B_2 = 0.00296$, so

$$v(l, t) = \frac{Pl^3}{EI} [-0.432 \sin \omega_1(t - T) - 0.0059 \sin \omega_2(t - T)$$

$$- 0.0066 \cos \omega_2(t - T) \cdots].$$

In this case the free vibration can be represented approximately by the first mode component; it is apparent from the above expression that the maxima during free vibration, $0.432 \, (Pl^3/EI)$, which in practice will decrease due to damping, are less than the maximum displacement occurring during the time 0 to T. The free vibration, neglecting the small components for $s \geq 2$, is shown in Fig. 3.10 for $1 \leq t/T \leq 1.8$. When the stresses at the root are investigated for $t/T > 1$, it is found that the maximum stress is given by: $|\sigma Z/Pl| = 1.517$, a value lower than the maximum stress occurring during the time $0 \leq t/T \leq 1$.

The above example shows that the dynamic response, expressed as the sum of a series of normal modes, converges more rapidly for displacement than for stress. Thus for an accurate stress–time curve more terms will have to be considered, in general, than for an accurate displacement–time curve. Now the rate at which the series converges depends also on the ratio of the pulse time t_0 to the period of the fundamental mode, T. In the above example a value of unity was assumed for this ratio. If t_0/T is small, many terms in the series giving the dynamic response will have to be considered.

As mentioned in Section 3.3, accurate values of the natural frequencies are obtained from equation (3.15) only if l/sK_r is large, where s is the mode number and K_r is the radius of gyration of the cross-section. Thus when t_0/T is small and many modes have to be included, the higher natural frequencies should be determined from the Timoshenko theory, which includes the effects of rotatory inertia and transverse shear deformation (Section 4.6). The problem of determining the initial response of a beam to pulses of short duration (i.e. $t_0/T \ll 1$) is outside the scope of this book, but various methods of approach are discussed by Flügge and Zajac[32].

If the dynamic response of a beam to a pulse of complex form is required and the contributions from a few modes have to be considered, lengthy calculations will be required to predict the maximum displacement or stress. Fung and Barton[35] suggest that if $t_m/T > 0.5$ for a single pulse of any shape, where t_m is the rise time for the pulse (i.e. the time for the force or pressure to increase from zero to its maximum value), a good approximation to the maximum response can be obtained from the algebraic sum of the peak responses of the individual modes. Thus it is assumed that the contribution from each mode reaches its peak at the same time. The integral in equation (3.46) for the dynamic response is identical to that in the transient response of a single-degree-of-freedom system [equation (1.36)]; thus as non-dimensional maxima for the response of single-degree-of-freedom systems have been plotted against t_m/T_s, where T_s is the period of the mode, these data[41] can be used to give the maximum response in each mode; then rapid approximations to the maximum dynamic response of a beam can be obtained using Fung and Barton's hypothesis.

3.7. Natural Frequencies of Non-uniform Beams by the Rayleigh–Ritz Method

In Section 3.3 the natural frequencies and mode shapes of uniform beams vibrating in flexure have been found for specified boundary conditions. For non-uniform beams for which the cross-sectional dimensions are functions of the axial coordinate x, equation (3.14) and the appropriate boundary conditions must be satisfied. However, with the exception of a few special cases, such solutions do not exist, and approximate methods are required to determine natural frequencies and their associated mode shapes. From the many approximate methods the Rayleigh and Rayleigh–Ritz methods will be given in this section and the finite element method will be considered in Section 3.9. These methods can be used to determine the natural frequencies of beams with (a) discrete changes in cross-section at one or more sections and (b) attached concentrated masses and/or springs. (If all segments of a beam have uniform, but not necessarily identical cross-sections, these problems can be solved, using the method of beam systems (Section 4.5).)

Rayleigh Method. The beam of Fig. 3.11 has cross-sectional dimensions which are continuous or discontinuous functions of the axial coordinate x. In addition to the distributed mass $\rho A(x)$ per unit length the beam carries concentrated masses $m_1, m_2, m_3, \ldots, m_J$ at distances $x_1, x_2, x_3, \ldots, x_J$ from the origin. Linear springs of stiffness $k_1, k_2, k_3, \ldots, k_N$ are attached to the beam at distances $x'_1, x'_2, x'_3, \ldots, x'_N$ from the origin; the axis of each spring is parallel to the axis OY and the spring lies between the beam and a fixed datum.

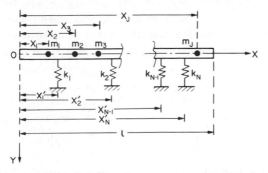

FIG. 3.11. General beam with attached masses and springs.

For free undamped vibrations the displacement of the beam is given by

$$v(x, t) = V(x) \sin (\omega t + \alpha).$$

The strain energy of the beam is

$$\mathfrak{S}_b = \frac{1}{2} \int_0^l \frac{M^2}{EI} \, dx$$

where the bending moment

$$M = -EI \frac{d^2 V}{dx^2} \sin (\omega t + \alpha)$$

using equation (3.13). Thus the total strain energy, including that of the springs, is

$$\mathfrak{S} = \frac{1}{2} \sin^2 (\omega t + \alpha) \left[\int_0^l EI \left(\frac{d^2 V}{dx^2} \right)^2 dx + \sum_{j=1}^N k_j \{V(x'_j)\}^2 \right]. \quad (3.57)$$

The kinetic energy

$$\mathfrak{T} = \frac{1}{2} \int_0^l \rho A \left(\frac{\partial v}{\partial t} \right)^2 dx + \frac{1}{2} \sum_{j=1}^J m_j \left[\left(\frac{\partial v}{\partial t} \right)_{x=x_j} \right]^2$$

$$= \frac{1}{2} \omega^2 \cos^2 (\omega t + \alpha) \left[\int_0^l \rho A V^2 \, dx + \sum_{j=1}^J m_j \{V(x_j)\}^2 \right]. \quad (3.58)$$

The sum of the strain energy and kinetic energy is a constant. Following Section 1.5, the kinetic energy in the mean position, where $\cos(\omega t + \alpha) = \pm 1$, equals the strain energy in a position of maximum displacement, where $\sin(\omega t + \alpha) = \pm 1$. Thus from equations (3.57) and (3.58)

$$\omega^2 = \frac{\int_0^l EI(d^2V/dx^2)^2\, dx + \sum_{j=1}^{N} k_j\{V(x_j')\}^2}{\int_0^l \rho AV^2\, dx + \sum_{j=1}^{J} m_j\{V(x_j)\}^2}. \tag{3.59}$$

If for a particular mode we know the correct form for $V(x)$, equation (3.59) gives the exact value of the natural frequency. The Rayleigh principle states that for a reasonable assumed mode shape, which must satisfy all geometric boundary conditions (i.e. displacement and slope conditions at the ends), a good approximation to the natural frequency will be obtained. If the difference between the assumed mode shape and the exact mode shape is a small quantity, the difference between the natural frequency from equation (3.59) and the exact value will be a very small quantity. Moreover, if the assumed mode shape is an approximation to the fundamental mode, the natural frequency determined from equation (3.59) will be slightly *higher* than the exact value[29], as the assumption is equivalent to introducing additional constraints.

Rayleigh–Ritz Method. The Rayleigh method can be extended, and its accuracy improved, by assuming for $V(x)$ a series

$$V(x) = \sum_{i=1}^{n} \Gamma_i V_i(x) \tag{3.60}$$

where each of the assumed functions $V_i(x)$ must satisfy the geometric boundary conditions; the parameters Γ_i are arbitrary. In order to reduce the effects of the additional constraints as much as possible, the parameters Γ_i are chosen to make the frequencies determined from equation (3.59) a minimum. Mathematically, the set of equations

$$\partial(\omega^2)/\partial\Gamma_i = 0, \qquad i = 1, 2, 3, \ldots, n \tag{3.61}$$

is formed, elimination of the parameters Γ_i leads to an nth order determinant in ω^2, of which the roots give approximate values for the first n natural frequencies of the system.

It will be shown now that the Rayleigh–Ritz method leads to the matrix equation for the eigenvalue problem, considered in Section 2.5. Substituting equation (3.60) into equation (3.59)

$$
\begin{aligned}
\omega^2 &= \frac{\int_0^l EI\left\{\sum_{i=1}^{n} \Gamma_i \, d^2 V_i/dx^2\right\}^2 dx + \sum_{s=1}^{N} k_s\left\{\sum_{i=1}^{n} \Gamma_i V_i(x_s')\right\}^2}{\int_0^l \rho A\left\{\sum_{i=1}^{n} \Gamma_i V_i\right\}^2 dx + \sum_{s=1}^{J} m_s\left\{\sum_{i=1}^{n} \Gamma_i V_i(x_s)\right\}^2} \\[2ex]
&= \frac{\sum_i \sum_j \Gamma_i \Gamma_j \left[\int_0^l EI \dfrac{d^2 V_i}{dx^2}\dfrac{d^2 V_j}{dx^2}\, dx + \sum_{s=1}^{N} k_s V_i(x_s')V_j(x_s')\right]}{\sum_i \sum_j \Gamma_i \Gamma_j \left[\int_0^l \rho A V_i V_j\, dx + \sum_{s=1}^{J} m_s V_i(x_s)V_j(x_s)\right]} \\[2ex]
&= \frac{\sum_i \sum_j k_{ij}\Gamma_i \Gamma_j}{\sum_i \sum_j m_{ij}\Gamma_i \Gamma_j}
\end{aligned}
\tag{3.62}
$$

where

$$
k_{ij} = \int_0^l EI \frac{d^2 V_i}{dx^2}\frac{d^2 V_j}{dx^2}\, dx + \sum_s k_s V_i(x_s')V_j(x_s')
\tag{3.63}
$$

and

$$
m_{ij} = \int_0^l \rho A V_i V_j\, dx + \sum_s m_s V_i(x_s)V_j(x_s).
$$

Substituting for ω^2 from equation (3.62) in equation (3.61) and differentiating with respect to Γ_i

$$
\frac{2\sum_j k_{ij}\Gamma_j}{\sum_i \sum_j m_{ij}\Gamma_i \Gamma_j} - \frac{2\left(\sum_i \sum_j k_{ij}\Gamma_i \Gamma_j\right)\left(\sum_j m_{ij}\Gamma_j\right)}{\left(\sum_i \sum_j m_{ij}\Gamma_i \Gamma_j\right)^2} = 0, \qquad i = 1, 2, \ldots, n.
$$

Using equation (3.62), this simplifies to

$$
\sum_j k_{ij}\Gamma_j - \omega^2 \sum_j m_{ij}\Gamma_j = 0, \qquad i = 1, 2, \ldots, n
$$

and can be written in matrix form as

$$(\mathbf{K} - \mathbf{M}\omega^2)\mathbf{g} = 0 \qquad (3.64)$$

where $\mathbf{g}$ is a vector of the parameters Γ_j. From the definitions of k_{ij} and m_{ij}, equations (3.63), the matrices $\mathbf{K}$ and $\mathbf{M}$ are symmetric. Equation (3.64) is of the same form as equation (2.47). The determination of natural frequencies ω_r and the corresponding modal values of the vector of parameters $_r\mathbf{g}$ from the latter equation has been discussed in Section 2.5.

If there are no concentrated masses m_s and orthogonal functions are used for $V_i(x)$, i.e.

$$\int_0^l \rho A V_i V_j \, dx = 0 \qquad \text{for } i \neq j,$$

the matrix $\mathbf{M}$ is diagonal.

Although the above analysis is for a beam undergoing flexural vibrations, the matrix equation (3.64) will be obtained if the Rayleigh–Ritz method is applied to other modes of deformation and other geometries, provided that the definitions of k_{ij} and m_{ij} are modified to conform with the appropriate strain and kinetic energy expressions respectively. It will be applied to rectangular plates in Section 5.2.

The Rayleigh and Rayleigh–Ritz methods will be illustrated in the following example.

Example. A cantilever beam of length l, flexural rigidity EI and constant mass per unit length ρA carries a single concentrated mass $2m$ at the tip, where $m = \rho A l$. Determine the first and second natural frequencies of this beam.

A simple example has been chosen so that it is possible to compare the approximate values of the natural frequencies with exact solutions from the beam equation. The latter gives $\omega_1 = 1{\cdot}1582(EI/ml^3)^{1/2}$ and $\omega_2 = 15{\cdot}861(EI/ml^3)^{1/2}$ [56].

First we will assume that

$$V(x) = B(3lx^2 - x^3)$$

and use the Rayleigh method to determine the fundamental natural frequency. The assumed function satisfies the essential conditions of zero deflection and slope at the root. It is of the form of the static deflection curve for a uniform light cantilever with a tip load. Now

$$\frac{d^2V}{dx^2} = 6B(l - x)$$

and

$$V(l) = 2Bl^{3'}.$$

Substituting in equation (3.59),

$$\omega_1^2 = \frac{36EIB^2 \int_0^l (l - x)^2 \, dx}{\rho AB^2 \int_0^l (3l - x)^2 x^4 \, dx + 2m(2Bl^3)^2}$$

and

$$\omega_1 = 1.1584(EI/ml^3)^{1/2}$$

after integrating and putting $\rho Al = m$. This approximate value is only 0.02 per cent higher than the true value. Static deflection curves are useful approximate forms for fundamental mode shapes. In this case the static deflection curve for a light cantilever with a tip load has been used, rather than that for a cantilever with distributed and tip loads. If the distributed load had been greater than the tip load, it would have been advisable to use for $V(x)$ the static deflection curve for a cantilever with a uniformly distributed load. However, in this case, where the tip mass/distributed mass = 2, use of the latter curve gives a value of ω_1 which is 2.2 per cent high.

The Rayleigh–Ritz method will now be applied to this problem, using two functions in the series for $V(x)$ so that approximate values of the first and second natural frequencies are obtained. We assume

$$V(x) = \Gamma_1 \phi_1(x) + \Gamma_2 \phi_2(x)$$

where $\phi_1(x)$ and $\phi_2(x)$ are the first and second normalized mode shapes of a uniform cantilever beam (i.e. the exact functions for no tip mass). Relevant numerical data[9] are:

$$\lambda_1 l = 1.8751, \ \lambda_2 l = 4.6941; \qquad \phi_1(l) = 2, \ \phi_2(l) = -2.$$

Substituting the above values in equations (3.62) and (3.63), integrating and using equations (A3.7), (A3.8) and

$$\int_0^l \left(\frac{d^2\phi_i}{dx^2}\right)^2 dx = \lambda_i^4 l,$$

$$k_{ii} = \lambda_i^4 EIl, \ k_{12} = 0, \qquad m_{ii} = \rho Al + 8m, \ m_{12} = -8m.$$

Applying the Rayleigh–Ritz minimization procedure,

$$[\lambda_1^4 EIl - (\rho Al + 8m)\omega^2]\Gamma_1 + 8m\omega^2\Gamma_2 = 0,$$

$$8m\omega^2\Gamma_1 + [\lambda_2^4 EIl - (\rho Al + 8m)\omega^2]\Gamma_2 = 0.$$

Putting $m = \rho Al$, $\Omega = ml^3\omega^2/EI$ and substituting for $\lambda_i l$, the frequency determinant is

$$\det \begin{vmatrix} 12 \cdot 362 - 9\Omega & 8\Omega \\ 8\Omega & 485 \cdot 519 - 9\Omega \end{vmatrix} = 0$$

with roots: $\Omega_1 = 1 \cdot 3463$ and $\Omega_2 = 262 \cdot 23$, i.e.

$$\omega_1 = 1 \cdot 1603(EI/ml^3)^{1/2}$$

and

$$\omega_2 = 16 \cdot 193(EI/ml^3)^{1/2}.$$

Thus in this example the Rayleigh–Ritz method gives approximate values of the first and second natural frequencies 0·18 and 2·1 per cent, respectively, higher than the exact values. From

$$\frac{\Gamma_2}{\Gamma_1} = -\frac{12 \cdot 362 - 9\Omega}{8\Omega}$$

$\Gamma_2/\Gamma_1 = -0 \cdot 0227$ and $1 \cdot 119$ for the first and second modes respectively.

If the mode shape $V(x) = \Gamma_1 \phi_1(x)$ had been assumed for the fundamental mode (i.e. assuming that the presence of tip mass does not alter the mode shape), it will be seen by putting $\Gamma_2 = 0$ in the above analysis that $9\Omega_1 = 12 \cdot 362$ and $\omega_1 = 1 \cdot 1719(EI/ml^3)^{1/2}$, an error of 1·2 per cent. Thus the use of the Rayleigh–Ritz method with only two terms leads to greater accuracy in the determination of the fundamental frequency and gives a good approximation for the second frequency. Better accuracy could be obtained by taking more terms in the series.

3.8. Response of Non-uniform Beams by the Rayleigh–Ritz Method

As mentioned in Section 3.6, the response of non-uniform beams can be obtained in theory from equations (3.46) or (3.56), but in practice the explicit determination of $\phi_s(x)$ for most non-uniform beams is either impossible or tedious. Thus approximate methods of determining response are essential; the approximate methods of determining natural frequencies can be extended to the determination of response. In this section the Rayleigh–Ritz method, applied to the determination of natural frequencies in Section 3.7, will be given; in Section 3.9 the finite element method will be considered.

It is assumed that the displacement can be represented by the series

$$v(x,\ t) = \sum_{i=1}^{n} \Gamma_i(t) V_i(x) \qquad (3.65)$$

where each of the assumed functions $V_i(x)$ must satisfy all geometric boundary conditions and the functions $\Gamma_i(t)$ have to be determined. We consider again the general beam of Fig. 3.11 with the addition of viscous damping. For a particular damper of constant c_p it is assumed that its line of action is parallel to axis OY, it is attached to the beam at coordinate x_p and it lies between the beam and a fixed datum. Springs and damping mechanisms which give rotational constraint at specific sections of the beam could, but will not, be added to the model. For this system the kinetic energy

$$\mathfrak{T} = \tfrac{1}{2}\dot{\mathbf{g}}^T \mathbf{M} \dot{\mathbf{g}} \qquad (3.66)$$

where m_{ij}, an element of the symmetric matrix $\mathbf{M}$, is defined by equations (3.63). The vector $\mathbf{g}$ contains the n functions Γ_i, i.e.

$$\mathbf{g}^T = [\Gamma_1, \Gamma_2, \ldots, \Gamma_n].$$

Also

$$\mathbf{\dot{g}}^T = [\dot{\Gamma}_1, \dot{\Gamma}_2, \ldots, \dot{\Gamma}_n].$$

The strain energy

$$\mathfrak{S} = \tfrac{1}{2}\mathbf{g}^T\mathbf{K}\mathbf{g} \qquad (3.67)$$

where k_{ij}, an element of the symmetric matrix $\mathbf{K}$, is defined in equations (3.63). The dissipation function

$$\begin{aligned}
\mathfrak{F} &= \frac{1}{2}\sum_p c_p[\partial v(x_p, t)/\partial t]^2 \\
&= \frac{1}{2}\sum_p c_p\left[\sum_i \dot{\Gamma}_i V_i(x_p)\right]^2 \\
&= \tfrac{1}{2}\mathbf{\dot{g}}^T\mathbf{C}\mathbf{\dot{g}} \qquad (3.68)
\end{aligned}$$

where

$$c_{ij} = \sum_p c_p V_i(x_p)V_j(x_p).$$

In order to apply the Lagrange equation to obtain a set of differential equations in terms of Γ_i and their derivatives with respect to time, we require the corresponding generalized forces Q_i. The virtual work associated with an increment $\delta\Gamma_i$, the other functions remaining constant, is by definition

$$\delta\mathfrak{W} = Q_i\,\partial\Gamma_i.$$

If the beam is subjected to an applied force of $p(x)f(t)$ per unit length,

$$\begin{aligned}
\delta\mathfrak{W} &= \int_0^l p(x)f(t)\,\delta v\,dx \\
&= \int_0^l p(x)f(t)V_i\,\delta\Gamma_i\,dx
\end{aligned}$$

as $\delta v = V_i\,\delta\Gamma_i$. Thus

$$Q_i = f(t)\int_0^l p(x)V_i(x)\,dx. \qquad (3.69)$$

Applying the Lagrange equation (A4.17) with Γ_i treated as independent coordinates,

$$\frac{d}{dt}\left(\frac{\partial \mathfrak{T}}{\partial \dot{\Gamma}_i}\right) - \frac{\partial \mathfrak{T}}{\partial \Gamma_i} + \frac{\partial \mathfrak{F}}{\partial \dot{\Gamma}_i} + \frac{\partial \mathfrak{S}}{\partial \Gamma_i} = Q_i, \qquad i = 1, 2, \ldots, n \qquad (3.70)$$

and using equations (3.66) to (3.69) for $\mathfrak{T}$, $\mathfrak{F}$, $\mathfrak{S}$ and Q_i

$$\mathbf{M}\ddot{\mathbf{g}} + \mathbf{C}\dot{\mathbf{g}} + \mathbf{K}\mathbf{g} = \mathbf{p}(t) \qquad (3.71)$$

where Q_i in equation (3.69) is the ith element of $\mathbf{p}(t)$. If internal damping is included, as defined by equation (3.34), E is effectively replaced by $E(1 + c\,\partial/\partial t)$, so a term $c\mathbf{K}\dot{\mathbf{g}}$ will be added to the left-hand side of equation (3.71). Apart from a change in nomenclature, equations (2.40) and (3.71) are identical. Thus the methods of Chapter 2 can be used to determine the response from equation (3.71). It will be noted that if $\mathbf{C}$ is non-zero it is unlikely to satisfy condition (2.72) and thus the normal mode method of Section 2.6 will not be applicable for this case; however, the inclusion of internal damping does not prohibit the use of the normal mode method.

Example. Determine the amplitude of vibration at the tip of the cantilever beam shown in Fig. 3.12, when the frequency of excitation equals the fundamental natural frequency of the system. Damping in the spring is included by using a complex stiffness $k(1 + 0\cdot01i)$; damping in the beam is neglected. $kl^3/EI = 50$.

As the response is required only at the fundamental resonant frequency, it is necessary to take sufficient and suitable terms in the series for v to model the first mode with reasonable accuracy. We assume that

$$v = \Gamma_1\phi_1 + \Gamma_2\phi_2$$

where ϕ_1 and ϕ_2 are the first and second mode shapes of a uniform

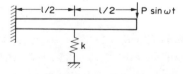

FIG. 3.12. Example beam.

cantilever (i.e. without the spring). Recalling that these mode shapes are normalized and orthogonal, the mass matrix

$$\mathbf{M} = \begin{bmatrix} m & 0 \\ 0 & m \end{bmatrix}$$

where m is the mass of the beam. Also, as

$$\int_0^l EI \frac{d^2\phi_i}{dx^2} \frac{d^2\phi_j}{dx^2} dx = 0 \qquad \text{for } i \neq j$$
$$= EI\lambda_i^4 l \qquad \text{for } i = j$$

from equations (A3.4) and (A3.8) and the normalizing condition for uniform beams (3.43), the stiffness matrix

$$\mathbf{K} = \begin{bmatrix} EI\lambda_1^4 l + k\{\phi_1(l/2)\}^2 & k\phi_1(l/2)\phi_2(l/2) \\ k\phi_1(l/2)\phi_2(l/2) & EI\lambda_2^4 l + k\{\phi_2(l/2)\}^2 \end{bmatrix}.$$

The damping can be included by replacing k by $k(1 + i \times 0.01)$ in $\mathbf{K}$ and using the matrix equation in the form

$$\mathbf{M\ddot{g}} + \mathbf{Kg} = \mathbf{p} \exp(i\omega t).$$

From equation (3.69)

$$\mathbf{p} = \begin{bmatrix} \phi_1(l) \\ \phi_2(l) \end{bmatrix} P.$$

The following properties relating to ϕ_i are obtained from Bishop and Johnson's tables[9]:

$$\phi_1(l/2) = 0.6790; \qquad \phi_1(l) = 2.0; \qquad (\lambda_1 l)^4 = 12.362,$$
$$\phi_2(l/2) = 1.4273; \qquad \phi_2(l) = -2.0; \qquad (\lambda_2 l)^4 = 485.52.$$

First the fundamental natural frequency is obtained by considering free vibrations, leading to

$$\begin{bmatrix} EI\lambda_1^4 l + k\{\phi_1(l/2)\}^2 - m\omega^2 & k\phi_1(l/2)\phi_2(l/2) \\ k\phi_1(l/2)\phi_2(l/2) & EI\lambda_2^4 l + k\{\phi_2(l/2)\}^2 - m\omega^2 \end{bmatrix} \begin{bmatrix} \Gamma_1 \\ \Gamma_2 \end{bmatrix} = 0,$$

i.e.

$$\det \begin{vmatrix} 12.362 + 0.6790^2\beta - \Omega & 0.6790(1.4273\beta) \\ 0.6790(1.4273\beta) & 485.52 + 1.4273^2\beta - \Omega \end{vmatrix} = 0$$

where $\beta = kl^3/EI = 50$ and $\Omega = m\omega^2 l^3/EI$.

Solving the resulting quadratic equation in Ω and taking the lower root

$$\Omega_1 = 31{\cdot}195, \qquad \omega_1 = 5{\cdot}5852(EI/ml^3)^{1/2}$$

and

$$\frac{\Gamma_2}{\Gamma_1} = -0{\cdot}0868.$$

To determine the response we return to equation (3.71) with damping included and put $\Omega = 31{\cdot}195$ to give

$$\frac{EI}{l^3} \begin{bmatrix} 12{\cdot}362 + 0{\cdot}6790^2 \times 50(1 + 0{\cdot}01i) - 31{\cdot}195 & 0{\cdot}6790 \times 1{\cdot}4273 \times 50(1 + 0{\cdot}01i) \\ 0{\cdot}6790 \times 1{\cdot}4273 \times 50(1 + 0{\cdot}01i) & 485{\cdot}52 + 1{\cdot}4273^2 \times 50(1 + 0{\cdot}01i) - 31{\cdot}195 \end{bmatrix} \begin{bmatrix} \Gamma_1 \\ \Gamma_2 \end{bmatrix} = \begin{bmatrix} 2 \\ -2 \end{bmatrix} P \exp{(i\omega t)}.$$

This pair of complex simultaneous equations is solved to give

$$\Gamma_1 = (0{\cdot}035 - 14{\cdot}13i)\frac{Pl^3}{EI}\exp{(i\omega t)},$$

$$\Gamma_2 = (-0{\cdot}017 + 1{\cdot}231i)\frac{Pl^3}{EI}\exp{(i\omega t)}.$$

At the tip

$$v(l,\, t) = 2\Gamma_1 - 2\Gamma_2 = (0{\cdot}104 - 30{\cdot}72i)\frac{Pl^3}{EI}\exp{(i\omega t)}.$$

Taking the imaginary part as the actual excitation is proportional to $\sin \omega t$,

$$v(l,\, t) = (0{\cdot}104 \sin \omega t - 30{\cdot}72 \cos \omega t)\frac{Pl^3}{EI}.$$

Thus the amplitude at the tip at the fundamental resonant frequency is $30{\cdot}72Pl^3/EI$. The above expression for $v(l,\, t)$ should be compared with the exact solution $v(l,\, t) = (0{\cdot}106 \sin \omega t - 30{\cdot}98 \cos \omega t)Pl^3/EI$, which is obtained in Section 4.5.

The geometry of this example differs from that in Section 3.3 only in the position of the spring and its stiffness. However, the change in spring position from the tip to mid-span means that the problem cannot be treated as a single beam by the method of Section 3.3, but has to be treated as a two-beam system by the method of Section 4.5 to obtain an exact solution. The change in spring position only affects the arithmetic when the Rayleigh–Ritz method is used. For the problem of Section 3.3 (Fig. 3.6) the fundamental natural frequency by the Rayleigh–Ritz method, following the procedure of this example, is given by $\Omega_1 = 31\cdot519$, which yields a natural frequency 0·22 per cent higher than the exact value found in Section 3.3.

3.9. Finite Element Method Applied to Beam Vibrations

In this method the structure is divided into a large number of small, but finite, parts. Each element is analysed, considering the deformations that occur in that section of the structure. The variation of displacement, e.g. linear, quadratic, etc., over the length of the element is assumed. (In this section one-dimensional elements for beams are considered; in Sections 5.4 and 5.5 two-dimensional elements for plate structures are introduced, and there the variation of the components of displacement and some derivatives over the area of the element is assumed.) This allows the displacement of any point in the element to be expressed in terms of the displacements at the ends of the element; the latter displacements are treated as unknowns. In finite element terminology these displacements (or for two-dimensional elements the displacements and some derivatives at the corners of the element) are called nodal variables. (It is unfortunate that the term "node" has different definitions in finite element and vibration theory.) By integrating over the length the strain and kinetic energies of the element are determined in terms of the nodal variables. By superposition of the energy contributions from the individual elements into which the structure has been divided the strain and kinetic energies of the structure or system are determined in terms of the nodal variables of the whole structure. Where two or more elements join at a node each displacement component at that node must have a single value. In general, the finite element method is based upon variational principles.

In this book, where only applications of the method to dynamic problems are considered, the differential equations for the nodal displacements are derived from the strain and kinetic energy functions of the structure by using the Lagrange equation. In this section the finite element method is applied to dynamic problems of beams in extension, torsion and flexure. Only a brief treatment can be included here; further details of the method, its mathematical background and its applications to other problems can be found in books on finite elements[14, 90].

In finite element analysis the nodal variables must include the appropriate components of displacement and their derivatives up to the order one less than that appearing in the relevant strain energy expression. For extensional or membrane deformations the strain energy expression contains first derivatives of the displacements [equations (3.73a) and (5.54)]; thus the nodal variables are the appropriate components of displacement [i.e. u for extensional vibrations of bars and u and v for membrane deformations of plates (Section 5.4)]. For flexural deformations this expression contains second derivatives of the displacement (equations (3.57) and (5.13) for beams and plates, respectively); thus the nodal variables are the displacement and its first derivatives with respect to the coordinates [i.e. v and $\partial v/\partial x$ for flexure of beams and w, $\partial w/\partial x$ and $\partial w/\partial y$ for flexure of plates (Section 5.5)]. (Higher order elements, which include additional higher derivatives as nodal variables, exist, but will not be discussed in this book.) The order of the assumed polynomial is controlled by the choice of nodal variables.

Extensional vibrations. Figure 3.13a shows an element of length l,

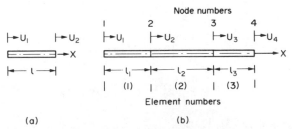

FIG. 3.13. Finite element method for extension. (a) Single element. (b) Beam consisting of three elements.

which undergoes extensional deformation. Only the axial displacement of the element is considered and this must be expressed in terms of the displacements u_1 and u_2, at the left- and right-hand ends of the element, i.e. u_1 and u_2 are the element nodal variables. Thus, assuming a linear variation of displacement with x, i.e.

$$u = a_1 + a_2 x = [1 \quad x]\begin{bmatrix} a_1 \\ a_2 \end{bmatrix} = \mathbf{ga}, \tag{3.72}$$

the nodal values are

$$u_1 = a_1$$

and

$$u_2 = a_1 + a_2 l,$$

or in matrix form $\mathbf{u}_e = \mathbf{Na}$ where $\mathbf{u}_e^T = [u_1 \quad u_2]$. Thus

$$\mathbf{a} = \mathbf{Bu}_e \tag{3.73}$$

where

$$\mathbf{B} = \mathbf{N}^{-1} = \begin{bmatrix} 1 & 0 \\ -1/l & 1/l \end{bmatrix}.$$

From equation (3.72)

$$\frac{\partial u}{\partial x} = \mathbf{g'a} \quad \text{with } \mathbf{g'} = [0 \quad 1].$$

The strain energy of the element

$$\mathfrak{S}_e = \frac{1}{2} \int_0^l AE\left(\frac{\partial u}{\partial x}\right)^2 dx \tag{3.73a}$$

$$= \frac{1}{2} \int_0^l [\mathbf{g'a}]^T AE[\mathbf{g'a}]\, dx$$

$$= \frac{1}{2} \int_0^l \mathbf{u}_e^T \mathbf{B}^T \mathbf{g'}^T AE\mathbf{g'Bu}_e\, dx$$

using equation (3.73). Thus

$$\mathfrak{S}_e = \tfrac{1}{2}\mathbf{u}_e^T \mathbf{K}_e\, \mathbf{u}_e \tag{3.74}$$

where the element stiffness matrix

$$\mathbf{K}_e = \mathbf{B}^T \left[\int_0^l \mathbf{g}'^T A E \mathbf{g}' \, dx \right] \mathbf{B} \tag{3.75}$$

$$= \frac{AE}{l} \begin{bmatrix} 1 & -1 \\ -1 & 1 \end{bmatrix} \tag{3.76}$$

substituting for $\mathbf{g}'$ and $\mathbf{B}$, integrating and performing the matrix multiplications.

The kinetic energy of the element

$$\mathfrak{T}_e = \frac{1}{2} \int_0^l \rho A \left(\frac{\partial u}{\partial t} \right)^2 dx$$

$$= \tfrac{1}{2} \dot{\mathbf{u}}_e^T \mathbf{B}^T \left[\int_0^l \mathbf{g}^T \rho A \mathbf{g} \, dx \right] \mathbf{B} \dot{\mathbf{u}}_e$$

$$= \tfrac{1}{2} \dot{\mathbf{u}}_e^T \mathbf{M}_e \, \dot{\mathbf{u}}_e \tag{3.77}$$

where the element mass matrix

$$\mathbf{M}_e = \mathbf{B}^T \left[\int_0^l \mathbf{g}^T \rho A \mathbf{g} \, dx \right] \mathbf{B}$$

$$= \rho A l \begin{bmatrix} \frac{1}{3} & \frac{1}{6} \\ \frac{1}{6} & \frac{1}{3} \end{bmatrix} \tag{3.78}$$

and $\dot{\mathbf{u}}_e^T = [\dot{u}_1 \quad \dot{u}_2]$.

The element mass matrix (3.78) is full and symmetric; this is the consistent mass matrix for the element and is the standard approach in finite element theory. The alternative formulation is to use a lumped mass matrix, where the distributed mass is replaced by lumped or concentrated masses at the nodes and the kinetic energy is defined in terms of contributions from these lumped masses. This procedure leads to a diagonal mass matrix; in this case the diagonal elements of $\mathbf{M}_e$ are $\frac{1}{2}\rho A l$. In this section and in Chapter 5, where finite elements are applied to plate and shell problems, consistent mass matrices will be used. The following points are noted:

(a) The general mass matrix of Chapter 2 [equation (2.44)] was defined as symmetric, rather than as diagonal, so that the eigenvalue and response analysis of that chapter is applicable to finite element models with consistent mass matrices.

(b) The effect on the accuracy of natural frequencies of using consistent and lumped mass matrices is illustrated in the example in Appendix 5.

(c) For the determination of response by the central difference method (Section 2.8) there are computational advantages if the mass matrix is diagonal.

We require now to assemble the strain and kinetic energy expressions of the structure from the corresponding expressions for the individual elements. Considering the beam consisting of three elements, shown in Fig. 3.13b, the strain and kinetic energy expressions for each element are obtained from equations (3.74) to (3.78), using appropriate subscripts to identify the elements. Continuity of displacements at nodes 2 and 3 is assured, if the nodal variables of the structure are u_1, u_2, u_3 and u_4, where the subscript refers to the node number; i.e. u_2 is the displacement at the right-hand end of element 1 and also at the left-hand end of element 2, etc. Thus the strain energy of the structure can be written

$$\mathfrak{S} = \tfrac{1}{2}\mathbf{u}^T\mathbf{K}\mathbf{u} \tag{3.79}$$

where

$$\mathbf{u}^T = [u_1, u_2, u_3, u_4]$$

and the structural stiffness matrix

$$\mathbf{K} = \begin{bmatrix} \gamma_1 & -\gamma_1 & 0 & 0 \\ -\gamma_1 & \gamma_1 + \gamma_2 & -\gamma_2 & 0 \\ 0 & -\gamma_2 & \gamma_2 + \gamma_3 & -\gamma_3 \\ 0 & 0 & -\gamma_3 & \gamma_3 \end{bmatrix} \tag{3.80}$$

where γ_j is the value of (AE/l) for element j. The kinetic energy of the structure is

$$\mathfrak{T} = \tfrac{1}{2}\dot{\mathbf{u}}^T\mathbf{M}\dot{\mathbf{u}} \tag{3.81}$$

where $\dot{\mathbf{u}}^T = [\dot{u}_1, \dot{u}_2, \dot{u}_3, \dot{u}_4]$ and the structural mass matrix

$$\mathbf{M} = \tfrac{1}{3}\begin{bmatrix} \beta_1 & \beta_1/2 & 0 & 0 \\ \beta_1/2 & \beta_1 + \beta_2 & \beta_2/2 & 0 \\ 0 & \beta_2/2 & \beta_2 + \beta_3 & \beta_3/2 \\ 0 & 0 & \beta_3/2 & \beta_3 \end{bmatrix} \tag{3.82}$$

where β_j is the value of (ρAl) for element j. The extensions to $\mathbf{u}$, $\mathbf{K}$ and $\mathbf{M}$ for a structure with a larger number of elements are apparent.

The equations of motion are obtained by applying the Lagrange equation (A4.17).

$$\frac{d}{dt}\left(\frac{\partial \mathfrak{T}}{\partial \dot{u}_j}\right) - \frac{\partial \mathfrak{T}}{\partial u_j} + \frac{\partial \mathfrak{F}}{\partial \dot{u}_j} + \frac{\partial \mathfrak{S}}{\partial u_j} = Q_j. \qquad (3.83)$$

In successive applications of equation (3.83) u_j takes each nodal displacement in turn, except those which boundary conditions prescribe to be zero. For example, if the left-hand end of the beam in Fig. 3.13b is fixed and the right-hand end is free, three equations are obtained, corresponding to differentiation with respect to u_2, u_3 and u_4. In practice, the stiffness and mass matrices of the structure are reduced by eliminating rows and columns associated with zero nodal variables. For free vibration problems, where $\ddot{\mathbf{u}} = -\omega^2 \mathbf{u}$, application of equation (3.83) leads to

$$[\mathbf{K} - \mathbf{M}\omega^2]\mathbf{u} = 0 \qquad (3.84)$$

where $\mathbf{K}$ and $\mathbf{M}$ are the stiffness and mass matrices for the structure after rows and columns associated with zero boundary displacements have been eliminated, ω is a natural frequency and $\mathbf{u}$ is the vector of nodal variables (with zero boundary values excluded). Equation (3.84) is of the same form as equation (2.47) and the determination of natural frequencies and normal modes from equation (2.47) and their properties have been discussed in Section 2.5.

Hysteretic damping can be included, following Section 3.5, by replacing γ_j in equation (3.80) by $\gamma_j(1 + i\mu_j)$, where μ_j is the nondimensional damping factor for element j. Damping mechanisms concentrated at nodal points can be introduced through the dissipation function $\mathfrak{F}$. The generalized forces Q_j can be expressed in terms of applied forces through the principle of virtual work; for example, if an axial force $P \sin \omega t$ acts at node 4 in Fig. 3.13b, $Q_4 = P \sin \omega t$ and $Q_j = 0$ for $j \neq 4$. Thus for response problems the finite element method yields equation (2.40), whose solution has been discussed in detail in Chapter 2.

Returning to free vibrations, if a beam is free at $x = 0$ and clamped at $x = L$ and is approximated by three finite elements each of length l, the reduced stiffness and mass matrices are obtained from equations (3.80) and (3.82) by omitting the fourth row and column. Noting that $\gamma_1 = \gamma_2 = \gamma_3$ and $\beta_1 = \beta_2 = \beta_3$, equation (3.84) becomes

$$\left\{ \begin{bmatrix} 1 & -1 & 0 \\ -1 & 2 & -1 \\ 0 & -1 & 2 \end{bmatrix} \frac{AE}{l} - \begin{bmatrix} \frac{1}{3} & \frac{1}{6} & 0 \\ \frac{1}{6} & \frac{2}{3} & \frac{1}{6} \\ 0 & \frac{1}{6} & \frac{2}{3} \end{bmatrix} \rho A l \omega^2 \right\} \begin{bmatrix} u_1 \\ u_2 \\ u_3 \end{bmatrix} = 0.$$

From the lowest root of the frequency determinant

$$\omega_1 L(\rho/E)^{1/2} = 1{\cdot}5819$$

which is 1·1 per cent higher than the correct value of $\pi/2$ obtained in Section 3.1. If the beam is clamped at both ends, the reduced matrices are obtained by deleting the first and fourth rows and columns in equations (3.80) and (3.82), and the frequency determinant is

$$\det \left| \begin{bmatrix} 2 & -1 \\ -1 & 2 \end{bmatrix} \frac{AE}{l} - \begin{bmatrix} \frac{2}{3} & \frac{1}{6} \\ \frac{1}{6} & \frac{2}{3} \end{bmatrix} \rho A l \omega^2 \right| = 0.$$

The lowest root yields $\omega_1 L(\rho/E)^{1/2} = 3{\cdot}286$ which is 4·6 per cent higher than the correct value of π. If four equal elements are used to idealize the beam, the error in ω_1 is reduced to 2·6 per cent. With a reasonable number of elements the natural frequencies of the fundamental and higher modes are obtained accurately. The natural frequencies of seg-mented beams, comprising lengths of different cross-sections, can be determined also; it is necessary only to ensure that changes in cross-section occur at nodes in the finite element idealization.

Torsional vibration. The analogy between torsional and extensional vibrations has been demonstrated in Section 3.2. A finite element analysis for torsional vibrations of bars of circular cross-section is obtained from the above equations by replacing: u_j by θ_j, the cross-sectional rotation at node j; A by J, the polar second moment of area; and E by G.

Flexural vibrations. Figure 3.14a shows an element of length l which undergoes flexural deformation. The nodal variables are the

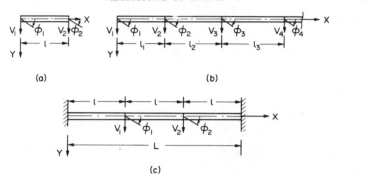

FIG. 3.14. Finite element method for flexure. (a) Single element. (b) Beam consisting of several elements. (c) Clamped–clamped beam divided into three elements.

transverse displacement v and the slope ϕ ($\equiv \partial v/\partial x$) at each node. Assuming a cubic variation of displacement with coordinate x, i.e.

$$v = a_1 + a_2 x + a_3 x^2 + a_4 x^3 = \begin{bmatrix} 1 & x & x^2 & x^3 \end{bmatrix} \begin{bmatrix} a_1 \\ a_2 \\ a_3 \\ a_4 \end{bmatrix} = \mathbf{ga}$$

(3.85)

the coefficients a_j can be related to the nodal values by

$$\mathbf{v}_e = \begin{bmatrix} v_1 \\ \phi_1 \\ v_2 \\ \phi_2 \end{bmatrix} = \begin{bmatrix} 1 & 0 & 0 & 0 \\ 0 & 1 & 0 & 0 \\ 1 & l & l^2 & l^3 \\ 0 & 1 & 2l & 3l^2 \end{bmatrix} \begin{bmatrix} a_1 \\ a_2 \\ a_3 \\ a_4 \end{bmatrix} = \mathbf{Na}.$$

Thus

$$\mathbf{a} = \mathbf{N}^{-1}\mathbf{v}_e = \mathbf{Bv}_e$$

(3.86)

where

$$\mathbf{B} = \begin{bmatrix} 1 & 0 & 0 & 0 \\ 0 & 1 & 0 & 0 \\ -3/l^2 & -2/l & 3/l^2 & -1/l \\ 2/l^3 & 1/l^2 & -2/l^3 & 1/l^2 \end{bmatrix}.$$

From equation (3.85)

$$\frac{\partial^2 v}{\partial x^2} = \mathbf{g}''\mathbf{a} \quad \text{with } \mathbf{g}'' = [0 \quad 0 \quad 2 \quad 6x]. \tag{3.87}$$

The strain energy of the element

$$\mathfrak{S}_e = \frac{1}{2}\int_0^l EI\left(\frac{\partial^2 v}{\partial x^2}\right)^2 dx = \tfrac{1}{2}\mathbf{v}_e^T \mathbf{B}^T \left[\int_0^l \mathbf{g}''^T EI\mathbf{g}'' \, dx\right]\mathbf{B}\mathbf{v}_e$$

using equations (3.86) and (3.87). Thus

$$\mathfrak{S}_e = \tfrac{1}{2}\mathbf{v}_e^T \mathbf{K}_e \, \mathbf{v}_e \tag{3.88}$$

where the element stiffness matrix

$$\mathbf{K}_e = \mathbf{B}^T \left[\int_0^l \mathbf{g}''^T EI\mathbf{g}'' \, dx\right]\mathbf{B}$$

$$= \frac{EI}{l^3}\begin{bmatrix} 12 & 6l & -12 & 6l \\ 6l & 4l^2 & -6l & 2l^2 \\ -12 & -6l & 12 & -6l \\ 6l & 2l^2 & -6l & 4l^2 \end{bmatrix} \tag{3.89}$$

after substitution for $\mathbf{g}''$ and $\mathbf{B}$, integration and matrix multiplication.

The kinetic energy of the element

$$\mathfrak{T}_e = \frac{1}{2}\int_0^l \rho A\left(\frac{\partial v}{\partial t}\right)^2 dx$$

$$= \tfrac{1}{2}\dot{\mathbf{v}}_e^T \mathbf{M}_e \, \dot{\mathbf{v}}_e \tag{3.90}$$

where the element mass matrix

$$\mathbf{M}_e = \mathbf{B}^T \left[\int_0^l \mathbf{g}^T \rho A\mathbf{g} \, dx\right]\mathbf{B}$$

$$= \frac{\rho Al}{420}\begin{bmatrix} 156 & 22l & 54 & -13l \\ 22l & 4l^2 & 13l & -3l^2 \\ 54 & 13l & 156 & -22l \\ -13l & -3l^2 & -22l & 4l^2 \end{bmatrix} \tag{3.91}$$

and

$$\dot{\mathbf{v}}_e^T = [\dot{v}_1,\ \dot{\phi}_1,\ \dot{v}_2,\ \dot{\phi}_2].$$

Considering the beam of Fig. 3.14b consisting of a number of elements, the nodal variables are v_1, ϕ_1, v_2, ϕ_2, v_3, ϕ_3, ... Each 4×4 element stiffness matrix is assembled to form the structure stiffness matrix as indicated by the dotted line boxes in $\mathbf{K}$ below; where two boxes overlap the contributions to $\mathbf{K}$ from the corresponding element matrices add; $\mathbf{K}_{ej}$ signifies the matrix for element j.

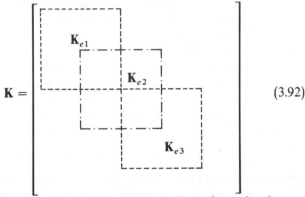

$$\text{(3.92)}$$

The structure mass matrix $\mathbf{M}$ is assembled similarly from the element mass matrices $\mathbf{M}_{ej}$. The matrices $\mathbf{K}$ and $\mathbf{M}$ are reduced by eliminating rows and columns associated with zero displacements and slopes at boundaries. Application of the Lagrange equation leads to equations (2.47) and (2.40) for eigenvalue and response problems respectively, as discussed earlier.

Figure 3.14c shows a uniform beam of length L with both ends clamped and divided into three equal elements, each of length l. The nodal variables are v_1, ϕ_1, v_2 and ϕ_2; displacement and slope are zero at both ends. On assembly of the three element matrices we obtain 8×8 structural matrices, but the first, second, seventh and eighth rows and columns are eliminated to account for the boundary conditions. Using equations (3.88) to (3.92), the Lagrange equation and putting $\ddot{\mathbf{v}} = -\omega^2\mathbf{v}$, the free vibration equations are

$$\left\{ \begin{bmatrix} 24 & 0 & -12 & 6l \\ 0 & 8l^2 & -6l & 2l^2 \\ -12 & -6l & 24 & 0 \\ 6l & 2l^2 & 0 & 8l^2 \end{bmatrix} \frac{EI}{l^3} - \begin{bmatrix} 312 & 0 & 54 & -13l \\ 0 & 8l^2 & 13l & -3l^2 \\ 54 & 13l & 312 & 0 \\ -13l & -3l^2 & 0 & 8l^2 \end{bmatrix} \frac{\rho Al\omega^2}{420} \right\} \begin{bmatrix} v_1 \\ \phi_1 \\ v_2 \\ \phi_2 \end{bmatrix} = 0$$

$$(3.93)$$

Natural frequencies ω_1 and ω_2 obtained from the determinant of equation (3.93) have errors of 0·4 and 2·0 per cent, respectively, compared with the solutions from the beam equation in Section 3.3. Using four equal elements for the same problem the errors reduce to 0·1 and 0·9 per cent. Leckie and Lindberg[46] showed that for this and other boundary conditions accurate natural frequencies can be obtained with a relatively small number of elements. Some further results for this problem are given in Appendix 5, where a method of reducing the size of the final matrices (a procedure that may be necessary when a large number of elements is used) is presented.

If the beam of Fig. 3.14 is subjected to transverse applied forces, the response, neglecting damping, can be found from the solution of the equation

$$\mathbf{M\ddot{v}} + \mathbf{Kv} = \mathbf{p}$$

where the matrices $\mathbf{K}$ and $\mathbf{M}$ can be inferred from equation (3.93). If the loading is a single force $Pf(t)$ at the node with displacement v_1 (Fig. 3.14c), the consistent force vector is

$$\mathbf{p}^T = [Pf(t) \quad 0 \quad 0 \quad 0].$$

If the loading is $p_0 f(t)$, per unit length, distributed over the middle element of the figure, the vector $\mathbf{p}$, which consists of the generalized forces corresponding to the nodal displacements $\mathbf{v}_e^T = [v_1, \phi_1, v_2, \phi_2]$, is found by virtual work, considering a virtual increment $\delta \mathbf{v}_e$ for the displacements of this element, from

$$\delta \mathfrak{W} = \mathbf{p}^T \, \delta \mathbf{v}_e = f(t) \int_0^l p_0 \, \delta v(x) \, dx$$

$$= f(t) \int_0^l p_0 \mathbf{gB} \, \delta \mathbf{v}_e \, dx$$

from equations (3.85) and (3.86). Thus

$$\mathbf{p}^T = f(t)\left[\int_0^l p_0 \mathbf{g} \, dx\right]\mathbf{B}$$

$$= f(t)[\tfrac{1}{2}p_0 l, \quad \tfrac{1}{12}p_0 l^2, \quad \tfrac{1}{2}p_0 l, \quad -\tfrac{1}{12}p_0 l^2]$$

after substitution for $\mathbf{g}$ and $\mathbf{B}$, integration and matrix multiplication.

In the problems discussed in this section all elements have had the same coordinate axes and flexural, extensional and torsional deformations have been considered separately. For a beam system for which the coordinate axes of individual beams are in different directions the element matrices are derived in terms of local coordinates and then transformed in terms of system or global axes before assembling the structural matrices. When an element is subjected simultaneously to flexure and torsion or extension, the appropriate 4 × 4 flexural and 2 × 2 torsional or extensional matrices can be merged to give a 6 × 6 element matrix.

For simple beams the finite element method may lead to more computation than other approximate methods, but its merit appears when applied to complex problems, such as the vibration of two- and three-dimensional frames, consisting of beams, and the vibration of stiffened structures when beam elements are used for the stiffeners in conjunction with plate or shell elements. Curved beam finite elements and deep beam elements, in which the effects of transverse shear deformation and rotatory inertia are included, have been developed[25–27].

Problems

1. A uniform steel bar of length 2 m is clamped at both ends. The cross-section is square of side 50 mm. Find the first three natural frequencies of extensional vibrations. (For steel $E = 207$ GN/m^2 and $\rho = 7800$ kg/m^3.)
2. A uniform bar of circular cross-section is clamped at the end $x = 0$. The end $x = l$ is given an initial angular rotation about the X-axis θ_0 and the bar is released from this position at time $t = 0$. Show that the resulting torque at the clamped end is

$$\frac{4GJ\theta_0}{\pi l} \sum_n \left[\frac{(-1)^{n+1}}{(2n-1)} \cos \frac{(2n-1)\pi C_2 t}{l}\right].$$

3. A uniform beam of length l is clamped at both ends. Show that the natural frequencies in flexure are given by the roots of the equation $\cos \lambda l \cosh \lambda l - 1 = 0$ and that the shape of the rth mode is given by

$$V_r(x) = B_r[\cosh \lambda_r x - \cos \lambda_r x - \eta_r(\sinh \lambda_r x - \sin \lambda_r x)]$$

where

$$\eta_r = \frac{\cosh \lambda_r l - \cos \lambda_r l}{\sinh \lambda_r l - \sin \lambda_r l}.$$

If the above beam has the dimensions and material properties given in Problem 1, find the first three natural frequencies in flexure.

4. A uniform beam of length l, flexural rigidity EI and mass per unit length ρA is clamped at one end and carries a concentrated mass $2m$ at the other end, where $m = \rho Al$. Without using an approximate method, show that the first and second natural frequencies of flexural vibration of this beam are given by $\omega_1 = 1 \cdot 1582(EI/ml^3)^{1/2}$ and $\omega_2 = 15 \cdot 861(EI/ml^3)^{1/2}$.

5. A uniform beam of length l and flexural rigidity EI is simply supported at one end and is supported at the other end by a spring of stiffness k; the spring exerts no directional restraint on the beam. Derive the frequency equation for flexural vibrations for the beam. Determine the first two values of the frequency parameter λl, if kl^3/EI equals (a) $0 \cdot 01$, (b) 1 and (c) ∞.

6. A transverse harmonic force $P \sin \omega t$ is applied at a quarter-span point of a uniform beam with simply supported ends. Find the steady-state amplitude of forced vibration at (a) the quarter-span point, (b) mid-span, if (i) $\omega/\omega_1 = 0 \cdot 75$, (ii) $\omega/\omega_1 = 1 \cdot 25$, where $\omega_1/2\pi$ is the fundamental natural frequency of flexural vibrations. Neglect damping.

7. A uniform cantilever beam of length l is subjected to a transverse harmonic force $P \sin \omega t$ at the tip. Determine the steady-state amplitude of displacement at the tip and the bending moment at the root-section in the form V/V_{st} and M/M_{st} respectively, where V_{st} and M_{st} are the corresponding static values, if (a) $\omega = \omega_1$, (b) $\omega = \omega_2$ and the internal damping in the beam is (i) hysteretic with $\mu = 0 \cdot 01$, (ii) viscous with $c\omega_1 = 0 \cdot 01$. In each case state the number of terms required in a normal mode solution to give a solution accurate to three significant figures. (The modal parameters required can be found in Section 3.6.)

8. A transverse force $P \sin \omega t$, $0 \le t \le \pi/\omega$, is applied at mid-span to a uniform beam with simply supported ends. If $\omega/\omega_1 = \frac{1}{2}$, where $\omega_1/2\pi$ is the fundamental natural frequency in flexure, find the displacement at mid-span when (a) $t = \pi/2\omega$, (b) $t = \pi/\omega$. Neglect damping.

If the applied force is zero for $t > \pi/\omega$, derive an expression for the resulting free vibrations of the beam measured at mid-span.

9. A transverse force:

$$P(t) = P_0, \qquad 0 \le t \le t_1,$$

$$P(t) = 0, \qquad t > t_1$$

is applied at the tip of a uniform cantilever. If $\omega_1 t_1 = \pi$, where $\omega_1/2\pi$ is the fundamental natural frequency in flexure, find the displacement at the tip when (a) $t = \frac{1}{2}t_1$, (b) $t = t_1$. Also find the maximum stress in the root section when $t = t_1$. Neglect damping.

10. If the transverse force in Problem 9 is replaced by a pressure $p(t)$ per unit length of beam, distributed over the whole length, and

$$p(t) = p_0, \qquad 0 \le t \le t_1,$$
$$p(t) = 0, \qquad t > t_1,$$

find the tip displacement for (a) $t = \frac{1}{2}t_1$ and (b) $t = t_1$, if $\omega_1 t_1 = \pi$. Neglect damping.

11. A transverse force:

$$P(t) = P_0(1 - \cos \omega t), \qquad 0 \le t \le 2\pi/\omega,$$
$$P(t) = 0, \qquad t > 2\pi/\omega$$

is applied at the tip of a uniform cantilever. If $\omega/\omega_1 = 2$, where $\omega_1/2\pi$ is the fundamental natural frequency in flexure, find the tip displacement when ωt equals (a) π, (b) 2π, and (c) 3π, if (i) damping is neglected and (ii) the internal damping is viscous with $c\omega_1 = 0.05$.

12. A cantilever beam of uniform width and length l tapers uniformly from a depth d at the fixed end to a depth $d/5$ at the free end. Using (a) the Rayleigh, or Rayleigh–Ritz, method and (b) the finite element method, obtain approximate values for the first and second natural frequencies in flexure of the beam; compare these values with the analytical solution, which can be expressed as $\omega_r = \beta_r(Ed^2/\rho l^4)^{1/2}$ with $\beta_1 = 1.239$ and $\beta_2 = 4.545$.

13. A simply supported beam of length l, flexural rigidity EI and constant mass per unit length ρA carries a concentrated mass m at mid-span, where $m = \rho Al$. Determine the fundamental natural frequency in flexure of the beam, (a) from the beam equation, considering one-half of the beam and using the symmetry properties, (b) by the Rayleigh–Ritz method and (c) by the finite element method.

14. A solid shaft of length l is supported at its ends so that it behaves as a beam with simply supported ends. A length of $2l/3$ from one end is of diameter D and the remaining length is of diameter $D/4$. Find the fundamental natural frequency in flexure of the shaft, using an approximate method, and use the solution from Problem 13 of Chapter 4 to assess its accuracy.

15. A uniform bar of cross-sectional area A and length l is clamped at one end and carries a concentrated mass m at the other end, where m is equal to the total distributed mass of the bar. Determine the fundamental natural frequency of extensional vibrations of the bar, (a) from a solution satisfying the governing equation and the boundary conditions and (b) by the finite element method.

16. Assuming the data of Problem 9, use the Rayleigh method to determine the tip displacement when (a) $t = \frac{1}{2}t_1$, (b) $t = t_1$.

CHAPTER 4

Vibrations of Beams—II

4.1. Response to Moving Loads

This section is an introduction to the complex practical problem of the vibration of bridges due to the passage of vehicles. The forces on the bridge due to a moving vehicle are of three types: (a) gravity forces; (b) inertia forces; (c) disturbing forces due to unbalanced parts in the vehicle. The lines of action of these forces move across the bridge at a certain velocity. Practical problems lie between the following extreme cases.

(1) The weight of the vehicle is large compared with that of the bridge, so that the system can be treated as a concentrated weight on a light elastic beam, i.e. as a single-degree-of-freedom system, but as the weight is moving across the beam, the stiffness of the system varies continuously. Timoshenko[75] gives an approximate solution of this problem.

(2) The weight of the vehicle is small compared with that of the bridge, so that the system can be idealized as an elastic beam subjected to a moving *force*. This will be considered, subject to the assumptions that the beam is uniform, its ends are simply supported (thus sinusoidal mode shapes can be used), and the velocity of the force is constant.

Harmonic force. Figure 4.1 shows a uniform beam with simply supported ends. The harmonic disturbing force $P \sin (\omega t + \alpha)$, which could be caused by unbalanced parts in a vehicle, moves uniformly across the beam from left to right with constant velocity U and it is assumed that the force is at the left-hand support at time $t = 0$. For a simply supported beam the response in terms of normal modes is

$$v = \sum_s 2^{1/2} \sin \frac{s\pi x}{l} q_s(t) \qquad (4.1)$$

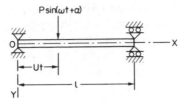

FIG. 4.1. Simply supported beam subjected to a harmonic force, which moves across the beam with uniform velocity U.

where the factor $2^{1/2}$ has been introduced to satisfy the normalizing condition (3.43). The force acts at $x = Ut$, so from equation (3.44), neglecting damping,

$$\ddot{q}_s + \omega_s^2 q_s = \frac{2^{1/2}P \sin(\omega t + \alpha)}{m} \sin\left(\frac{s\pi Ut}{l}\right)$$

$$= \frac{P}{2^{1/2}m}\left[\cos\left(\omega t + \alpha - \frac{s\pi Ut}{l}\right) - \cos\left(\omega t + \alpha + \frac{s\pi Ut}{l}\right)\right].$$

$$(4.2)$$

The complete solution is

$$q_s = B_s \sin \omega_s t + C_s \cos \omega_s t$$

$$+ \frac{P}{2^{1/2}m}\left\{\frac{\cos[\omega t + \alpha - (s\pi Ut/l)]}{\omega_s^2 - [\omega - (s\pi U/l)]^2} - \frac{\cos[\omega t + \alpha + (s\pi Ut/l)]}{\omega_s^2 - [\omega + (s\pi U/l)]^2}\right\}.$$

$$(4.3)$$

If the beam is at rest in its equilibrium position at $t = 0$, then $q_s(0) = 0 = \dot{q}_s(0)$, and the constants B_s and C_s are evaluated from these conditions. Inspection of equation (4.3) shows that large amplitudes will occur when

$$\omega_s = \omega \pm s\pi U/l. \qquad (4.4)$$

If either of conditions (4.4) is satisfied, one of the terms in equation (4.2) can be rewritten to give

$$\ddot{q}_s' + \omega_s^2 q_s' = K \cos(\omega_s t + \alpha). \qquad (4.5)$$

If the positive sign in equation (4.4) is taken, the particular integral q_s' from equation (4.5) gives the part of q_s corresponding to the cos $(\omega t + \alpha + s\pi Ut/l)$ term in equation (4.3); the other part of the particular integral of q_s is as given in equation (4.3). It can be shown by substitution that the particular integral of equation (4.5) is

$$q_s' = \frac{Kt}{2\omega_s} \sin(\omega_s t + \alpha). \tag{4.6}$$

All the above equations apply only to $t \leq l/U$, i.e. when the force is on the beam. Thus considering the form of equation (4.6) and the various modes, large finite amplitudes are possible if the forcing frequency

$$\omega = \omega_s \pm s\pi U/l, \qquad s = 1, 2, 3 \ldots .$$

Constant force. Next the maximum deflection due to a single constant force P, which comes on to the beam at one end at time $t = 0$ and moves across the beam at constant velocity U, will be determined. Assuming that the beam is simply supported at both ends, the response in terms of normal modes is given by equation (4.1). At time t the force is at $x = Ut$, so in equation (3.44),

$$\phi_s(a) = 2^{1/2} \sin \frac{s\pi Ut}{l}. \tag{4.7}$$

The solution is

$$q_s = B_s \sin \omega_s t + C_s \cos \omega_s t + 2^{1/2} \frac{P}{m} \frac{\sin s\pi Ut/l}{[\omega_s^2 - (s\pi U/l)^2]}.$$

If the beam is at rest in its equilibrium position at $t = 0$, then $q(0) = 0 = \dot{q}(0)$, giving B_s and C_s, and the solution becomes

$$q_s = \frac{2^{1/2}P}{m} \frac{[\sin s\pi Ut/l - (s\pi U/l\omega_s) \sin \omega_s t]}{\omega_s^2 - (s\pi U/l)^2}. \tag{4.8}$$

Equations (4.1) and (4.8) give the response of the beam, provided that $\omega_s \neq s\pi U/l$. If $\omega_s = s\pi U/l$, equation (4.7) is replaced by

$$\ddot{q}_s + \omega_s^2 q_s = \frac{2^{1/2}P}{m} \sin \omega_s t$$

with the solution

$$q_s = B_s \sin \omega_s t + C_s \cos \omega_s t - \frac{2^{1/2}P}{m} \frac{t}{2\omega_s} \cos \omega_s t.$$

Applying the conditions for $t = 0$,

$$q_s = \frac{P}{2^{1/2}m\omega_s^2} (\sin \omega_s t - \omega_s t \cos \omega_s t). \qquad (4.9)$$

Considering first the case: $\omega_1 = \pi U/l$, the maximum value of q_1 from equation (4.9) occurs when $\omega_1 t = \pi$, i.e. when $t = l/U$ or the force is just leaving the beam, and neglecting the higher modes the maximum displacement, which is at $x = l/2$, is given by

$$v(l/2, \pi/\omega_1) = \frac{P}{m\omega_1^2} (0 + \pi).$$

Now $m\omega_1^2 = \pi^4 EI/l^3$, so

$$v_{\max} = \pi \left(\frac{Pl^3}{\pi^4 EI} \right).$$

It can be seen from equation (4.8) that the contribution of the higher modes, $s > 1$, to the displacement when $t = \pi/\omega_1$ is zero. The maximum static deflection for a force P is $Pl^3/48EI$. Thus if $\omega_1 = \pi U/l$, the maximum displacement occurs at mid-span when the force is just leaving the beam and is $1\cdot548 \times$ the static deflection.

Considering the range of speeds, $0 < (\pi U/l\omega_1) < 1$, and assuming initially that the displacement is dominated by the contribution from the fundamental mode, the maximum displacement occurs at mid-span and is obtained from equations (4.1) and (4.8) as

$$v(l/2, t) = \frac{2Pl^3}{\pi^4 EI} \left[\frac{\sin \psi\omega_1 t - \psi \sin \omega_1 t}{1 - \psi^2} \right] \qquad (4.10)$$

where $\psi = \pi U/l\omega_1$. Equation (4.10) applies while the force is on the beam, i.e. for $0 \le t \le l/U$. For $t > l/U$ free vibrations of the form

$$v(l/2, t) = B \sin \omega_1(t - l/U) + C \cos \omega_1(t - l/U)$$

occur; the constants B and C are found from the continuity conditions for v and $\dot{v}$ at $t = l/U$. However, even neglecting damping, the amplitude of free vibrations is always less than the maximum displacement of the beam during the period of application of the force if $\psi < 1$. Thus maximum values of v from equation (4.10) are required; these will occur when

$$\cos \psi \omega_1 t - \cos \omega_1 t = 0. \tag{4.11}$$

In Fig. 4.2 the displacement at mid-span has been plotted against t [from equation (4.10)] for the time interval $0 \leq t \leq l/U$ for the values of ψ: 0·1, 0·2 and 0·3; the corresponding curve for $\psi = 1$ from equation (4.9) has been added for comparison. For $\psi = 0·3$ there are two mathematical maximum displacements, but the first is considerably greater

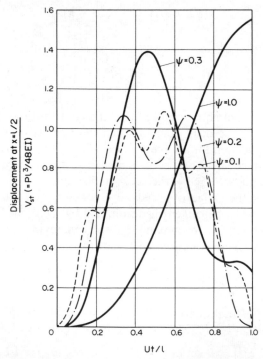

FIG. 4.2. Response at mid-span of a simply supported beam to a constant force P, moving across the beam with uniform velocity U; $\psi = \pi U/\omega_1 l$.

than the second; for $\psi = 0.2$ the first two mathematical maximum displacements are equal in magnitude; for $\psi = 0.1$ the third mathematical maximum is the greatest in magnitude.

Thus, solving equation (4.11), the maximum displacement occurs when $\omega_1 t = 2\pi/(1 + \psi)$ for $0.2 < \psi < 1.0$ and is

$$v_{max} = \frac{2Pl^3}{\pi^4 EI} \frac{\sin\left[2\pi\psi/(1 + \psi)\right]}{1 - \psi}. \tag{4.12}$$

In Fig. 4.3 the dynamic magnification factor, defined as the maximum displacement at $x = \frac{1}{2}l/v_{st}$, where v_{st} is the static deflection at $x = l/2$ for a force P at that point (i.e. $v_{st} = Pl^3/48EI$), is plotted against the velocity parameter ψ. The curve for $0.2 < \psi < 1.0$ is obtained from equation (4.12); the numbers marked against the different parts of the curve refer to the mathematical maximum which is greatest in magnitude. There are discontinuities in slope at the values of ψ for which two mathematical maxima are of equal magnitude, i.e. at $\psi = \frac{1}{5}, \frac{1}{9}, \frac{1}{13} \ldots$. It can be shown that the maximum value of expression (4.12) occurs when $\psi = 0.617$ and that the maximum dynamic magnification factor

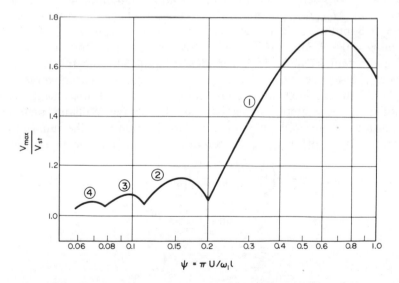

FIG. 4.3. Variation of maximum amplitude at mid-span with velocity parameter, ψ.

is 1·743. It should be noted that some writers have given particular attention to the condition $\psi = 1$, for which the dynamic magnification factor is 1·548, and have inferred incorrectly that this represents the worst possible condition. Figure 4.3 shows that there is a significant dynamic magnification factor for a wide range of speeds.

In the above discussion the effect of the modes, corresponding to $s = 2, 3, 4 \ldots$, has been neglected. However, their effect on displacement is relatively small. If the first five modes of vibration are considered, the maximum dynamic magnification is 1·738 and occurs when $\psi = 0\cdot625$ and $x/l = 0\cdot53$.

The bending moment is given by

$$M(x, t) = -EI \frac{\partial^2 v}{\partial x^2}$$

$$= EI \sum_s \frac{2^{1/2}s^2\pi^2}{l^2} (\sin s\pi x/l)q_s$$

with q_s given by equation (4.8) for $\psi < 1\cdot0$. If only the fundamental mode is considered, a curve of the dynamic magnification factor for moment, M_{max}/M_{st}, where $M_{st} = Pl/4$, would have the same form as Fig. 4.3, but the ordinates would differ. For $x = l/2$, $\omega_1 t = 2\pi/(1 + \psi)$ and $\psi = 0\cdot617$, $M_{max}/M_{st} = 1\cdot434$. However, the contribution to bending moment from the higher modes is significant and the series for $M(x, t)$ converges slowly. (A comparable deterioration in convergence, when bending moment is considered instead of displacement, was noted in the example of Section 3.6.) Using the Williams method (Section 3.6) and obtaining a closed form solution for the sum of the terms outside the integral, the bending moment can be expressed as

$$\frac{M(x, t)}{M_{st}} = 4\left(1 - \frac{Ut}{l}\right)\frac{x}{l} - 4\left|\frac{x}{l} - \frac{Ut}{l}\right|$$

$$+ \frac{8}{\pi^2}\sum_s \frac{\sin\frac{s\pi x}{l}\left(\sin\frac{s\pi Ut}{l} - \frac{s}{\psi}\sin\omega_s t\right)}{s^2\left(\frac{s^2}{\psi^2} - 1\right)} \quad (4.13)$$

where the term $\{\cdots\}$ is included only if $x \geq Ut$. Investigation of equation (4.13), including seven modes in the series expression, shows that (a) the maximum value of $M(x, t)/M_{st}$ is 1·550, occurring when $\psi = 0·525$, $x/l = 0·636$ and $Ut/l = 0·636$; (b) $M(x, t)/M_{st}$ exceeds 1·45 for the speed range $0·36 \leq \psi \leq 0·7$; and (c) for this speed range the maximum value of $M(x, t)$ occurs when $x \simeq Ut$, i.e. at the section where the force is instantaneous, but this value of Ut/l increases as ψ increases.

In the above problem of a constant force moving uniformly across a beam, there is a paradox as the force apparently performs no net work (its net displacement in the transverse direction is zero), but the beam is left in a state of oscillation. If the force is considered to be applied to the beam through a circular disc of negligible mass, which rolls on the beam, the point of contact having the prescribed horizontal velocity U, then a couple must be applied to the disc to maintain this motion, and it is this couple which does work equal to the energy stored by the beam[58].

Frýba[34] has made an extensive study of the vibrations of beams and plates due to moving loads.

4.2. Response of Uniform Beams to Time-dependent Boundary Conditions

In many practical problems a disturbance is transmitted to a beam through its supports, so that a method of determining the response of the beam when a displacement, linear or angular, $f(t)$ is imposed on one end is required. This problem is more complex than that of determining the response to a transient force applied at some point on the beam. The method given here is a simplification of the general case considered by Mindlin and Goodman[61].

It will be assumed that only one of the boundary conditions is time-dependent. Figure 4.4 shows four examples, where a time-dependent displacement (linear or angular) is imposed at a boundary: (a) a displacement $f(t)$ imposed at the root of a cantilever; (b) a dependent displacement (linear or angular) is imposed at a boundary: $f(t)$ imposed on the end $x = 0$ of a simply supported beam; and (d) a rotation $f(t)$ imposed on the end of a simply supported beam. Figure 4.4 shows also two examples of excitation by force or moment applied

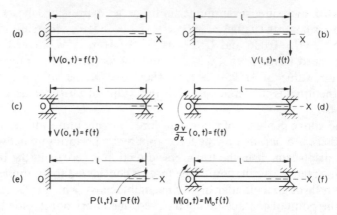

FIG. 4.4. Beams with prescribed time-dependent boundary conditions. (a) Cantilever with root displacement. (b) Cantilever with tip displacement. (c) Simply supported beam with displacement at a support. (d) Simply supported beam with rotation imposed at a support. (e) Cantilever with tip force. (f) Simply supported beam with moment applied at a support.

at a boundary; (e) a force $Pf(t)$ at the tip of a cantilever; and (f) a moment $M_0 f(t)$ at the end $x = 0$ of a simply supported beam. The method to be developed in this section is intended primarily for cases (a) to (d); it is applicable to cases (e) and (f), but the method of Section 3.6 can be used for these cases and is preferable. However, (e) and (f) are included here in order to illustrate the relationship between the methods of this section and Section 3.6.

In each case a solution of the form

$$v(x, t) = v_1(x, t) + v_2(x, t)$$

where

$$v_1(x, t) = f(t)g(x)$$

and

$$v_2(x, t) = \sum_s \phi_s(x)q_s(t)$$

is sought.

The solution $v(x, t)$ must satisfy the boundary conditions, including the time-dependent condition, the initial conditions and the governing equation for vibrations of a uniform beam,

$$EI \frac{\partial^4 v}{\partial x^4} + \rho A \frac{\partial^2 v}{\partial t^2} = 0. \tag{4.14}$$

The function $v_1(x, t)$ must satisfy the complete set of boundary conditions; thus the appropriate influence function is chosen for $g(x)$. For a prescribed displacement (or rotation) at a boundary the influence function gives the displacement at x due to unit translation (or rotation) at that boundary and satisfies the other boundary conditions. The function $v_2(x, t)$ must satisfy the *reduced* boundary conditions, which are obtained by replacing $f(t)$, when it occurs in a geometric boundary condition, by zero.

Boundary Conditions and Functions $g(x)$

Figure 4.4a. The boundary conditions

$$v = f(t) \quad \text{and} \quad \partial v/\partial x = 0 \quad \text{at } x = 0$$

and

$$\partial^2 v/\partial x^2 = 0 \quad \text{and} \quad \partial^3 v/\partial x^3 = 0 \quad \text{at } x = l$$

are satisfied by $v_1(x, t)$, if $g(x) = 1$ (i.e. a rigid body translation). Then $v_2(x, t)$ must satisfy

$$v_2 = 0 \quad \text{and} \quad \partial v_2/\partial x = 0 \quad \text{at } x = 0$$

and

$$\partial^2 v_2/\partial x^2 = 0 \quad \text{and} \quad \partial^3 v_2/\partial x^3 = 0 \quad \text{at } x = l.$$

These are the boundary conditions for a cantilever beam; thus the functions $\phi_s(x)$ are the orthogonal normal modes of a uniform cantilever beam.

Figure 4.4b. The boundary conditions

$$v = 0 \quad \text{and} \quad \partial v/\partial x = 0 \quad \text{at } x = 0$$

and

$$v = f(t) \quad \text{and} \quad \partial^2 v/\partial x^2 = 0 \quad \text{at } x = l$$

are satisfied by $v_1(x, t)$, if

$$g(x) = \frac{3}{2}\left(\frac{x}{l}\right)^2 - \frac{1}{2}\left(\frac{x}{l}\right)^3.$$

(This influence function has the same shape as that for a cantilever beam subjected to a tip force.) The reduced boundary conditions to be satisfied by $v_2(x, t)$ are

$$v_2 = 0 \quad \text{and} \quad \partial v_2/\partial x = 0 \qquad \text{at } x = 0,$$

$$v_2 = 0 \quad \text{and} \quad \partial^2 v_2/\partial x^2 = 0 \qquad \text{at } x = l.$$

As these are the boundary conditions for a beam, which is clamped at $x = 0$ and simply supported at $x = l$, the functions $\phi_s(x)$ are the normal modes for this type of beam.

The influence functions $g(x)$ and the reduced boundary conditions, which govern the normal modes $\phi_s(x)$, are given in Table 4.1 for the various cases shown in Fig. 4.4.

TABLE 4.1

Figure	$g(x)$	Reduced boundary conditions for $\phi_s(x)$ at	
		$x = 0$	$x = l$
4.4a	1	Clamped	Free
4.4b	$\dfrac{3}{2}\left(\dfrac{x}{l}\right)^2 - \dfrac{1}{2}\left(\dfrac{x}{l}\right)^3$	Clamped	Simply supported
4.4c	$1 - x/l$	Simply supported	Simply supported
4.4d	$\dfrac{l}{2}\left[2\left(\dfrac{x}{l}\right) - 3\left(\dfrac{x}{l}\right)^2 + \left(\dfrac{x}{l}\right)^3\right]$	Clamped	Simply supported
4.4e	$\dfrac{Pl^3}{3EI}\left[\dfrac{3}{2}\left(\dfrac{x}{l}\right)^2 - \dfrac{1}{2}\left(\dfrac{x}{l}\right)^3\right]$	Clamped	Free
4.4f	$\dfrac{M_0 l^2}{6EI}\left[2\left(\dfrac{x}{l}\right) - 3\left(\dfrac{x}{l}\right)^2 + \left(\dfrac{x}{l}\right)^3\right]$	Simply supported	Simply supported

General solution. Having determined $\phi_s(x)$ and $g(x)$, the complete expression $v = g(x)f(t) + v_2(x, t)$ is substituted in equation (4.14), giving

$$EI\frac{\partial^4 v_2}{\partial x^4} + \rho A\frac{\partial^2 v_2}{\partial t^2} = -\rho Ag(x)\ddot{f}(t). \qquad (4.15)$$

As the influence functions $g(x)$ do not contain powers higher than x^3, $d^4g/dx^4 = 0$. Substituting $v_2(x, t) = \sum_s \phi_s(x)q_s(t)$ in equation (4.15),

$$\sum_s \phi_s(x)\ddot{q}_s(t) + \frac{EI}{\rho A} \sum_s \frac{d^4\phi_s}{dx^4} q_s(t) = -g(x)\ddot{f}(t). \qquad (4.16)$$

Multiplying equation (4.16) by $\phi_s(x)$, integrating with respect to x from 0 to l and using the orthogonality conditions (A3.2) and (A3.9), the normalizing condition (3.43) and equation (A3.14),

$$\ddot{q}_s + \omega_s^2 q_s = -\ddot{f}(t)(1/l)\int_0^l g(x)\phi_s(x)\,dx = K_s\ddot{f}(t) \qquad (4.17)$$

where

$$K_s = -(1/l)\int_0^l g(x)\phi_s(x)\,dx. \qquad (4.18)$$

The complete solution of equation (4.17), using Duhamel's integral, is

$$q_s = \frac{K_s}{\omega_s}\int_0^t \ddot{f}(\tau)\sin\omega_s(t-\tau)\,.\,d s\tau + B_s\sin\omega_s t + C_s\cos\omega_s t$$

$$= K_s F_s(t) + B_s\sin\omega_s t + C_s\cos\omega_s t \qquad (4.19)$$

where

$$F_s(t) = \frac{1}{\omega_s}\int_0^t \ddot{f}(\tau)\sin\omega_s(t-\tau)\,d\tau. \qquad (4.20)$$

The response is given by

$$v(x, t) = g(x)f(t) + \sum_s [\phi_s(x)\{K_s F_s(t) + B_s\sin\omega_s t + C_s\cos\omega_s t\}]. \qquad (4.21)$$

The constants B_s and C_s are determined from the initial conditions, which will be assumed to be $v(x, 0) = 0$, $\dot{v}(x, 0) = 0$. If the function $g(x)$ is expanded as a series of normal modes, i.e.

$$g(x) = \sum_s D_s \phi_s(x),$$

the coefficients D_s are obtained from

$$\int_0^l g(x)\phi_s(x)\, dx = l \cdot D_s,$$

using the orthogonality and normalizing conditions.

Thus $D_s = -K_s$ and

$$v(x, t) = \sum_s [\phi_s(x)\{K_s F_s(t) - K_s f(t) + B_s \sin \omega_s t + C_s \cos \omega_s t\}].$$

For the assumed initial conditions to be satisfied at all values of x,

$$K_s[F_s(0) - f(0)] + C_s = 0,$$

$$K_s[\dot{F}_s(0) - \dot{f}(0)] + \omega_s B_s = 0. \tag{4.22}$$

Provided that $\ddot{f}(\tau)$ is well behaved in the vicinity of the origin (i.e. $\ddot{f}(\tau)$ can be expressed as $a_0 + a_1 \tau + a_2 \tau^2 + \cdots$ for $0 \le \tau \le \varepsilon$, where ε is small), $F_s(0)$ and $\dot{F}_s(0)$ are zero. Then from equations (4.21) and (4.22) the dynamic response is given by

$$v(x, t) = g(x)f(t)$$

$$+ \sum_s \left[\phi_s(x)K_s \left\{ F_s(t) + f(0) \cos \omega_s t + \frac{\dot{f}(0)}{\omega_s} \sin \omega_s t \right\} \right].$$

$$\tag{4.23}$$

(The necessity to expand $g(x)$ as a series of normal modes is avoided when $f(0) = 0$ and $\dot{f}(0) = 0$; in this case $B_s = 0$ and $C_s = 0$ by inspection.)

It is of interest to compare equation (4.23) with that obtained by the method of Section 3.6 for the force and moment excitation given in Fig. 4.4e and f. From equation (3.46), neglecting damping, the response of the cantilever of Fig. 4.4e is

$$v(x, t) = \sum_s \left[\frac{P\phi_s(l)\phi_s(x)}{m\omega_s} \int_0^t f(\tau) \sin \omega_s(t - \tau)\, d\tau \right]. \tag{4.24}$$

For the simply supported beam of Fig. 4.4f $M_0[d\phi_s/dx]_{x=0}$ replaces $P\phi_s(l)$ in equation (4.24). Integrating equation (4.24) by parts twice, using the appropriate definition of $g(x)$ and noting that, after writing K_s in the form

$$K_s = -\frac{1}{\lambda_s^4 l} \int_0^l g(x) \frac{d^4\phi_s}{dx^4} dx,$$

K_s can be evaluated for each case by integrating by parts and using the relevant boundary conditions, equation (4.23) is obtained. As mentioned in Section 3.6, integration by parts can have beneficial effects on convergence; it is necessary for $f(0)$ and $\dot{f}(0)$ to be zero for equation (4.23) to have advantages over equations (3.46) and (3.56) when force excitation is considered.

When a displacement is prescribed at a support, as in Fig. 4.4 a–d, it may be preferable for the response to be defined in terms of the displacement $f(\tau)$, rather than in terms of the acceleration $\ddot{f}(\tau)$ as in equations (4.20) and (4.23). Using an integral equation method, the author[13] obtained the response in terms of $f(\tau)$ as

$$v(x, t) = -\sum_s \phi_s(x) K_s \omega_s \int_0^t f(\tau) \sin \omega_s(t - \tau) \, d\tau. \qquad (4.25)$$

Equation (4.25) can be modified to treat multi-span beams, non-uniform beams and excitation at more than one support. If excitations $f_A(t)$ and $f_B(t)$ are prescribed at boundaries A and B, respectively, the influence function $g_A(x)$ must give unit displacement at A and zero displacement at B, $g_B(x)$ must give unit displacement at B and zero displacement at A, and the reduced boundary conditions, which govern the normal modes $\phi_s(x)$, include zero displacement at A and at B. With this proviso the method of this section can be extended to multi-support excitation.

4.3. Effect of an Axial Force on Flexural Vibrations

The effect of a constant axial compressive force on the natural frequencies of flexural vibrations of uniform beams is of practical importance. Figure 4.5 shows a simply supported beam, subjected at

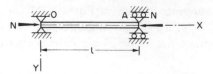

FIG. 4.5. Simply supported beam subjected to axial forces.

the ends to axial compressive forces N. At any section x there are two contributions to the bending moment $M(x)$: (i) $M_1(x)$ due to the distributed transverse inertia force of $\rho A\, \partial^2 v/\partial t^2$ per unit length; and (ii) $M_2(x) = Nv$ due to the axial force; v is the displacement of section x. As

$$\frac{\partial^2 M_1(x)}{\partial x^2} = \rho A \frac{\partial^2 v}{\partial t^2}$$

and the bending moment–curvature relation is

$$-EI\, \partial^2 v/\partial x^2 = M(x),$$

the equation of motion for a uniform beam is

$$-EI \frac{\partial^4 v}{\partial x^4} = \rho A \frac{\partial^2 v}{\partial t^2} + N \frac{\partial^2 v}{\partial x^2}$$

or

$$EI \frac{\partial^4 v}{\partial x^4} + N \frac{\partial^2 v}{\partial x^2} + \rho A \frac{\partial^2 v}{\partial t^2} = 0. \tag{4.26}$$

Considering a beam with simply supported ends, a solution which satisfies the end conditions is

$$v = B \sin \frac{n\pi x}{l} \sin (\omega_n t + \alpha). \tag{4.27}$$

This satisfies equation (4.26), provided that

$$EI \left(\frac{n\pi}{l}\right)^4 - N \left(\frac{n\pi}{l}\right)^2 - \rho A \omega_n^2 = 0.$$

Thus the natural frequencies are given by

$$\omega_n = \left(\frac{n\pi}{l}\right)^2 \left[\frac{EI}{\rho A}\left(1 - \frac{Nl^2}{n^2\pi^2 EI}\right)\right]^{1/2}, \qquad n = 1, 2, 3, \ldots.$$

Considering the fundamental mode, $n = 1$,

$$\omega_1 = \left(\frac{\pi}{l}\right)^2 \left(\frac{EI}{\rho A}\right)^{1/2} \left(1 - \frac{N}{N_c}\right)^{1/2} \tag{4.28}$$

where N_c ($= \pi^2 EI/l^2$) is the Euler critical load for this strut. Buckling of the strut, occurring when $N = N_c$ and represented by equation (4.27) with $\omega_n = 0$ and $\alpha = \pi/2$, is thus a limiting case.

Constant axial tensile forces, applied at the ends of the beam, will raise the natural frequencies. As shown in Fig. 4.5, it is assumed that the point A can move in the axial direction, although the movement is very small provided that $v \ll l$. This assumption is made for beams with simply supported ends, when there is no axial force acting (see Figs. 3.3 and 3.7). If the distance between the ends is kept constant, non-linear vibrations occur, and the natural frequencies depend on the amplitude, but provided that the ratio (maximum amplitude of vibration/radius of gyration) is small, the increase in the fundamental natural frequency is very small[89].

From equation (4.28)

$$(\omega_1/\omega_0)^2 = 1 - (N/N_c)$$

where ω_0 is the fundamental natural frequency of the simply supported beam with no axial force ($N = 0$). If a linear relation between square of frequency and axial force existed for practical systems, it would enable the buckling load to be predicted from vibration tests, which gave the natural frequencies corresponding to two values of N. However, for general end conditions substitution of $v = B \exp(rx) \sin(\omega t + \alpha)$ in equation (4.26) gives the auxiliary equation

$$r^4 + \frac{N}{EI}r^2 - \frac{\rho A\omega^2}{EI} = 0. \tag{4.29}$$

Solving equation (4.29),

$$r^2 = -\frac{N}{2EI} \pm \left[\left(\frac{N}{2EI}\right)^2 + \frac{\rho A\omega^2}{EI}\right]^{1/2}.$$

Thus

$$v = [B_1 \sinh r_1 x + B_2 \cosh r_1 x + B_3 \sin \lambda x + B_4 \cos \lambda x] \sin (\omega t + \alpha)$$

$$(4.30)$$

where

$$r_1 = \left[\left\{ \left(\frac{N}{2EI} \right)^2 + \frac{\rho A \omega^2}{EI} \right\}^{1/2} - \frac{N}{2EI} \right]^{1/2}$$

and

$$\lambda = \left[\left\{ \left(\frac{N}{2EI} \right)^2 + \frac{\rho A \omega^2}{EI} \right\}^{1/2} + \frac{N}{2EI} \right]^{1/2}.$$

Using the four boundary conditions and eliminating B_j, a frequency equation is obtained. In general, the mode shape, represented by the terms in the square brackets in equation (4.30), depends upon the force N. As shown by Massonet[57], the relation between ω^2 and N is linear if the mode shapes for free vibration without axial force and for buckling are identical; this is true only for beams with simply supported ends, but for other standard end conditions the relation is very nearly linear.

4.4. Beams on Elastic Foundations

A uniform beam supported along its length l by a foundation of modulus C will be considered, i.e. C equals the load on unit length of the foundation required to produce unit deflection. Thus considering an element of the beam of length dx, which at any time t has a displacement v, the foundation will exert an upward force on the element $Cv\,dx$. The equation of motion for this element is (from Fig. 3.2 with the force $Cv\,dx$ added)

$$\frac{\partial S}{\partial x} - Cv = \rho A \frac{\partial^2 v}{\partial t^2}.$$

As $M = -EI\,\partial^2 v/\partial x^2$ and $\partial M/\partial x = S$,

$$EI\,\partial^4 v/\partial x^4 + Cv + \rho A\,\partial^2 v/\partial t^2 = 0.$$

For free vibrations $v(x, t) = V(x) \sin (\omega t + \alpha)$. Hence

$$EI \frac{d^4 V}{dx^4} + CV - \rho A \omega^2 V = 0. \tag{4.31}$$

The general solution of equation (4.31) is

$$V = B_1 \sin \lambda x + B_2 \cos \lambda x + B_3 \sinh \lambda x + B_4 \cosh \lambda x$$

provided that

$$EI\lambda^4 + C = \rho A \omega^2. \tag{4.32}$$

As usual, the constants B_r are determined from the end conditions. It will be assumed that the ends of the beam are free, i.e. the beam is supported solely by the elastic foundation. Thus at $x = 0$ and at $x = l$, $d^2V/dx^2 = 0 = d^3V/dx^3$; substituting and eliminating the constants leads to the frequency equation (as in Section 3.3) for a beam with free ends,

$$\cos \lambda l \cosh \lambda l = 1. \tag{4.33}$$

However, the existence of the foundation modifies the relation (4.32) between ω and λ. The roots of equation (4.33) are $\lambda l = 0$, 4·730, 7·853, $(r + \frac{1}{2})\pi$ for $r - 3, 4, 5, \ldots$ (The frequency equation is the same as that for a beam with fixed ends.) On substituting the roots λl in equation (4.32) the natural frequencies are determined. A beam with free ends supported on an elastic foundation can vibrate as a rigid body in translation and rotation. In translation, where each element of the beam has the same displacement, the total restoring force is Cvl and thus the equation of motion is $\rho Al(d^2v/dt^2) + Cvl = 0$. Thus the natural frequency is given by $\omega^2 = C/\rho A$, which corresponds to the root $\lambda l = 0$. In rotation the displacement of an element, distant x from the centre of gravity, is $v(t) = x\theta(t)$, the restoring force on an element dx is $Cx\theta(t)\,dx$ and the restoring moment about the centre of gravity is $\int_{-l/2}^{l/2} Cx^2\theta(t)\,dx$. Thus the equation of motion is

$$\frac{d^2\theta}{dt^2} \int_{-l/2}^{l/2} \rho Ax^2\,dx + \theta \int_{-l/2}^{l/2} Cx^2\,dx = 0$$

and the natural frequency is given by $\omega^2 = C/\rho A$. Thus the two frequencies of vibration as a rigid body are equal and are given by the first root, $\lambda l = 0$, of equation (4.33).

4.5. Vibrations of Beam Systems

Figure 4.6 shows four simple beam systems; in each case natural frequencies and the response to harmonic excitation can be determined by treating the structures as two simple beams of lengths l_1 and l_2 for Fig. 4.6 a, b and c and as three simple beams of lengths l_1, l_2 and l_3 for Fig. 4.6d. Hysteretic damping can be included for the beam material; dashpots, representing concentrated damping forces, can be added at junctions between beams. The structures of Fig. 4.6 a, b and d are assumed to vibrate in the plane of the diagram; points on the beams of Fig. 4.6c are assumed to vibrate in the direction perpendicular to the plane of the diagram and a harmonic force acts at point D in this perpendicular direction. Thus for Fig. 4.6c the individual beams will undergo flexural and torsional deformations. For Fig. 4.6d flexural and extensional (or longitudinal) deformations of the individual beams will be considered, although in practice the significance of extensional deformations depends upon the geometry and frequency of excitation. It is possible to extend the method to three-dimensional frames where all three types of deformation could be significant. The axis of each individual beam must be parallel to one of the coordinate directions OX, OY or OZ. The cross-section of an individual beam does not vary along its length; thus changes in cross-section can be allowed only at

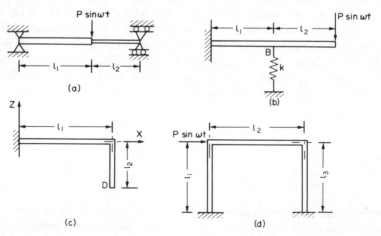

FIG. 4.6. Four simple beam systems.

joints. In this method the dynamic stiffness matrix is determined, using the solutions for extensional, torsional and flexural vibrations of uniform beams, developed in Sections 3.1, 3.2 and 3.3, respectively. The dynamic stiffness method is recommended for simple beam systems, although the finite element method (Section 3.9) could be used. However, for systems consisting of non-uniform component beams and for large systems of beams it is necessary to use the finite element method. A comparison of the methods has been given by the author[39].

The method of this section can be used to determine a closed form solution for the steady-state amplitude of a single uniform beam subjected to a harmonic force or moment—a problem for which a series solution based on normal modes has been given in Section 3.5. If the force is applied at $x = a$ to a beam of length l, the two sections of the beam, $0 \leq x \leq a$ and $a \leq x \leq l$, must be treated separately and appropriate continuity and equilibrium conditions applied at the "joint" $x = a$; this is illustrated by the example based on Fig. 4.6a.

Before considering the specific problems of Fig. 4.6 the dynamic stiffness matrices will be established for a single beam for each type of deformation.

Flexure

The displacements, slopes, applied forces and moments at the ends A and B of a uniform beam of length l are shown in Fig. 4.7. The sign convention for displacement and slope is that of Section 3.3; applied forces and moments are positive in the direction of positive displacement and slope respectively. All displacements, slopes, forces and moments are proportional to $\sin (\omega t + \alpha)$, (or to $\exp (i\omega t)$ if damping is included and the analysis is in terms of complex quantities) but this

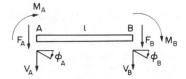

FIG. 4.7. Applied end forces and displacements for a beam in flexure.

term is omitted from all the equations. From equation (3.18) for a uniform beam vibrating at frequency ω in flexure

$$V(x) = B_1 \sin \lambda x + B_2 \cos \lambda x + B_3 \sinh \lambda x + B_4 \cosh \lambda x \quad (4.34)$$

where $\lambda^4 = \rho A \omega^2 / EI$. The coefficients B_i can be found in terms of the end displacements and slopes by solving the equations

$$V_A = B_2 + B_4,$$

$$\phi_A = \lambda B_1 + \lambda B_3,$$

$$V_B = B_1 \sin \lambda l + B_2 \cos \lambda l + B_3 \sinh \lambda l + B_4 \cosh \lambda l$$

and

$$\phi_B = \lambda B_1 \cos \lambda l - \lambda B_2 \sin \lambda l + \lambda B_3 \cosh \lambda l + \lambda B_4 \sinh \lambda l$$

to give

$$\mathbf{b} = \mathbf{D}_1 \mathbf{v} \qquad (4.35)$$

where

$$\mathbf{b}^T = [B_1, B_2, B_3, B_4]$$

and

$$\mathbf{v}^T = [V_A, V_B, \phi_A, \phi_B].$$

The elements of matrix $\mathbf{D}_1$ depend upon λl. The applied end forces and moments are given in terms of displacements by

$$M_A = -EI\left(\frac{d^2 V}{dx^2}\right)_{x=0}, \qquad M_B = EI\left(\frac{d^2 V}{dx^2}\right)_{x=l},$$

$$F_A = EI\left(\frac{d^3 V}{dx^3}\right)_{x=0}, \qquad F_B = -EI\left(\frac{d^3 V}{dx^3}\right)_{x=l}. \qquad (4.36)$$

Substituting for V from equation (4.34), equations (4.36) can be written

$$\mathbf{f} = \mathbf{D}_2 \mathbf{b} \qquad (4.37)$$

where

$$\mathbf{f}^T = [F_A, F_B, M_A, M_B].$$

From equations (4.35) and (4.37)

$$\mathbf{f} = \mathbf{Jv} \qquad (4.38)$$

where the dynamic stiffness matrix

$$\mathbf{J} = \mathbf{D}_2\mathbf{D}_1 = \begin{bmatrix} j_a & j_b & j_c & j_d \\ j_b & j_a & -j_d & -j_c \\ j_c & -j_d & j_e & j_f \\ j_d & -j_c & j_f & j_e \end{bmatrix}, \qquad (4.39)$$

$$j_a = -K\lambda^3(\cos \lambda l \sinh \lambda l + \sin \lambda l \cosh \lambda l),$$

$$j_b = K\lambda^3(\sin \lambda l + \sinh \lambda l),$$

$$j_c = -K\lambda^2 \sin \lambda l \sinh \lambda l,$$

$$j_d = K\lambda^2(\cos \lambda l - \cosh \lambda l),$$

$$j_e = K\lambda(\cos \lambda l \sinh \lambda l - \sin \lambda l \cosh \lambda l),$$

$$j_f = K\lambda(\sin \lambda l - \sinh \lambda l)$$

and

$$K = EI/(\cos \lambda l \cosh \lambda l - 1).$$

Extension

The sign convention for displacements and applied axial forces at the ends A and B of a uniform bar is shown in Fig. 4.8a. From equation (3.4) for a beam vibrating at frequency ω in extension (i.e. axially or longitudinally)

$$U(x) = B_5 \sin (\omega x/C_1) + B_6 \cos (\omega x/C_1)$$

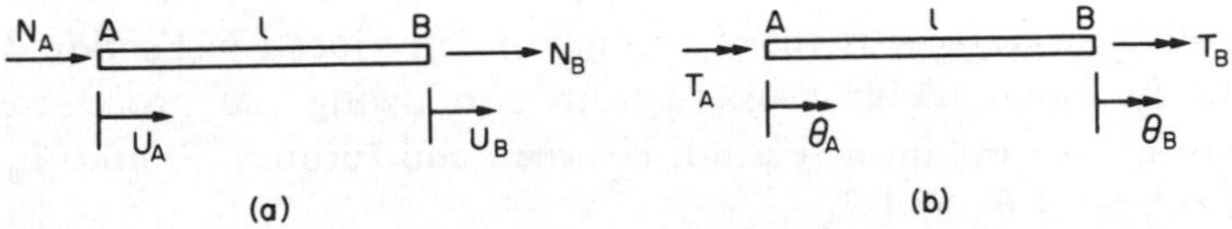

(a)   (b)

FIG. 4.8. Applied end forces and displacements for a beam in (a) extension, (b) torsion.

where $C_1^2 = E/\rho$. The coefficients B_5 and B_6 can be found in terms of U_A and U_B from

$$U_A = B_6,$$

$$U_B = B_5 \sin (\omega l/C_1) + B_6 \cos (\omega l/C_1).$$

The applied forces at the ends are expressed in terms of displacements by

$$N_A = -AE\left(\frac{dU}{dx}\right)_{x=0} \quad \text{and} \quad N_B = AE\left(\frac{dU}{dx}\right)_{x=l}.$$

Expressing N_A and N_B in terms of B_5 and B_6 and the latter in terms of U_A and U_B, we obtain

$$\mathbf{f'} = \mathbf{J'v'} \tag{4.40}$$

or

$$\begin{bmatrix} N_A \\ N_B \end{bmatrix} = \frac{EA\omega}{C_1} \begin{bmatrix} \cot (\omega l/C_1) & -\operatorname{cosec} (\omega l/C_1) \\ -\operatorname{cosec} (\omega l/C_1) & \cot (\omega l/C_1) \end{bmatrix} \begin{bmatrix} U_A \\ U_B \end{bmatrix}. \tag{4.41}$$

Torsion

The sign convention is shown in Fig. 4.8b. For clarity the torques T and the angular rotations θ, which act about the longitudinal axis of the beam, are denoted by double-headed arrows in the direction of this axis; application of the right-hand screw rule indicates the positive sense of T and θ. From equation (3.11) for a beam vibrating at frequency ω in torsion

$$\theta(x) = B_7 \sin (\omega x/C_2) + B_8 \cos (\omega x/C_2)$$

where $C_2^2 = \kappa G/\rho$; κ is the shape factor introduced in Section 3.1 to allow for non-circular cross-sections. Following the procedure for extension, we obtain a relation between end torques T_A and T_B and end rotations θ_A and θ_B

$$\mathbf{f''} = \mathbf{J''v''} \tag{4.42}$$

or

$$\begin{bmatrix} T_A \\ T_B \end{bmatrix} = \frac{GJ\omega}{C_2} \begin{bmatrix} \cot(\omega l/C_2) & -\csc(\omega l/C_2) \\ -\csc(\omega l/C_2) & \cot(\omega l/C_2) \end{bmatrix} \begin{bmatrix} \theta_A \\ \theta_B \end{bmatrix}. \quad (4.43)$$

Systems, which consist of several beams, may be analysed by matching shear and axial forces and bending and twisting moments at the joints between beams. The forces and moments are written in terms of joint displacements and rotations, using equations (4.38), (4.40) and (4.42), and are then eliminated by considerations of equilibrium. Thus, if there are n independent non-zero displacements at the joints and boundaries of the system, we obtain n equilibrium equations, which can be written in terms of displacements as

$$Sv = p \quad (4.44)$$

where S is an $n \times n$ symmetric matrix composed of terms derived from the J matrices of the component beams in the system. The terms of S depend upon frequency ω and the stiffnesses and masses of the system. Vector v lists the n independent non-zero displacements of the system; vector p lists any external forces which are exciting the system at frequency ω. For free vibration problems p is zero and the natural frequencies are determined from

$$\det |S| = 0. \quad (4.45)$$

The response can be obtained from equation (4.44) as

$$v = S^{-1}p. \quad (4.46)$$

So far, damping has been neglected; thus all displacements and force quantities are proportional to $\sin(\omega t + \alpha)$ and all terms in equation (4.46) are real. With damping included all displacement and force quantities are proportional to $\exp(i\omega t)$; the response can be obtained from equation (4.46), but some or all of the terms in S and v are complex. Two methods of adding damping exist, namely discrete damping forces applied at joints of the system through dashpots, etc., and internal damping in the beam material, and will be considered separately, although obviously they can exist together.

A viscous or a hysteretic damper acting at joint B, in parallel with the spring k, in Fig. 4.6b is an example of the first type of damping. In the equilibrium equation for forces at B the stiffness k is replaced by the complex quantity $(k + i\omega c)$ for viscous damping or $k(1 + i\mu)$ for hysteretic damping, recalling that all quantities are proportional to exp $(i\omega t)$. For each damping mechanism added one term in the matrix S becomes complex, as illustrated later for the example of Fig. 4.6b.

When hysteretic damping for the material of a beam in flexure is included, Young's modulus E, occurring in the undamped beam equations, is replaced by the complex quantity $E(1 + i\mu)$, where μ is the hysteretic damping constant, as in Section 3.5. For an undamped beam the auxiliary equation (3.17) leads to four roots, of which two are real and two are imaginary. For a damped beam the corresponding four roots are complex; the general solution, replacing equation (3.18), is

$$V = B_1 \sin \lambda^+ x + B_2 \cos \lambda^+ x + B_3 \sinh \lambda^+ x + B_4 \cosh \lambda^+ x$$

$$(4.47)$$

where

$$(\lambda^+)^4 = \frac{\rho A \omega^2}{EI(1 + i\mu)} = \frac{\lambda^4}{1 + i\mu}.$$

For small damping

$$\lambda^+ = \lambda(1 - \mu/4). \tag{4.48}$$

The analysis, which leads to equation (4.44), is unchanged except that λ^+ replaces λ in the terms in the matrix J [equation (4.39)]. This is a significant change, as these terms include trigonometric and hyperbolic functions with the complex argument $\lambda^+ l$. This method of including damping is treated in detail by Snowdon[69]; its general principles have been given here and its application to the example of Fig. 4.6a will be considered later. If hysteretic damping is included for extensional motion, ω/C_1 in equation (4.41) is replaced by $\omega/C_1(1 + i\mu)^{1/2}$; for small damping the latter is equal to $\omega(1 - i\mu/2)/C_1$. For torsional motion it is assumed that the shear modulus G is replaced by $G(1 + i\mu)$. Thus for small damping $\omega(1 - i\mu/2)/C_2$ replaces ω/C_2 in the matrix J'' [equations (4.42) and (4.43)].

The four examples of Fig. 4.6 will be considered now in some detail.
Figure 4.6a. The two component beams for this structure are shown

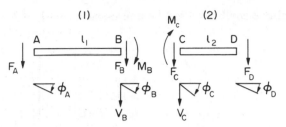

FIG. 4.9. The two component beams for the system of Fig. 4.6a.

separated in Fig. 4.9 and numbered 1 and 2, respectively. Only flexural deformation is considered. The matrix equation for beam 1 is

$$\mathbf{f}_1 = \mathbf{J}_1 \mathbf{v}_1$$

where

$$\mathbf{f}_1^T = [F_A \quad F_B \quad 0 \quad M_B] \quad \text{and} \quad \mathbf{v}_1^T = [0 \quad v_B \quad \phi_A \quad \phi_B].$$

The terms in $\mathbf{J}_1$ are obtained from equation (4.39) with the addition of subscripts 1, noting that $\lambda_1^4 = \rho A_1 \omega^2 / EI_1$. Similarly for beam 2

$$\mathbf{f}_2 = \mathbf{J}_2 \mathbf{v}_2$$

where the terms in $\mathbf{J}_2$ include the subscript 2 and $\lambda_2^4 = \rho A_2 \omega^2 / EI_2$. As $V_B = V_C$ and $\phi_B = \phi_C$ for continuity and $V_A = 0$ and $V_D = 0$ from the boundary conditions, the independent non-zero displacements are ϕ_A, V_B, ϕ_B and ϕ_D. The equilibrium conditions for the joint B–C and the remaining boundary conditions are

$$M_A = 0,$$

$$F_B + F_C = P,$$

$$M_B + M_C = 0,$$

$$M_D = 0.$$

If $j_{a1}, j_{b1}, \ldots$ are the values of $j_a, j_b, \ldots$, defined in equation (4.39), in terms of $\lambda_1 l_1$ and $j_{a2}, j_{b2}, \ldots$ are the corresponding values in terms of

$\lambda_2 l_2$, the above conditions can be expressed, using $\mathbf{f}_i = \mathbf{J}_i \mathbf{v}_i$, $i = 1, 2$, as

$$
\begin{bmatrix}
j_{e1} & -j_{d1} & j_{f1} & 0 \\
-j_{d1} & (j_{a1} + j_{a2}) & (j_{c2} - j_{c1}) & j_{d2} \\
j_{f1} & (j_{c2} - j_{c1}) & (j_{e1} + j_{e2}) & j_{f2} \\
0 & j_{d2} & j_{f2} & j_{e2}
\end{bmatrix}
\begin{bmatrix}
\phi_A \\
V_B \\
\phi_B \\
\phi_D
\end{bmatrix}
=
\begin{bmatrix}
0 \\
P \\
0 \\
0
\end{bmatrix}.
\tag{4.49}
$$

Natural frequencies are obtained by putting $P = 0$ and finding the values of ω, and thus of λ_1 and λ_2, for which the determinant of equation (4.49) is zero. As the terms in the determinant are complicated functions of frequency, this is best achieved by an iterative solution starting from a trial value of ω.

For a prescribed excitation frequency ω the coefficients j_{a1}, j_{a2}, etc., can be computed and equation (4.49) solved to give the response. If hysteretic damping is included, j_{ai}, j_{ci}, etc., are defined in terms of λ_i^+; for example,

$$
j_{a1} = -EI(1 + i\mu)(\lambda_1^+)^3 (\cos \lambda_1^+ l_1 \sinh \lambda_1^+ l_1
$$
$$
+ \sin \lambda_1^+ l_1 \cosh \lambda_1^+ l_1)/(\cos \lambda_1^+ l_1 \cosh \lambda_1^+ l_1 - 1).
$$

If the two component beams are of the same material and dimensions, i.e. $l_1 = l_2$ and λ_1 and λ_2, and we consider symmetric modes of vibration, $\phi_D = -\phi_A$ and $\phi_B = 0$. Thus, omitting the subscript i in j_{ai}, etc., equations (4.49) reduce to

$$
j_e \phi_A - j_d V_B = 0
$$

and

$$
-2j_d \phi_A + 2j_a V_B = P
$$

with the solution

$$
V_B = \frac{P}{2(j_a - j_d^2/j_e)}.
$$

Substituting for j_a, j_d and j_e from equation (4.39) and simplifying,

$$
V_B = \frac{P[\sin \lambda^+ l \cosh \lambda^+ l - \cos \lambda^+ l \sinh \lambda^+ l]}{4EI(1 + i\mu)(\lambda^+)^3 \cos \lambda^+ l \cosh \lambda^+ l}.
\tag{4.50}
$$

Equation (4.50) has been given in terms of λ^+. Temporarily neglecting damping, $\lambda^+ \to \lambda$ and $\mu = 0$. The resonant condition, for which $V_B \to \infty$, is $\cos \lambda l = 0$, i.e. $\lambda l = (2n - 1)\pi/2$, $n = 1, 2, 3, \ldots$. As expected, these values of λl coincide with the natural frequencies for the symmetric modes of a uniform simply supported beam. The response with damping included is obtained by using the following expressions, valid for small damping:

$$\lambda^+ = \lambda(1 - i\mu/4);$$

$$\sin \lambda^+ l = \sin \lambda l - \tfrac{1}{4}i\mu\lambda l \cos \lambda l;$$

$$\cos \lambda^+ l = \cos \lambda l + \tfrac{1}{4}i\mu\lambda l \sin \lambda l;$$

$$\sinh \lambda^+ l = \sinh \lambda l - \tfrac{1}{4}i\mu\lambda l \cosh \lambda l;$$

$$\cosh \lambda^+ l = \cosh \lambda l - \tfrac{1}{4}i\mu\lambda l \sinh \lambda l.$$

Substituting in equation (4.50), the response at resonance, obtained by putting $\cos \lambda l = 0$ and $\sin \lambda l = (-1)^{n+1}$, is

$$V_B = \frac{Pl^3[1 - \tfrac{1}{2}i\mu\lambda l \tanh \lambda l]}{4EI(\lambda l)^3(1 + \tfrac{1}{4}i\mu)(\tfrac{1}{4}i\mu\lambda l)(1 - \tfrac{1}{4}i\mu\lambda l \tanh \lambda l)}.$$

The first approximation to the resonant amplitude, valid for small damping, is

$$V_B = \frac{Pl^3}{\mu EI(\lambda l)^4}.$$

Putting the span

$$L = 2l \quad \text{and} \quad \lambda l = (2n - 1)\pi/2,$$

$$V_B = \frac{2PL^3}{\pi^4 \mu EI(2n - 1)^4}.$$

(4.51)

Equation (4.51), giving the amplitude at mid-span at successive resonances, agrees with the corresponding expression from the normal mode method of Section 3.5, if only the dominant term in that series solution is included.

For this symmetric problem there are resonant peaks in the response curve when the excitation frequency $\omega = \omega_1, \omega_3, \omega_5, \ldots$ where ω_s is the sth natural frequency of the beam. Considering the response at midspan when the excitation frequency is not near to any resonant frequency, specifically when $\omega = \omega_2$ and ω_4, $\lambda l = \pi$ and 2π, respectively; using equation (4.50) and the approximate expressions for λ^+, $\sin \lambda^+ l$, etc., the amplitude at B, provided that the damping is small, is

$$V_B = \frac{Pl^3}{4EI} \frac{\tanh (s\pi/2)}{(s\pi/2)^3}, \qquad s = 2, 4,$$

i.e. $V_B = 0{\cdot}001004 PL^3/EI$ and $0{\cdot}0001260 PL^3/EI$ for $\omega = \omega_2$ and ω_4, respectively. Using the normal mode method of Section 3.5 [equation (3.53)], there are contributions to V_B only from terms in the series corresponding to odd values of s, as $\phi_s(L/2) = 0$ for s even. For these off-resonance responses convergence is slow. When $\omega = \omega_2$, $EIV_B/PL^3 = 0{\cdot}001019$, $0{\cdot}001006$ and $0{\cdot}001005$ if the terms for $s = 1$ to 5, $s = 1$ to 11 and $s = 1$ to 15, respectively, are used. When $\omega = \omega_4$, $EIV_B/PL^3 = 0{\cdot}0001422$, $0{\cdot}0001279$ and $0{\cdot}0001268$ for the three ranges of s above. Both methods show that at these off-resonance frequencies the amplitude is not affected significantly by the level of damping provided that this level is low.

Figure 4.6b. The two component beams, 1 and 2, are shown separately in Fig. 4.10. Only flexural deformation is considered. The matrix equations are $\mathbf{f}_i = \mathbf{J}_i \mathbf{v}_i$, $i = 1, 2$. As $V_B = V_C$ and $\phi_B = \phi_c$ for continuity and $V_A = 0$ and $\phi_A = 0$, the independent non-zero displace-

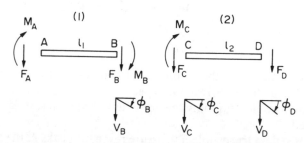

FIG. 4.10. The two component beams for the system of Fig. 4.6b.

ments are V_B, ϕ_B, V_D and ϕ_D. The equilibrium conditions for the joint $B - C$ and the boundary conditions at D are

$$F_B + F_C = -kV_B,$$

$$M_B + M_C = 0,$$

$$F_D = P,$$

$$M_D = 0.$$

With the notation of the previous example, these equations become

$$\begin{bmatrix} (j_{a1} + j_{a2} + k) & (j_{c2} - j_{c1}) & j_{b2} & j_{d2} \\ (j_{c2} - j_{c1}) & (j_{e1} + j_{e2}) & -j_{d2} & j_{f2} \\ j_{b2} & -j_{d2} & j_{a2} & -j_{c2} \\ j_{d2} & j_{f2} & -j_{c2} & j_{e2} \end{bmatrix} \begin{bmatrix} V_B \\ \phi_B \\ V_D \\ \phi_D \end{bmatrix} = \begin{bmatrix} 0 \\ 0 \\ P \\ 0 \end{bmatrix}.$$

If the two component beams are of the same material and dimensions, $j_{a1} = j_{a2}$, etc., and the subscripts 1 and 2 may be omitted. The natural frequencies of the system are obtained from

$$\det \begin{vmatrix} (2j_a + k) & 0 & j_b & j_d \\ 0 & 2j_e & -j_d & j_f \\ j_b & -j_d & j_a & -j_c \\ j_d & j_f & -j_c & j_e \end{vmatrix} = 0. \qquad (4.52)$$

If hysteretic damping is included in parallel with the spring, k is replaced by $k(1 + i\mu)$, one element of the matrix S becomes complex and the response is obtained for a specified excitation frequency from the solution of

$$\begin{bmatrix} 2j_a + k(1 + i\mu) & 0 & j_b & j_d \\ 0 & 2j_e & -j_d & j_f \\ j_b & -j_d & j_a & -j_c \\ j_d & j_f & -j_c & j_e \end{bmatrix} \begin{bmatrix} V_B \\ \phi_B \\ V_D \\ \phi_D \end{bmatrix} = \begin{bmatrix} 0 \\ 0 \\ P \\ 0 \end{bmatrix}. \qquad (4.53)$$

Response quantities are determined as complex numbers, for example $V_B = a + ib$; if the excitation force is $P \sin \omega t$, the displacement at B

$$v_B(t) = \text{Im} \left[(a + ib) \exp (i\omega t) \right]$$

$$= a \sin \omega t + b \cos \omega t$$

$$= (a^2 + b^2)^{1/2} \sin (\omega t + \alpha)$$

where

$$\tan \alpha = b/a.$$

If $kl^3/EI = 6\cdot25$, the lowest root of equation (4.52) is $\lambda l = 1\cdot1804$, corresponding to $\omega_1 = 1\cdot3934(EI/\rho Al^4)^{1/2}$. If $\mu = 0\cdot01$, the response at the tip at the fundamental resonant frequency, obtained by putting $\lambda l = 1\cdot1804$ and solving equation (4.53), is

$$v_D = \frac{Pl^3}{EI}(0\cdot850 \sin \omega t - 247\cdot8 \cos \omega t)$$

and the amplitude $V_D = 247\cdot8Pl^3/EI$. This problem has been solved by the Rayleigh–Ritz method in Section 3.8; allowing for the different definitions of l adopted in the two methods, the errors in the Rayleigh–Ritz method are $0\cdot21$ and $-0\cdot84$ per cent for the fundamental natural frequency and the tip amplitude for excitation at this frequency, respectively.

Figure 4.6c. The two component beams, 1 and 2, are shown separately in Fig. 4.11. From Fig. 4.6c for a force at D in the Y-direction beam no. 1 bends in the XY-plane and beam no. 2 bends in the ZY-plane. Each component beam is drawn in its own plane of bending in Fig. 4.11. The system or global axes XYZ are defined in Fig. 4.6c with the Y-axis into the plane of the diagram to give a right-hand set of axes and are shown also in Fig. 4.11. Flexural and torsional deformations

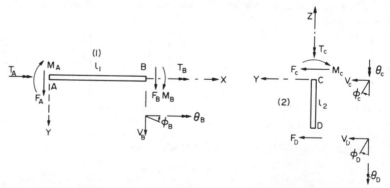

Fig. 4.11. The two component beams for the system of Fig. 4.6c.

are considered. For each beam a 6×6 symmetric matrix $\mathbf{J}_i'''$ is assembled from equations (4.38) and (4.42) and can be expressed in partitioned form as

$$\mathbf{J}_i''' = \begin{bmatrix} \mathbf{J}_i & \vdots & \mathbf{0} \\ \cdots & \vdots & \cdots \\ \mathbf{0} & \vdots & \mathbf{J}_i'' \end{bmatrix} \tag{4.54}$$

Then

$$\mathbf{f}_i = \mathbf{J}_i''' \mathbf{v}_i, \qquad i = 1, 2$$

where, for example,

$$\mathbf{f}_1^T = [F_A \quad F_B \quad M_A \quad M_B \quad T_A \quad T_B]$$

and

$$\mathbf{v}_1^T = [0 \quad V_B \quad 0 \quad \phi_B \quad 0 \quad \theta_B].$$

The continuity conditions are $V_B = V_C$, $\theta_B = \phi_C$ and $\phi_B = -\theta_C$. The independent displacements are V_B, ϕ_B, θ_B, V_D, ϕ_D and θ_D. The equilibrium equations for the joint $B - C$ and the boundary conditions at D are:

$$\begin{aligned} F_B + F_C &= 0, \\ M_B - T_C &= 0, \\ T_B + M_C &= 0, \\ F_D &= P, \\ M_D &= 0, \end{aligned} \tag{4.55}$$

and

$$T_D = 0.$$

Natural frequencies and the response to harmonic excitation can be obtained from equations (4.55) by the methods given for the previous examples.

Figure 4.6d. The three component beams, no. 1, 2 and 3, are shown separately in Fig. 4.12. Flexural and axial deformations are considered.

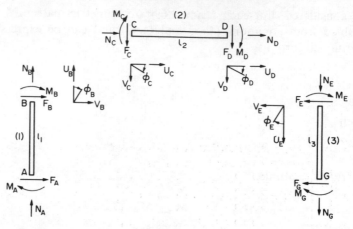

FIG. 4.12. The three component beams for the system of Fig. 4.6d.

The matrix equations for the three beams are

$$\mathbf{f}_i = \mathbf{J}_i''' \mathbf{v}_i, \qquad i = 1, 2 \text{ and } 3$$

where

$$\mathbf{f}_1^T = [F_A \quad F_B \quad M_A \quad M_B \quad N_A \quad N_B]$$

and

$$\mathbf{v}_1^T = [0 \quad V_B \quad 0 \quad \phi_B \quad 0 \quad U_B].$$

The 6×6 symmetric matrices $\mathbf{J}_i'''$ are assembled from equations (4.38) and (4.40); in partitioned form they are similar to equation (4.54) with $\mathbf{J}_i'$ replacing $\mathbf{J}_i''$.

The continuity conditions at joint $B - C$ are:

$$V_B = U_C,$$
$$\phi_B = \phi_C,$$
$$U_B = -V_C.$$

The equilibrium equations for joint $B - C$ are:

$$F_B + N_C = P,$$

$$N_B - F_C = 0,$$

$$M_B + M_C = 0.$$

With the equilibrium equations for joint $D - E$ six equations exist for the six independent coordinates, namely the values of V, ϕ and U at the two joints.

4.6. Shear Deformation and Rotatory Inertia

In the preceding sections of Chapters 3 and 4 the analysis has been based on the elementary, or Euler–Bernoulli, theory of beams, for which dynamic equilibrium of an element in flexure is governed by equation (3.14). In this section an introduction to a higher-order theory is given by including the separate effects of transverse shear deformation and rotatory inertia. The former is included by relaxing the assumption that plane sections, initially normal to the neutral axis, remain plane and normal to that axis during deformation. It will be shown that the influence of these effects upon natural frequencies increases as the ratio of the axial wavelength to the radius of gyration of the cross-section decreases.

Two general points regarding higher-order theories should be noted. The theories for the vibration of thin plates and shells, to be given in Chapter 5, neglect the effects of transverse shear deformation and rotatory inertia. Higher-order theories for plates and shells, comparable to that for beams to be given here, exist, but will not be included. In general, results from a higher-order theory may be used to assess the accuracy and range of applicability of the corresponding elementary theory. For a limited range of problems for the vibration of beams, plates and shells solutions from three-dimensional elasticity theory exist. The latter can be used to assess the accuracy of the appropriate higher-order theory.

The element $ABCD$ of a beam, shown in Fig. 4.13, deforms due to shear stresses (denoted by unlabelled arrows) into position $AB'C'D$. If the faces AD and BC form parts of cross-sections perpendicular to the

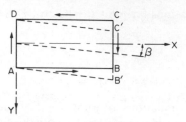

FIG. 4.13. Deformation of an element of a beam due to shear.

neutral axis, the shear stresses on these faces, for a given shear force, are not uniform. It follows that elements at different distances from the neutral axis will undergo different deformations, and thus, if shear deformation is considered, plane sections, initially perpendicular to the neutral axis, will not remain plane. Let β be the angle of shear at the neutral axis (Fig. 4.13) and ψ be the slope of the deflection curve when the shear deformation is neglected. Then the total slope

$$\frac{\partial v}{\partial x} = \psi + \beta. \tag{4.56}$$

The moment–curvature relation, replacing equation (3.13), is

$$M = -EI\frac{\partial \psi}{\partial x}. \tag{4.57}$$

The shear force

$$S = \kappa AG\beta \tag{4.58}$$

where κ is a numerical factor, A is the cross-sectional area and G is the shear modulus. The factor κ is usually defined as the ratio of the average shear strain over the cross-section (S/AG) to the shear strain at the centroid β; it depends upon the shape of the cross-section and strictly upon the mode of vibration. Even for a simple shape such as a rectangle several slightly different values are given in the literature. Cowper[23] gives a good discussion, derivation and numerical values of κ.

Considering the element of Fig. 3.2, equilibrium in the Y-direction gives equation (3.12), i.e.

$$\frac{\partial S}{\partial x} = \rho A \frac{\partial^2 v}{\partial t^2}. \tag{4.59}$$

Taking moments about the centre line of the element and including rotatory inertia,

$$S - \frac{\partial M}{\partial x} = \rho I \frac{\partial^2 \psi}{\partial t^2}. \tag{4.60}$$

Substituting from equations (4.56), (4.57) and (4.58) in equation (4.60)

$$EI \frac{\partial^2 \psi}{\partial x^2} + \kappa A G \left(\frac{\partial v}{\partial x} - \psi \right) = \rho I \frac{\partial^2 \psi}{\partial t^2}. \tag{4.61}$$

Substituting from equations (4.56) and (4.58) in equation (4.59)

$$\kappa A G \left(\frac{\partial^2 v}{\partial x^2} - \frac{\partial \psi}{\partial x} \right) = \rho A \frac{\partial^2 v}{\partial t^2}. \tag{4.62}$$

From equation (4.61)

$$\left[EI \frac{\partial^2}{\partial x^2} - \kappa A G - \rho I \frac{\partial^2}{\partial t^2} \right] \frac{\partial \psi}{\partial x} = -\kappa A G \frac{\partial^2 v}{\partial x^2}. \tag{4.63}$$

Eliminating $\partial \psi / \partial x$ between equations (4.62) and (4.63)

$$EI \frac{\partial^4 v}{\partial x^4} - \left(\rho \frac{EI}{\kappa G} + \rho I \right) \frac{\partial^4 v}{\partial x^2 \partial t^2} + \rho A \frac{\partial^2 v}{\partial t^2} + \frac{\rho^2 I}{\kappa G} \frac{\partial^4 v}{\partial t^4} = 0. \tag{4.64}$$

If K_r is the radius of gyration, i.e. $I = AK_r^2$,

$$\frac{\partial^4 v}{\partial x^4} - \left(\frac{\rho}{\kappa G} + \frac{\rho}{E} \right) \frac{\partial^4 v}{\partial x^2 \, dt^2} + \frac{\rho}{EK_r^2} \frac{\partial^2 v}{\partial t^2} + \frac{\rho^2}{\kappa EG} \frac{\partial^4 v}{\partial t^4} = 0. \tag{4.65}$$

Equation (4.65) is the governing equation for free vibrations of a beam, including shear deformation and rotatory inertia; it is usually known as the Timoshenko beam equation.

For free vibrations we put $v(x, t) = B \exp (\lambda x) \sin (\omega t + \alpha)$. This satisfies equation (4.65), provided that

$$\lambda^4 + \lambda^2 (\eta + 1)\Omega - \Omega/K_r^2 + \Omega^2 \eta = 0 \tag{4.66}$$

where $\Omega = \rho\omega^2/E$ and $\eta = E/\kappa G$. For $\Omega\eta K_r^2 < 1$, equation (4.66) yields one negative and one positive value of λ^2; i.e. the roots are $\pm i\lambda_1, \pm\lambda_2$. (High-frequency solutions, for which $\Omega\eta K_r^2 > 1$, will not be considered here.) Thus

$$v(x, t) = V(x) \sin (\omega t + \alpha)$$

and

$$V(x) = B_1 \sin \lambda_1 x + B_2 \cos \lambda_1 x + B_3 \sinh \lambda_2 x + B_4 \cosh \lambda_2 x.$$
$$(4.67)$$

The constants B_j are eliminated and a frequency equation obtained by applying the boundary conditions. Consider a beam simply supported at $x = 0$ and $x = l$. If $M(x, t) = M(x) \sin (\omega t + \alpha)$,

$$M(x) = -EI\left(\frac{d^2 V}{dx^2} + \frac{\rho\omega^2}{\kappa G} V\right)$$

using equations (4.57) and (4.62). Hence, at a simply supported end the conditions reduce to

$$V = 0 \quad \text{and} \quad d^2 V/dx^2 = 0.$$

Substituting in equation (4.67), $B_2 = 0$ and $B_4 = 0$.
Also

$$B_1 \sin \lambda_1 l + B_3 \sinh \lambda_2 l = 0$$

and

$$-\lambda_1^2 B_1 \sin \lambda_1 l + \lambda_2^2 B_3 \sinh \lambda_2 l = 0.$$

Thus

$$\sin \lambda_1 l = 0 \quad \text{and} \quad B_3 = 0.$$

Substituting $\lambda_1 = n\pi/l$ in equation (4.66)

$$\eta\Omega^2 - \left[(\eta + 1)\left(\frac{n\pi}{l}\right)^2 + \frac{1}{K_r^2}\right]\Omega + \left(\frac{n\pi}{l}\right)^4 = 0. \qquad (4.68)$$

As l/nK_r becomes large, the solution of equation (4.68) reduces to $\Omega = K_r^2(n\pi/l)^4$, which is the value given by the Euler–Bernoulli theory. In general, equation (4.68) gives two natural frequencies, but only the lower value will be considered here. For specified values of l/nK_r and η non-dimensional frequency factors can be obtained from equation (4.68) and compared with values from the Euler–Bernoulli equation.

Taking $\eta = 3$, which corresponds to $v = 0\cdot25$ and $\kappa = 5/6$ (a standard value for a rectangular cross-section), natural frequencies from the Euler–Bernoulli equation exceed those from the Timoshenko equation by $1\cdot2$, $4\cdot7$, $17\cdot2$ and $54\cdot4$ per cent for $l/nK_r = 40$, 20, 10 and 5, respectively.

For other boundary conditions the derivation of a frequency equation is more complicated (as it is for the Euler–Bernoulli equation, Section 3.3), as terms in λ_1 and λ_2 in equation (4.67) coexist. Carr[16] gives the frequency equations and numerical results for six sets of boundary conditions. These results show that, although differences in frequency from the Euler–Bernoulli and Timoshenko equations depend upon the boundary conditions, these differences depend primarily upon the value of l/nK_r.

Results from three-dimensional elasticity theory are limited to the case when the axial waveform is harmonic and the cross-section of the beam is circular. There is close agreement between these results and natural frequencies from the Timoshenko beam equation for all values of l/nK_r[1, 44].

4.7. Dynamic Response of Rigid-plastic Beams

This section is an introduction to a complex subject, which has been the subject of considerable study. The assumptions that are required for the solution of a relatively simple problem and the possible limitations on the applicability of the results are discussed. In general, the appropriate method of determining the dynamic response of plastic beams depends on the nature of the problem and of the results required; although only one method is given here, Goldsmith[36] reviews available methods.

In the previous sections on the vibration of beams displacements were limited to the elastic region, so that the relation between bending moment and curvature was linear. Figure 4.14 shows three idealized relations between bending moment and curvature for plastic beams: (a) represents an elastic-work-hardening material, (b) an elastic-perfectly plastic material, and (c) a rigid-perfectly plastic material.

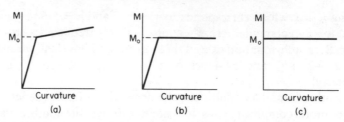

FIG. 4.14. Moment–curvature relations for: (a) elastic-work-hardening material; (b) elastic-perfectly plastic material; (c) rigid-perfectly plastic material.

Here a rigid-perfectly plastic material will be assumed; the validity of neglecting work-hardening depends upon the material of the beam; the limitations imposed by neglecting the elastic curvature will be discussed later. In this rigid-plastic type of analysis it is assumed that infinite curvature occurs at a cross-section where the bending moment is M_0, the limit moment or fully plastic moment, used in limit analysis of structures under static loads; i.e. a plastic hinge forms. The stress distribution across a rectangular cross-section where the limit moment M_0 is acting is shown in Fig. 4.15. If σ_0 is the yield stress, then for this cross-section

$$M_0 = \sigma_0(\tfrac{1}{2}bd)(d/2)$$
$$= \tfrac{1}{4}\sigma_0 bd^2. \tag{4.69}$$

It is assumed that deformation occurs locally at the plastic hinge sections at which the bending moment equals the limit moment M_0.

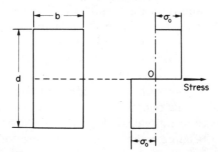

FIG. 4.15. Stress distribution across a rectangular cross-section corresponding to the limit moment M_0.

The dynamic response is required for impact problems, in which some cross-section (or sections) of the beam is subjected to a sudden change of velocity or to a transient force. A uniform beam of length $2l$ (Fig. 4.16a), subjected to a central transverse force $P(t)$, will be considered; the ends of the beam are assumed to be free. Assuming the beam material to be rigid-perfectly plastic the initial response of the beam as the force is applied will be rigid-body motion. This will continue as the force increases until the bending moment due to the loading (applied force and inertia forces) reaches the value M_0, when a plastic hinge will be formed at the central section. In the next

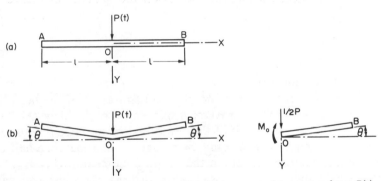

FIG. 4.16. (a) Beam with free ends subjected to a central transverse force $P(t)$. (b) Configuration of the beam after development of a central, plastic hinge.

phase of the motion the two halves of the beam move as rigid bodies, connected by the plastic hinge at the centre, where there are discontinuities in slope and curvature, i.e. a "kink" is formed. This phase continues until further increase of the applied force causes the moment at other sections to reach the value M_0, when further plastic hinges are formed. The initial phases will now be considered in detail.

Phase I. Rigid-body motion. The acceleration of the beam is given by

$$\frac{d^2v}{dt^2} = \frac{P}{2\rho Al}.$$

The inertia force on an element of length dx at a distance x from the central section is $(P/2l)\, dx$ and the bending moment at section x is

$$M(x) = \int_x^l \frac{P}{2l}(l - x)dx$$

$$= \frac{P}{4l}(l - x)^2.$$

The maximum bending moment occurs at $x = 0$; thus the first phase is completed when $M(0) = M_0$, i.e. when

$$\frac{Pl}{4} = M_0.$$

Introducing the non-dimensional parameter

$$\zeta = \frac{Pl}{M_0}, \tag{4.70}$$

the first phase ends when $\zeta = 4$.

Phase II. Central plastic hinge. The configuration of the beam is shown in Fig. 4.16b. The transverse displacement of the hinge O in the direction OY is v; each half of the beam makes an angle θ with the X-axis. By symmetry it is only necessary to consider one-half of the beam, say OB; by definition the bending moment at O is M_0; by symmetry the shear force at O is $\frac{1}{2}P$. Provided that the angle θ is small enough to replace $\cos \theta$ by unity, the transverse acceleration of the centre of gravity of OB is $\ddot{v} - \frac{1}{2}l\ddot{\theta}$. Thus considering the transverse forces on OB and taking moments about O for OB, the two equations of motion are

$$\tfrac{1}{2}P = \rho Al(\ddot{v} - \tfrac{1}{2}l\ddot{\theta})$$

and

$$M_0 = \tfrac{1}{2}\rho Al^2(\ddot{v} - \tfrac{1}{2}l\ddot{\theta}) - \tfrac{1}{12}\rho Al^3\ddot{\theta}.$$

Solving these two equations and using $Pl = \zeta M_0$,

$$\ddot{v} = 2(\zeta - 3)M_0/\rho Al^2$$

and

$$\ddot{\theta} = 3(\zeta - 4)M_0/\rho Al^3. \tag{4.71}$$

The bending moment in OB at a distance x from O is

$$M(x) = M_0 - \tfrac{1}{2}Px + \int_0^x \rho A(\ddot{v} - s\ddot{\theta})(x - s)\, ds$$

$$= M_0 - \tfrac{1}{2}Px + \tfrac{1}{2}\rho A\ddot{v}x^2 - \tfrac{1}{6}\rho A\ddot{\theta}x^3.$$

Substituting from equations (4.71)

$$M(x) = M_0\left[1 - \tfrac{1}{2}r\,\frac{x}{l} + (\zeta - 3)\frac{x^2}{l^2} - \tfrac{1}{2}(\zeta - 4)\frac{x^3}{l^3}\right]. \qquad (4.72)$$

Equation (4.72) is valid, provided that for

$$0 < x \leq l, \qquad |M(x)| < M_0,$$

i.e. provided a second plastic hinge has not been formed. The bending moment is a maximum (or minimum) where $dM(x)/dx = 0$; from equation (4.72) this leads to a quadratic equation in x with roots:

$$x/l = 1$$

and

$$x/l = \zeta/[3(\zeta - 4)]. \qquad (4.73)$$

At $x/l = 1$ the bending moment and shear force are both zero from physical considerations; for the second root only values of ζ which make $x/l = \zeta/[3(\zeta - 4)]$ less than unity are of interest; this corresponds to $\zeta > 6$. The variation of $M(x)$ along the beam OB is shown approximately for a few values of ζ in Fig. 4.17. Substituting $x/l = \zeta/[3(\zeta - 4)]$ into equation (4.72) gives M_{min} and equating this to $-M_0$ gives the value of ζ at which a second plastic hinge occurs. (This follows from the form of the curves of Fig. 4.17.) This is the condition for the end of phase II. The condition $M_{min} = -M_0$ leads to

$$-1 = 1 - \frac{\zeta^2}{6(\zeta - 4)} + \frac{\zeta^2(\zeta - 3)}{9(\zeta - 4)^2} - \frac{\zeta^3}{54(\zeta - 4)^2},$$

i.e.

$$\zeta^3 - 31 \cdot 5\zeta^2 + 216\zeta - 432 = 0. \qquad (4.74)$$

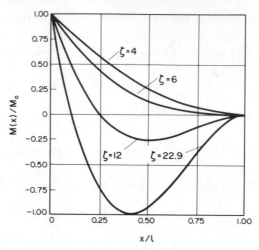

FIG. 4.17. Variation of bending moment along the beam; $\zeta = Pl/M_0$.

The only real root of equation (4.74) is $\zeta = 22\cdot9$. Substituting in equation (4.73), $x/l = 0\cdot404$. Thus there is a plastic hinge at the centre only, provided that

$$4 \leq \zeta \leq 22\cdot9. \tag{4.75}$$

Phase III. When $\zeta = 22\cdot9$, a second plastic hinge is also formed at $x/l = 0\cdot404$. By symmetry a plastic hinge is also formed in bar OA at $x/l = -0\cdot404$. For further increase of the applied force, i.e. $\zeta > 22\cdot9$, the new plastic hinges move out towards the ends of the beam. Analysis for this phase requires the consideration of the motion of bars OC and CB, linked by a plastic hinge at C, where C is the instantaneous position of the hinge. For this analysis the reader is referred to Lee and Symonds[47] or Goldsmith[36]. Here the final deformation of the beam if the applied force is removed during phase II will be considered.

Considering an applied force in the form of a rectangular pulse of duration τ and magnitude P_0 (Fig. 4.18a) such that $4 \leq (P_0 l/M_0) \leq 22\cdot9$, then due to the instantaneous increase of the force to its terminal

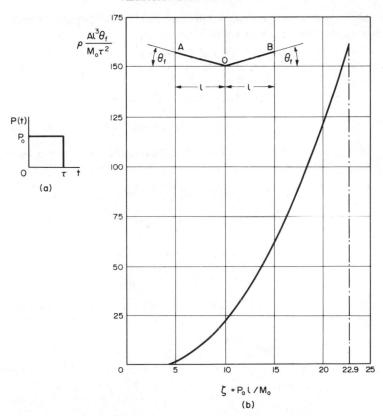

FIG. 4.18. (a) Rectangular pulse applied to beam of Fig. 4.16a. (b) Variation of permanent hinge angle θ_f with ζ.

value P_0 there is no initial rigid-body motion (no phase I); due to the limitation on the values of P_0, no second hinges are formed (no phase III). Thus the motion consists of: (a) phase II motion, with accelerations given by equations (4.71) for $0 \leq t \leq \tau$; (b) further permanent deformation, which occurs with no applied force until the angular velocity $\dot{\theta}$ is reduced to zero; assuming that $t = t_f$ and $\theta = \theta_f$ when $\dot{\theta} = 0$, this phase occurs for $\tau \leq t \leq t_f$, and the angle θ_f gives the final permanent hinge angle for the bar; (c) rigid-body motion for $t \geq t_f$.

(a) $0 \leq t \leq \tau$. Integrating equations (4.71), the displacements and velocities at time $t = \tau$ are given by:

$$\dot{v}_\tau = 2(\zeta - 3)M_0\tau/\rho Al^2,$$

$$v_\tau = (\zeta - 3)M_0\tau^2/\rho Al^2,$$

$$\dot{\theta}_\tau = 3(\zeta - 4)M_0\tau/\rho Al^3,$$

$$\theta_\tau = \tfrac{3}{2}(\zeta - 4)M_0\tau^2/\rho Al^3. \tag{4.76}$$

(b) $\tau \leq t \leq t_f$. As there is no applied force, the equations of motion for bar OB become:

$$0 = \rho Al(\ddot{v} - \tfrac{1}{2}l\ddot{\theta}),$$

$$M_0 = \tfrac{1}{2}\rho Al^2(\ddot{v} - \tfrac{1}{2}l\ddot{\theta}) - \tfrac{1}{12}A\rho l^3\ddot{\theta}$$

with solutions:

$$\ddot{\theta} = \frac{-12M_0}{\rho Al^3} \quad \text{and} \quad \ddot{v} = \frac{-6M_0}{\rho Al^2}. \tag{4.77}$$

Integrating equations (4.77) and using the initial conditions for this phase (4.76),

$$\dot{\theta} = \frac{3(\zeta - 4)M_0\tau}{\rho Al^3} - \frac{12M_0(t - \tau)}{\rho Al^3},$$

$$\theta = \frac{3(\zeta - 4)M_0\tau(t - \tau)}{\rho Al^3} - \frac{6M_0(t - \tau)^2}{\rho Al^3} + \frac{3(\zeta - 4)M_0\tau^2}{2\rho Al^3}$$

and

$$\dot{v} = \frac{2(\zeta - 3)M_0\tau}{\rho Al^2} - \frac{6M_0(t - \tau)}{\rho Al^2}. \tag{4.78}$$

This phase ends when $\dot{\theta} = 0$, when the conditions are defined as $t = t_f$, $\theta = \theta_f$, and $\dot{v} = \dot{v}_f$. From equations (4.78) this leads to

$$t_f - \tau = \frac{(\zeta - 4)}{4}\tau \quad \text{or} \quad t_f = \frac{\zeta\tau}{4},$$

$$\theta_f = \frac{3\zeta(\zeta - 4)M_0\tau^2}{8\rho Al^3} \tag{4.79}$$

and

$$\dot{v}_f = \frac{\zeta M_0 \tau}{2\rho A l^2}.$$

(4.80)

Equation (4.79) gives the final value of the kink or hinge angle, defined in Fig. 4.18b, where this angle is plotted non-dimensionally against $\zeta = P_0 l/M_0$.

(c) $t \geq t_f$. The deformed beam moves as a rigid body with velocity $\dot{v}_f$, given by equation (4.80).

As a check on the above expressions, it is instructive to write down the energy equation in the form:

Work done by the applied force = Energy dissipated by
the hinge moment + residual kinetic energy.

For the half-beam OB,

$$\frac{P}{2}v_\tau = M_0 \theta_f + \tfrac{1}{2}\rho A l \dot{v}_f^2.$$

Substituting for v_τ, θ_f and $\dot{v}_f$ from equations (4.76), (4.79) and (4.80), respectively, it is found that the energy equation is satisfied.

The limitations imposed by the assumptions will now be considered; the assumptions were:

(1) Rigid-perfectly plastic material. Thus elastic deformations have been neglected compared with plastic deformations; this is reasonable if the total energy absorbed in plastic deformation greatly exceeds the maximum possible elastic strain energy that could be stored in the beam.

(2) In deriving the equations of motion it has been assumed that $\cos \theta = 1$.

(3) As phase III of the motion has not been considered, $P_0 l/M_0 \leq 22.9$.

Considering the first assumption, the energy absorbed in plastic deformation of the beam OB is $M_0 \theta_f$, and the maximum elastic strain energy that can be stored in OB is $(M_e^2 l/2EI)$, where M_e is the maximum bending moment that the cross-section can withstand without exceeding the elastic limit. For a rectangular cross-section, $b \times d$,

$M_e = (1/6)\sigma_0 bd^2$, where σ_0 is the stress at the elastic limit. Comparing this expression with equation (4.69) $M_0 = \beta M_e$, where β is a "shape factor" equal to 1·5 for a rectangular cross-section. In general, β will not differ greatly from unity, and will be unity for an ideal I-section in which the total cross-sectional area is assumed to be divided equally between the two flanges. Thus for any cross-section the criterion for the validity of this assumption is

$$\theta_f \gg \frac{M_0 l}{2EI\beta^2}.$$

Using

$$M_0 = \beta M_e = \beta\left(\frac{2\sigma_0 I}{d}\right),$$

where d = depth of beam,

$$\theta_f \gg \frac{\sigma_0}{E}\frac{l}{d\beta}.$$

For steels, $\sigma_0/E \simeq 10^{-3}$. Thus

$$\theta_f \gg 10^{-3}\frac{l}{d\beta}. \tag{4.81}$$

The second assumption, $\cos\theta = 1$, is satisfied approximately if $\theta < 15°$. If the work done at the plastic hinge is at least ten times the maximum possible elastic strain energy (to satisfy the first assumption), then the bounds for θ_f are

$$0·262 \geq \theta_f \geq 10^{-2}\frac{l}{d\beta}. \tag{4.82}$$

The limits on ζ in Fig. 4.18b, corresponding to the above limits on θ_f (4.82), will be evaluated. The ordinate in Fig. 4.18b is $\theta_f \rho A l^3/M_0 \tau^2$; if T is the fundamental period of vibrations of the elastic free-free beam, then

$$T^2 = \frac{4\pi^2}{(4·730)^4}\frac{\rho A(2l)^4}{EI} \quad \text{or} \quad \frac{\rho A l^4}{EI} = 0·7925 T^2.$$

Using

$$M_0 = \beta\left(\frac{2\sigma_0 I}{d}\right), \quad \text{and} \quad \frac{E}{\sigma_0} = 10^3,$$

$$\frac{\theta_f \rho A l^3}{M_0 \tau^2} = \theta_f \frac{0.7925}{2\beta} 10^3 \left(\frac{T}{\tau}\right)^2 \frac{d}{l}.$$

Substituting in equation (4.82) and putting $\beta = 1.5$ for a rectangular cross-section,

$$69.2\left(\frac{T}{\tau}\right)^2 \frac{d}{l} \geq \theta_f\left(\frac{\rho A l^3}{M_0 \tau^2}\right) \geq 1.76\left(\frac{T}{\tau}\right)^2. \tag{4.83}$$

For specific values of T/τ and d/l, the above limits can be evaluated, and corresponding maximum and minimum values of ζ found from Fig. 4.18b. Some examples are tabulated below; it will be seen that the above assumptions impose in most cases close limits on the allowable values of $P_0 l/M_0$.

TABLE 4.2. Limits on ζ $(= P_0 l/M_0)$

	(1)	(2)	(3)	(4)	(5)	(6)	(7)
l/d	5	5	5	5	5	10	20
T/τ	0·5	1	2	4	6	4	4
ζ_{min}	4·0	5·0	6·8	10·9	15·2	10·9	10·9
ζ_{max}	5·5	8·4	14·2	22·9	22·9	19·2	14·2

The upper limit ζ_{max} in the fourth and fifth columns is imposed by the restricted theory presented here (neglect of phase III), rather than by the assumptions. Even if allowance is made for additional moving hinges[36, 47] the results can be applied only to beams with $l/d \leq 39.3$ [from equation (4.83)], and for the higher values of l/d in this range they can be used only for a very small range of values of $P_0 l/M_0$. Limits on ζ for values of $l/d < 5$ have not been included, because the general theory of this chapter which neglects rotatory inertia and shear deformation is applicable only to $2l/K_r > 20$ where K_r is the radius of gyration of the cross-section.

The above, relatively simple, example has been treated in detail as an introduction to the dynamic response of beams in the plastic range, and in order to show the effects of the assumptions on the applicability of the results. For simplicity the response to a rectangular pulse has been determined, but other pulse shapes give similar results[36].

Problems

1. A bridge can be idealized as a uniform simply supported beam of 10 m span, flexural rigidity 4 GN m^2 and mass per unit length 15,000 kg/m. A transverse harmonic force $P \sin 2\pi f t$ moves across the bridge with uniform velocity U. Find the values of U, up to a limit of 40 m/s, at which vibrations of large amplitude will tend to build up, if (a) $f = 8$ Hz, (b) $f = 10$ Hz and (c) $f = 12$ Hz. Find the range of values of f for which large amplitudes may occur for the speed range $0 \leq U \leq 40$ m/s.

2. A constant transverse force of 10 kN moves with uniform velocity of 40 m/s across the bridge of Problem 1. Plot the displacement at mid-span against time for the period when the force is on the bridge. How would this curve be altered if the span of the beam is increased to 50 m and other conditions are unchanged?

3. (a) One support of a uniform simply supported beam is given a transverse displacement

$$v(t) = a \sin \omega t, \qquad 0 \leq t \leq \pi/\omega,$$
$$v(t) = 0, \qquad\qquad t > \pi/\omega.$$

Derive an expression for the displacement at any point on the beam during the time interval $t = 0$ to $t = \pi/\omega$.

(b) If both supports are subjected simultaneously to the above displacement obtain the corresponding expression for the displacement at any point on the beam.

(c) If one support of this beam is subjected to a moment

$$M(t) = M_0 \sin \omega t, \qquad 0 \leq t \leq \pi/\omega,$$
$$M(t) = 0, \qquad\qquad t > \pi/\omega,$$

derive the general expression for the displacement for $t \leq \pi/\omega$.

4. A transverse displacement $v(t) = a(1 - \cos \omega t), 0 \leq t \leq 2\pi/\omega$, is applied to one end of a uniform cantilever beam. Derive general expressions for the displacement of the beam during this time interval, if the prescribed displacement occurs at (a) the root, (b) the tip.

5. It is observed that when axial forces N are applied to the ends of a uniform simply supported beam the fundamental natural frequency in flexure is reduced by 5 per cent. If the length of the beam is 5 m and the flexural rigidity is 20 MN m^2, determine the value of N. What will be the percentage change in the second natural frequency due to applying these axial forces?

6. The fundamental natural frequency in flexure of a uniform simply supported beam is observed to be 55·0 and 45·3 Hz when subjected to compressive axial forces of 100 N and 1 kN, respectively. Estimate the Euler critical load for this beam.

7. A uniform beam of length 10 m, mass per unit length 100 kg/m and flexural rigidity 10 MN m^2 rests on an elastic foundation of modulus 100 kN/m run/m deflection. There is no other constraint on the beam. Find the natural frequency of the rigid-body modes and the first and second natural frequencies of flexural vibration.

8. A uniform beam is clamped at the end $x = 0$ and simply supported at $x = l$. If a harmonic bending moment $M_0 \sin \omega t$ is applied at $x = l$, and damping is neglected, show that the slope at $x = l$ is given by

$$\phi = \frac{M_0(\cos \lambda l \cosh \lambda l - 1) \sin \omega t}{EI\lambda(\cos \lambda l \sinh \lambda l - \sin \lambda l \cosh \lambda l)}.$$

9. A uniform beam of length $2l$ is clamped at one end and simply supported at the other; there is an intermediate support at mid-span. Show that the natural frequencies of flexural vibration are given by the roots of the equation:

$$2(\cos \lambda l \sinh \lambda l - \sin \lambda l \cosh \lambda l)^2 - (\sin \lambda l - \sinh \lambda l)^2 = 0.$$

10. A single-storey plane frame consists of three identical beams AB, BC and CD. Each beam is of length l, mass per unit length ρA and flexural rigidity EI. The beams AB and CD are vertical, BC is horizontal. The ends A and D are clamped; there is no relative rotation at either of the joints B and C. Considering flexural vibrations in the plane of the frame, show that the fundamental natural frequency is given by $\omega_1 = 3 \cdot 17(EI/\rho Al^4)^{1/2}$.

11. A uniform cantilever beam is subjected to a transverse harmonic force $P \sin \omega t$ at the tip.

    (a) Neglecting damping, show that the resulting displacement at the tip is

$$\frac{P(\sin \lambda l \cosh \lambda l - \cos \lambda l \sinh \lambda l) \sin \omega t}{EI\lambda^3(1 + \cos \lambda l \cosh \lambda l)}$$

where the symbols have their usual meanings.

    (b) Allowing for hysteretic damping in the beam material, where the damping coefficient μ is small, show that the amplitude of vibrations at the tip, when the excitation frequency is equal to the rth natural frequency ω_r, is approximately $4P/\mu m\omega_r^2$, where m is the mass of the beam. Also show that the same value is obtained for this amplitude by the normal mode method of Section 3.5, if only the dominant term in the series is considered.

12. A uniform cantilever beam of length l and rigid-plastic material (limit moment M_0) is subjected to a transverse force at the tip

$$P(t) = P_0 \qquad 0 \leq t \leq t_1,$$

$$P(t) = 0 \qquad t > t_1.$$

    (a) Determine the limiting values of $P_0 l/M_0$ for which there will be a single plastic hinge. (b) If P_0 lies within the limits obtained in (a), find the tip displacement when $t = t_1$ and the final tip displacement. (c) Investigate the limits imposed on $P_0 l/M_0$ by the assumptions made in the theory, if the beam is of steel, the cross-section is rectangular, $l/d = 10$ where d = depth of beam, and $T/t_1 = 1$ where T = fundamental period of the equivalent elastic beam.

13. Considering the shaft described in Problem 14 of Chapter 3, but treating it as a system of two beams, determine the fundamental natural frequency in flexure.

CHAPTER 5

Vibrations of Plates and Shells

5.1. General Theory

In this chapter an introduction to the vibration of rectangular and circular plates and of cylindrical and shallow shells will be given. It should be recalled that if a structure is subjected to a harmonic applied force, large amplitudes will occur if the frequency of this force is equal, or nearly equal, to one of the natural frequencies, and that the natural frequencies may be required to evaluate response, using an extension of the method of Section 3.5. Thus in some sections only natural frequencies and mode shapes will be considered.

First the necessary relations between stress, strain and displacement will be derived in Cartesian coordinates for use when considering rectangular plates in Sections 5.2, 5.4 and 5.5. Then corresponding relations in cylindrical polar coordinates will be given for use when considering cylindrical shells in Section 5.7.

Cartesian Coordinates

In three-dimensional elasticity theory the stress at a point is specified by the six quantities:

$$\sigma_x \sigma_y \sigma_z \quad \text{the components of direct stress;}$$
$$\sigma_{xy} \sigma_{yz} \sigma_{zx} \quad \text{the components of shear stress.}$$

The components of stress on the face *ABCD* of an element are shown in Fig. 5.1, from which the sign conventions are seen to be: direct

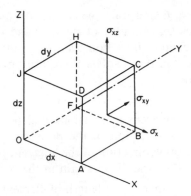

FIG. 5.1. Stress components on the face *ABCD* of an element.

stresses are positive when tensile; the shear stress σ_{xz} acts on a face perpendicular to the X-axis in a direction parallel to the Z-axis and is positive if it acts in the positive direction of the Z-axis on a face for which the positive direct stress is in the direction $\overrightarrow{OX}$. If the lengths of the sides of the element are dx, dy and dz, the shear stresses σ_{xz} on the faces *ABCD* and *OFHJ* form a couple about the Y-axis of magnitude $(\sigma_{xz}\, dy\, dz)\, dx$. Considering the other components of shear stress, only the components σ_{zx} acting on the faces *DCHJ* and *OABF* form a couple about the Y-axis. Thus taking moments about the Y-axis for the equilibrium of the element, $\sigma_{xz} = \sigma_{zx}$. Similarly, $\sigma_{xy} = \sigma_{yx}$ and $\sigma_{yz} = \sigma_{zy}$. The strain at a point is defined similarly by:

$\varepsilon_x \varepsilon_y \varepsilon_z$ the components of direct strain;

$\varepsilon_{xy} \varepsilon_{yz} \varepsilon_{zx}$ the components of shear strain.

The components of displacement at any point (x, y, z) are u, v and w, positive in the directions $\overrightarrow{OX}$, $\overrightarrow{OY}$ and $\overrightarrow{OZ}$, respectively.

The components of direct stress and strain are related by Hooke's law, extended to include Poisson's ratio effects,

$$\varepsilon_x = (1/E)[\sigma_x - v(\sigma_y + \sigma_z)] \qquad (5.1)$$

with analogous expressions for ε_y and ε_z. The components of shear stress and strain are related by

$$\varepsilon_{xy} = (1/G)\sigma_{xy}, \text{ etc.} \qquad (5.2)$$

In these equations E, G and v are the elastic constants, Young's modulus, shear modulus (or modulus of rigidity) and Poisson's ratio, respectively. In this chapter only homogeneous isotropic elastic solids will be considered; for these solids there are only two *independent* elastic constants and

$$E = 2G(1 + v). \tag{5.3}$$

Next the strain–displacement relations will be established. If the displacement in the X-direction of the point (x, y, z) is u, then the displacement in the same direction of the adjacent point $(x + dx, y, z)$ is $[u + (\partial u/\partial x)\, dx]$. Thus the direct strain in the X-direction

$$\varepsilon_x = \frac{\text{Increase in length of element}}{\text{Initial length of element}}$$

$$= \frac{[dx + (\partial u/\partial x)\,.\, dx] - dx}{dx}$$

$$= \partial u/\partial x. \tag{5.4}$$

Similarly, $\varepsilon_y = \partial v/\partial y$ and $\varepsilon_z = \partial w/\partial z$.

Considering an element $ABCD$, initially rectangular and having sides of length dx and dy parallel to the X- and Y-axes, the corner A, initially at the point (x, y), is displaced to A_1 in the plane OXY, with components of displacement u and v in the X- and Y-directions, respectively (Fig. 5.2). The point B is displaced to B_1, with a displacement in the Y-direction of $[v + (\partial v/\partial x)\, dx]$; the point D is displaced to

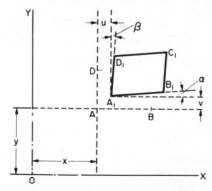

FIG. 5.2. Displacements in the plane OXY.

D_1, with a displacement in the X-direction of $[u + (\partial u/\partial y)\, dy]$. The deformed shape of the element is the parallelogram $A_1\, B_1\, C_1\, D_1$ and the shear strain is $(\alpha + \beta)$. Thus for small angles

$$\varepsilon_{xy} = \alpha + \beta = \frac{\partial v}{\partial x} + \frac{\partial u}{\partial y}. \tag{5.5}$$

Similarly,

$$\varepsilon_{yz} = \frac{\partial w}{\partial y} + \frac{\partial v}{\partial z} \quad \text{and} \quad \varepsilon_{zx} = \frac{\partial u}{\partial z} + \frac{\partial w}{\partial x}.$$

For an element of sides dx, dy and dz (Fig. 5.1) the force due to the stress σ_x is $\sigma_x\, dy\, dz$; if the corresponding strain is ε_x, the extension in the X-direction is $\varepsilon_x\, dx$ and the work done on the element is $\frac{1}{2}(\sigma_x\, dy\, dz)\,(\varepsilon_x\, dx) = \frac{1}{2}\sigma_x\varepsilon_x\, dV$, where dV is the volume of the element. Considering the element $ABCD$ of Fig. 5.2, of thickness dz in the Z-direction, the force corresponding to the shear stress σ_{xy} on the face BC is $\sigma_{xy}\, dy\, dz$; there is an equal and opposite force on the face AD, the two forces forming a couple of magnitude $\sigma_{xy}\, dx\, dy\, dz$; the rotation due to this couple is β. Due to the complementary shear stress on the faces AB and CD, there is a couple of equal magnitude with corresponding rotation α. Using $\varepsilon_{xy} = \alpha + \beta$, the work done on the element is $\frac{1}{2}\sigma_{xy}\varepsilon_{xy}\, dV$. Generalizing for a three-dimensional state of stress, the strain energy in an elastic body

$$\mathfrak{S} = \int_V \tfrac{1}{2}\left(\sigma_x\varepsilon_x + \sigma_y\varepsilon_y + \sigma_z\varepsilon_z + \sigma_{xy}\varepsilon_{xy} + \sigma_{yz}\varepsilon_{yz} + \sigma_{zx}\varepsilon_{zx}\right) dV. \tag{5.6}$$

Using equations (5.1) and (5.2) the strain energy may be expressed in terms of strain only or of stress only.

Cylindrical Polar Coordinates

Equations (5.4) and (5.5) give the strain–displacement relations in Cartesian coordinates, but for application to cylindrical shells these relations are required in cylindrical polar coordinates. A point A on a cylindrical shell is defined by the coordinates (x, θ, r), where x, θ and r are the axial, tangential (angular) and radial coordinates, respectively

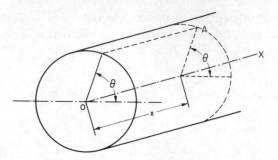

Fig. 5.3. Cylindrical polar coordinates.

(Fig. 5.3). The components of displacement of the point A are u, v and w in the x-, θ- and r-directions, respectively. In Fig. 5.4a the line AB is displaced to $A_1 B_1$, remaining in the same plane perpendicular to the X-axis; the components of displacement of A are v and w in the θ- and r-directions, respectively. The initial length AB is $r\, d\theta$. In the displaced position

$$A_1 B_1 = (r + w)\, d\theta + \left(v + \frac{\partial v}{\partial \theta}\, d\theta\right) - v.$$

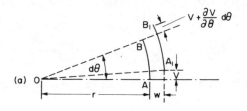

(a)

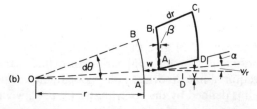

(b)

Fig. 5.4. Displacements in cylindrical polar coordinates.

The component of direct strain in the tangential direction

$$\varepsilon_\theta = \frac{A_1 B_1 - AB}{AB}$$

$$= \left(w \, d\theta + \frac{\partial v}{\partial \theta} \, d\theta \right) \Big/ r \, d\theta$$

$$= \frac{w}{r} + \frac{1}{r} \frac{\partial v}{\partial \theta}. \tag{5.7}$$

From Fig. 5.4b the shear strain in the $r - \theta$ plane $\varepsilon_{r\theta} = \alpha + \beta$ (being the change in the angle BAD). The tangential displacement of the point D to D_1 is:

$$v + \frac{v}{r} \, dr + \alpha \, dr = v + \frac{\partial v}{\partial r} \, dr$$

(by definition). Thus

$$\alpha = \frac{\partial v}{\partial r} - \frac{v}{r}.$$

Now

$$\beta = \frac{w + (\partial w / \partial \theta) \, d\theta - w}{r \, d\theta} = \frac{1}{r} \frac{\partial w}{\partial \theta}.$$

Hence

$$\varepsilon_{r\theta} = \frac{1}{r} \frac{\partial w}{\partial \theta} + \frac{\partial v}{\partial r} - \frac{v}{r}. \tag{5.8}$$

The other components of strain can be found from the corresponding expressions in Cartesian coordinates, (5.4) and (5.5), replacing ∂y by $r \, \partial \theta$; for example,

$$\varepsilon_{x\theta} = \frac{\partial v}{\partial x} + \frac{1}{r} \frac{\partial u}{\partial \theta}. \tag{5.9}$$

5.2. Transverse Vibrations of Rectangular Plates

The undeformed middle plane of the plate is defined as OXY, with the X- and Y-axes parallel to the edges of the plate; the Z-axis will be taken as positive upwards (Fig. 5.7). The following assumptions are made:

1. The plate is thin and of uniform thickness h; thus the free surfaces of the plate are the planes $z = \pm\frac{1}{2}h$.
2. The direct stress in the transverse direction, σ_z, is zero. This stress must be zero at the free surfaces, and provided that the plate is thin, it is reasonable to assume that it is zero at any section z.
3. Stresses in the middle plane of the plate (membrane stresses) are neglected, i.e. transverse forces are supported by bending stresses, as in flexure of a beam. For membrane action not to occur, the displacements must be small compared with the thickness of the plate, as shown by Jaeger[42].
4. Plane sections that are initially normal to the middle plane remain plane and normal to it. A similar assumption was made in the elementary theory for beams (used in the whole of Chapters 3 and 4, except Section 4.6), and implies that deformation due to transverse shear is neglected. Thus with this assumption the shear strains ε_{xz} and ε_{yz} are zero.
5. Only the transverse displacement w (in the Z-direction) has to be considered.

Figure 5.5a shows an element of the plate of length dx in the unstrained state and Fig. 5.5b the corresponding element in the strained state. If OA represents the middle surface of the plate, then $OA = O_1 A_1$ from the third assumption; also $O_1 A_1 = R_x\, d\theta$, where R_x is the radius of curvature of the deformed middle surface. Thus the strain in BC, at a distance z from the middle surface,

$$
\begin{aligned}
\varepsilon_x &= \frac{B_1 C_1 - BC}{BC} \\
&= \frac{(R_x + z) - R_x}{R_x} \\
&= z/R_x .
\end{aligned}
$$

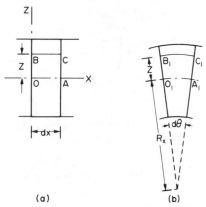

(a)   (b)

FIG. 5.5. Element of plate. (a) Unstrained. (b) Strained.

As in Chapter 3, the relation between the curvature and the displacement of the middle surface, w, is:

$$\frac{1}{R_x} = -\frac{\partial^2 w}{\partial x^2}.$$

Thus

$$\varepsilon_x = -z\frac{\partial^2 w}{\partial x^2}. \tag{5.10}$$

Similarly

$$\varepsilon_y = -z\frac{\partial^2 w}{\partial y^2}.$$

From equation (5.5) the shear strain ε_{xy} at a distance z from the middle surface is $[(\partial u/\partial y) + (\partial v/\partial x)]$, where u and v are the displacements at depth z in the X- and Y-directions, respectively. Using the assumption that sections normal to the middle plane remain normal to it,

$$u = -z\frac{\partial w}{\partial x} \quad \text{(Fig. 5.6)}.$$

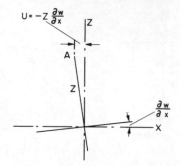

FIG. 5.6. Relation between u and w.

Similarly

$$v = -z \frac{\partial w}{\partial y}.$$

Thus

$$\varepsilon_{xy} = -2z \frac{\partial^2 w}{\partial x \, \partial y}. \tag{5.11}$$

(In equation (5.11) the term $(\partial^2 w/\partial x \, \partial y)$ is the twist of the surface.)

From the second assumption and equations (5.1) and (5.2), the stress–strain relations for a thin plate are

$$E\varepsilon_x = \sigma_x - v\sigma_y$$

and

$$E\varepsilon_y = \sigma_y - v\sigma_x,$$

which can be arranged to give

$$\sigma_x = \frac{E}{1 - v^2}(\varepsilon_x + v\varepsilon_y),$$

$$\sigma_y = \frac{E}{1 - v^2}(\varepsilon_y + v\varepsilon_x). \tag{5.12}$$

Also

$$\sigma_{xy} = \frac{E}{2(1 + v)}\varepsilon_{xy}.$$

Substituting from equations (5.10), (5.11) and (5.12) in the strain energy expression (5.6),

$$
\mathfrak{S} = \int_0^a \int_0^b \int_{-h/2}^{h/2} \frac{E}{2(1-v^2)} \left[\varepsilon_x^2 + \varepsilon_y^2 + 2v\varepsilon_x \varepsilon_y \right.
$$

$$
\left. + \frac{1}{2}(1-v)\varepsilon_{xy}^2 \right] dz \, dy \, dx
$$

$$
= \frac{D}{2} \int_0^a \int_0^b \left[\left(\frac{\partial^2 w}{\partial x^2} \right)^2 + \left(\frac{\partial^2 w}{\partial y^2} \right)^2 + 2v \left(\frac{\partial^2 w}{\partial x^2} \right) \left(\frac{\partial^2 w}{\partial y^2} \right) \right.
$$

$$
\left. + 2(1-v) \left(\frac{\partial^2 w}{\partial x \, \partial y} \right)^2 \right] dy \, dx \tag{5.13}
$$

after integrating with respect to z, where

$$
D = \frac{Eh^3}{12(1-v^2)} .
$$

The limits in the above integrals imply that the plate is bounded by the lines $x = 0$, $x = a$, $y = 0$ and $y = b$.

If ρ is the density of the plate, the kinetic energy

$$
\mathfrak{T} = \int_0^a \int_0^b \int_{-h/2}^{h/2} \frac{1}{2} \rho \left(\frac{\partial w}{\partial t} \right)^2 dz \, dy \, dx
$$

$$
= \frac{1}{2} \rho h \int_0^a \int_0^b \left(\frac{\partial w}{\partial t} \right)^2 dy \, dx \tag{5.14}
$$

after integrating with respect to z. Expressions (5.13) and (5.14) will be used with the Rayleigh–Ritz and finite element methods.

The element of the plate with sides dx and dy and thickness h, shown in Fig. 5.7, is subjected to a bending moment M_x, a twisting moment M_{xy} and a transverse shear force S_x per unit length on the face OB; on the face OA there are, per unit length, a bending moment M_y, a twisting moment M_{yx} and a shear force S_y. Each moment is denoted by a double-headed arrow in the direction of its axis (following the convention adopted in Section 4.5). The bending moments M_x and M_y are the resultant moments due to the direct stresses σ_x and σ_y, respectively,

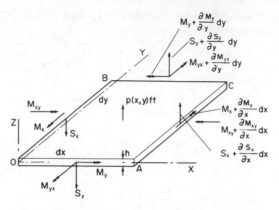

FIG. 5.7. Forces and moments on an element of a plate.

after integrating through the thickness of the plate. Similarly, the twisting moments are the resultants due to the shear stress σ_{xy}. Maintaining a consistent sign convention between the definitions for stresses in Fig. 5.1 and the moments shown in Fig. 5.7, we have

$$M_x = \int_{-h/2}^{h/2} \sigma_x z \, dz, \qquad M_y = \int_{-h/2}^{h/2} \sigma_y z \, dz$$

and

$$M_{xy} = M_{yx} = \int_{-h/2}^{h/2} \sigma_{xy} z \, dz. \tag{5.15}$$

Figure 5.7 shows also the incremental quantities acting on the faces AC and BC and the applied force per unit area, $p(x, y)f(t)$, in the Z-direction. In addition there is an inertia force per unit area, $\rho h \, \partial^2 w/\partial t^2$, in the Z-direction. The equilibrium equations, obtained by resolving in the Z-direction and taking moments about the Y- and X-axes, are, after dividing by $dx \, dy$:

$$\frac{\partial S_x}{\partial x} + \frac{\partial S_y}{\partial y} + p(x, y)f(t) = \rho h \frac{\partial^2 w}{\partial t^2},$$

$$\frac{\partial M_x}{\partial x} + \frac{\partial M_{yx}}{\partial y} - S_x = 0$$

and

$$-\frac{\partial M_y}{\partial y} - \frac{\partial M_{xy}}{\partial x} + S_y = 0.$$

Eliminating S_x and S_y,

$$\frac{\partial^2 M_x}{\partial x^2} + 2\frac{\partial^2 M_{xy}}{\partial x\,\partial y} + \frac{\partial^2 M_y}{\partial y^2} + p(x,\,y)f(t) = \rho h\frac{\partial^2 w}{\partial t^2}. \qquad (5.16)$$

Substituting from equations (5.10) to (5.12) in equations (5.15) and integrating with respect to z,

$$M_x = -D\left(\frac{\partial^2 w}{\partial x^2} + v\frac{\partial^2 w}{\partial y^2}\right), \qquad M_y = -D\left(\frac{\partial^2 w}{\partial y^2} + v\frac{\partial^2 w}{\partial x^2}\right)$$

and

$$M_{xy} = -D(1-v)\frac{\partial^2 w}{\partial x\,\partial y}. \qquad (5.17)$$

Substituting from equations (5.17) in equation (5.16) gives the equilibrium equation for an element of the plate in terms of w and its derivatives,

$$D\left[\frac{\partial^4 w}{\partial x^4} + 2\frac{\partial^4 w}{\partial x^2\,\partial y^2} + \frac{\partial^4 w}{\partial y^4}\right] + \rho h\frac{\partial^2 w}{\partial t^2} = p(x,\,y)f(t). \qquad (5.18)$$

For a dynamic problem $w(x,\,y,\,t)$ must satisfy equation (5.18) together with the boundary conditions.

The standard simple boundary conditions are simply supported, clamped and free. Considering the edges OB and AC of Fig. 5.7, the boundary conditions are:

Simply supported. $w = 0$ and $M_x = 0$, i.e.

$$w = 0 \quad \text{and} \quad \frac{\partial^2 w}{\partial x^2} + v\frac{\partial^2 w}{\partial y^2} = 0. \qquad (5.19)$$

Clamped.

$$w = 0 \quad \text{and} \quad \frac{\partial w}{\partial x} = 0. \qquad (5.20)$$

Free. For the edge to be free from force and moment M_x, M_{xy} and S_x should be zero. The latter two quantities are combined into a single condition following Kirchhoff. Consider two adjacent elements of the face OB, H_1H_2 and H_2H_3, each of length dy. The twisting moment $M_{xy}\,dy$ on H_1H_2 is replaced by equal and opposite forces M_{xy}; the twisting moment

$$\left(M_{xy} + \frac{\partial M_{xy}}{\partial y}\,dy\right)dy$$

on H_2H_3 is replaced by equal and opposite forces

$$M_{xy} + \frac{\partial M_{xy}}{\partial y}\,dy$$

(Fig. 5.8). Considering equilibrium in the vertical direction at H_2,

$$\frac{\partial M_{xy}}{\partial y}\,dy + S_x\,dy = 0.$$

Thus, if

$$V_x \equiv S_x + \frac{\partial M_{xy}}{\partial y}$$

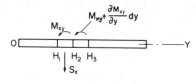

FIG. 5.8. Derivation of effective shear force.

is the effective shear force, the free edge conditions are $M_x = 0$ and $V_x = 0$, i.e.

$$\frac{\partial^2 w}{\partial x^2} + v \frac{\partial^2 w}{\partial y^2} = 0$$

and

$$\frac{\partial^3 w}{\partial x^3} + (2 - v) \frac{\partial^3 w}{\partial x \, \partial y^2} = 0. \qquad (5.21)$$

Corresponding conditions for faces parallel to the X-axis are found from equations (5.19) to (5.21) by interchanging x and y.

Natural Frequencies

If a rectangular plate is simply supported on the four edges $x = 0$, $x = a$, $y = 0$ and $y = b$, the expression

$$w(x, y, t) = A_{ij} \sin \frac{i\pi x}{a} \sin \frac{j\pi y}{b} \sin (\omega_{ij} t + \alpha), \qquad (5.22)$$

where i and j are integers, satisfies all the boundary conditions and, on substituting in equation (5.18) with $p(x, y)f(t) = 0$ for free vibrations, leads to

$$\rho h \omega_{ij}^2 = D \left[\left(\frac{\pi i}{a} \right)^4 + \frac{2\pi^4 i^2 j^2}{a^2 b^2} + \left(\frac{\pi j}{b} \right)^4 \right],$$

i.e.

$$\left(\frac{\rho h a^4}{\pi^4 D} \right)^{1/2} \omega_{ij} = i^2 + \frac{j^2 a^2}{b^2}, \qquad i, j = 1, 2, 3, \dots . \qquad (5.23)$$

The integers i and j are equal to the number of half-waves in the X- and Y-directions respectively. The nodal lines, i.e. lines for which w is zero for all values of t, are parallel to the edges of the plate; there are $(i - 1)$ and $(j - 1)$ nodal lines parallel to the Y- and X-axes, respectively. For the fundamental mode $i = 1$ and $j = 1$. The second natural frequency is given by $i = 2$ and $j = 1$ if $a/b > 1$ and by $i = 1$ and $j = 2$ if $a/b < 1$; for a square plate the natural frequencies associated with $i = 2, j = 1$ and $i = 1, j = 2$ are equal.

In Section 3.3 it is shown that the frequency expression for a uniform beam in flexure is considerably simpler if the ends are simply supported than for any other set of boundary conditions. For rectangular plates a simple frequency expression is obtained only when all edges are simply supported. If two parallel edges are simply supported, say $y = 0$ and $y = b$, the expression

$$w(x, y, t) = F(x) \sin \frac{j\pi y}{b} \sin (\omega t + \alpha) \tag{5.24}$$

satisfies the boundary conditions on $y = 0$ and $y = b$. Substituting equation (5.24) into equation (5.18) with $p(x, y)f(t) = 0$, a fourth-order ordinary differential equation in $F(x)$ is obtained. The solution of this equation contains four constants, which can be eliminated by application of the boundary conditions at $x = 0$ and $x = a$ to give a complicated frequency equation. Details will not be given here. The frequency equations and numerical results for all possible combinations of clamped, free and simply supported conditions on $x = 0$ and $x = a$ are given by Leissa[50].

When the plate does not have two parallel edges simply supported, no single expression of the form

$$w(x, y, t) = F(x)f(y) \sin (\omega t + \alpha)$$

satisfies the plate equation and all the boundary conditions. In order to be able to determine natural frequencies of rectangular plates with general boundary conditions and to present methods applicable to nonuniform plates and plates of complex geometry, the Rayleigh–Ritz and finite element methods will be given here and in Section 5.5, respectively.

Rayleigh–Ritz Method

The general method is described in Section 3.7.

The strain and kinetic energy expressions for a uniform rectangular plate are given by equations (5.13) and (5.14). It is assumed that

$$w(x, y, t) = \sum_{i=1}^{I} \sum_{j=1}^{J} \phi_i(x)\psi_j(y)\Gamma_s(t) \tag{5.25}$$

where $s = j + (i - 1)J$; thus s takes all integer values from 1 to IJ. For natural frequency calculations the functions $\Gamma_s(t)$ will be assumed subsequently to be harmonic. Each function $\phi_i(x)$ must satisfy any geometric boundary conditions on the edges $x = 0$ and $x = a$; each function $\psi_j(y)$ must satisfy any geometric boundary conditions on the edges $y = 0$ and $y = b$. It is assumed that the functions are normalized and orthogonal; in practice, appropriate beam mode shapes are often used, as they satisfy all these conditions. Thus

$$\int_0^a \phi_i(x)\phi_n(x)\,dx = \delta_{in}a$$

and

$$\int_0^b \psi_j(y)\psi_n(y)\,dy = \delta_{jn}b \tag{5.26}$$

where the Kronecker delta $\delta_{ii} = \delta_{jj} = 1$, $\delta_{in} = 0$ and $\delta_{jn} - 0$ for $i \neq n$ and $j \neq n$, respectively. Also for beam modal functions [equation (A3.8)]

$$\int_0^a \frac{d^2\phi_i}{dx^2}\frac{d^2\phi_n}{dx^2}\,dx = 0 \quad \text{if} \quad i \neq n$$

and

$$\int_0^b \frac{d^2\psi_j}{dy^2}\frac{d^2\psi_n}{dy^2}\,dy = 0 \quad \text{if} \quad j \neq n. \tag{5.27}$$

Substituting from equation (5.25) in equations (5.13) and (5.14) and using conditions (5.26) and (5.27), the energy expressions become

$$\mathfrak{T} = \tfrac{1}{2}\dot{\mathbf{g}}^T\mathbf{M}\dot{\mathbf{g}}$$

where

$$\dot{\mathbf{g}}^T = [\dot{\Gamma}_1, \dot{\Gamma}_2, ..., \dot{\Gamma}_{IJ}] \tag{5.28}$$

and the matrix $\mathbf{M}$ is diagonal with each diagonal term equal to ρhab, i.e.

$$\mathbf{M} = \rho hab\mathbf{I}. \tag{5.29}$$

(If a concentrated mass m_c is attached to the plate at (x_c, y_c), $\mathbf{M}$ is symmetric and contains additional terms, which depend upon $\phi_i(x_c)$ and $\psi_j(y_c)$.)

The strain energy

$$\mathfrak{S} = \tfrac{1}{2}\mathbf{g}^T\mathbf{K}\mathbf{g} \tag{5.30}$$

where a typical diagonal term in the matrix $\mathbf{K}$

$$k_{ss} = D\left[\frac{b}{a^3}A_i + \frac{a}{b^3}B_j + \frac{2v}{ab}C_{i,\,i}F_{j,\,j} + \frac{2(1-v)}{ab}G_{i,\,i}H_{j,\,j}\right] \tag{5.31}$$

and a typical off-diagonal term $(s \neq n)$

$$k_{sn} = D\left[\frac{v}{ab}(C_{i,\,i'}F_{j,\,j'} + C_{i',\,i}F_{j',\,j}) + \frac{2(1-v)}{ab}G_{i,\,i'}H_{j,\,j'}\right] \tag{5.32}$$

where $n = j' + (i'-1)J$ and the coefficients A_i, B_j, $C_{i,\,i'}$, etc., are defined by the following integrals:

$$A_i = a^3\int_0^a \left(\frac{d^2\phi_i}{dx^2}\right)^2 dx,$$

$$C_{i,\,i'} = a\int_0^a \frac{d^2\phi_i}{dx^2}\phi_{i'}\,dx,$$

$$G_{i,\,i'} = a\int_0^a \frac{d\phi_i}{dx}\frac{d\phi_{i'}}{dx}\,dx. \tag{5.33}$$

B_j, $F_{j,\,j'}$ and $H_{j,\,j'}$ are obtained from the expressions for A_i, $C_{i,\,i'}$ and $G_{i,\,i'}$, respectively, by replacing ϕ, x, a, i and i' by ψ, y, b, j and j', respectively.

For free vibrations, applying the Lagrange equation and putting $\mathbf{g}(t) = \mathbf{g}\sin(\omega t + \alpha)$, where $\mathbf{g}$ is a vector of constants, the standard eigenvalue equation is obtained:

$$[\mathbf{K} - \mathbf{M}\omega^2]\mathbf{g} = 0 \tag{5.34}$$

with $\mathbf{K}$ and $\mathbf{M}$ defined by equations (5.29), (5.31) and (5.32). Leissa[50] has used the Rayleigh–Ritz method to determine natural frequencies for several modes for all combinations of clamped, simply supported and free edges for which solutions are not available from the plate equation by the method based on equation (5.24). He used appropriate beam modal functions for $\phi_i(x)$ and $\psi_j(y)$ with $I = 6$ and $J = 6$. For example, for a plate with $x = 0$ and $y = 0$ clamped, $x = a$ simply supported and $y = b$ free, $\phi_i(x)$ are the modal functions of a clamped–simply supported beam and $\psi_j(y)$ are those of a clamped–free beam.

Earlier, the author[81] considered a single-term (Rayleigh) approximation for w, i.e.

$$w(x, y, t) = \phi_i(x)\psi_j(y)\Gamma_s(t) \tag{5.35}$$

is an approximation to the mode i/j. With $\phi_i(x)$ and $\psi_j(y)$ represented by beam modal functions, satisfying conditions (5.26) and (5.27), substitution of equation (5.35) in equations (5.13) and (5.14) and application of the Rayleigh method lead to

$$\rho\frac{ha^4}{D}\,\omega_{ij}^2 = A_i + \left(\frac{a}{b}\right)^4 B_j + 2\left(\frac{a}{b}\right)^2 [vC_{i,i}F_{j,j} + (1 - v)G_{i,i}H_{j,j}]$$

$$\tag{5.36}$$

where A_i, $C_{i,i}$, etc., are defined by equations (5.33). For higher values of i or j relatively simple approximate expressions can be used for the integrals A_i, $C_{i,i}$, etc. Leissa[50] has compared the lowest six natural frequencies from equation (5.36) with his values from the thirty-six-term Rayleigh–Ritz solution for aspect ratios a/b from 0·4 to 2·5 and fifteen sets of boundary conditions. The Rayleigh–Ritz values are upper bounds, i.e. they always exceed the true values, although the errors are expected to be small. If all the edges are simply supported or clamped, frequencies from equation (5.36) exceed the Rayleigh–Ritz values by less than 0·8 per cent. In these cases the assumed functions satisfy all the boundary conditions. When there is one free edge or more but no free corner, frequencies from equation (5.36) exceed the Rayleigh–Ritz values by less than 2·0 per cent. When there is one free corner or more, the percentage differences may be significant and occasionally a frequency from equation (5.36), which does not necessarily give an upper bound except for the fundamental mode, is slightly less than the corresponding value from the Rayleigh–Ritz solution.

Response. When two parallel edges of the plate are simply supported, the response can be determined in terms of normal modes by an extension of the method of Section 3.4. For general boundary conditions the response can be determined from an extension of the Rayleigh–Ritz method of Section 3.8. The two cases will be considered separately.

For a rectangular plate, which is simply supported on the edges $y = 0$ and $y = b$, we require the normal modes [equation (5.24)]

$$w(x, y, t) = F_{ij}(x)\psi_j(y) \sin (\omega_{ij}t + \alpha) \tag{5.37}$$

where $\psi_j(y) = 2^{1/2} \sin (j\pi y/b)$. Equation (5.37) satisfies the equilibrium equation (5.18) with $p(x, y)f(t) = 0$ and all the boundary conditions and yields the natural frequency ω_{ij} for mode i/j. $F_{ij}(x)$ signifies the ith function of x which satisfies these conditions for a specified value of j. By analysis similar to that of Appendix 3 it can be shown that for any combination of homogeneous boundary conditions on $x = 0$ and $x = a$

$$\int_0^a F_{ij}(x)F_{nj}(x) \, dx = 0, \qquad i \neq n. \tag{5.38}$$

Also, from equation (5.18), the modal functions must satisfy

$$D\left[\frac{d^4F_{ij}}{dx^4}\psi_j + 2\frac{d^2F_{ij}}{dx^2}\frac{d^2\psi_j}{dy^2} + F_{ij}\frac{d^4\psi_j}{dy^4}\right] = \rho h\omega_{ij}^2 F_{ij}\psi_j. \tag{5.39}$$

Considering the response problem with the plate subjected to a force per unit area in the Z-direction of $p(x, y)f(t)$, we seek a solution of the form

$$w(x, y, t) = \sum_i \sum_j F_{ij}(x)\psi_j(y)q_{ij}(t). \tag{5.40}$$

Substituting equation (5.40) into equation (5.18) and using equation (5.39),

$$\sum_i \sum_j \rho h F_{ij}\psi_j[\omega_{ij}^2 q_{ij} + \ddot{q}_{ij}] = p(x, y)f(t). \tag{5.41}$$

Multiplying equation (5.41) by $F_{ns}(x)\psi_s(y)$, integrating over the area of the plate and using equation (5.38) and the orthogonal property of functions $\psi_j(y)$, a set of uncoupled equations is obtained:

$$\ddot{q}_{ij} + \omega_{ij}^2 q_{ij} = K_{ij}f(t)/m_{ij} \tag{5.42}$$

where

$$K_{ij} = \int_0^a \int_0^b p(x, y) F_{ij}(x) \psi_j(y) \, dy \, dx$$

and

$$m_{ij} = \rho h \int_0^a \int_0^b F_{ij}^2 \psi_j^2 \, dy \, dx.$$

Damping can be included in equation (5.42), as in Section 3.4. The solution of equation (5.42) is obtained from the Duhamel integral and then substitution for $q_{ij}(t)$ in equation (5.40) gives the response. Snowdon[70] plots steady-state amplitudes against excitation frequency for rectangular plates, which are subjected to concentrated harmonic forces, have all edges simply supported and are hysteretically damped.

For general boundary conditions we use the Rayleigh–Ritz method, using equations (5.25) to (5.33), apply the Lagrange equation (A4.17) and obtain

$$\mathbf{M}\ddot{\mathbf{g}} + \mathbf{K}\mathbf{g} = \mathbf{p} \qquad (5.43)$$

where the matrices $\mathbf{M}$ and $\mathbf{K}$ are defined by equations (5.29) to (5.33) and $\mathbf{p}$ is a vector of generalized forces, which can be evaluated using the principle of virtual work. A typical term in $\mathbf{p}$ is

$$p_s = f(t) \int_0^a \int_0^b p(x, y) \phi_i(x) \psi_j(y) \, dy \, dx$$

where $s = j + (i - 1)J$ as before. Equation (5.43) does not represent a set of uncoupled equations, because of the presence of off-diagonal terms in $\mathbf{K}$ [equation (5.32)]. The response is obtained by solving equation (5.43), using an appropriate method from Chapter 2. If the normal mode method is used, the transformation $\mathbf{g} = \mathbf{Z}\mathbf{q}$ leads to a set of uncoupled equations, if $\mathbf{Z}$ consists of the normalized eigenvectors, found by the Rayleigh–Ritz method when solving equation (5.34).

As mentioned in Section 4.6, higher-order theories for the flexural vibrations of rectangular plates exist. Such theories are necessary when the thickness is not small compared with the other plate dimensions;

they include Mindlin's theory[60], which includes the effects of shear deformation and rotatory inertia [analogous to the Timoshenko equation (4.65) for beam vibrations], and solutions from three-dimensional elasticity theory. Srinivas *et al.*[71] show that for a simply supported rectangular plate with $v = 0.3$ the natural frequency of mode i/j from thin plate theory [equation (5.23)] exceeds the corresponding value from three-dimensional elasticity theory by 2·1, 3·4, 8·2, 15·5 and 28·2 per cent for values of $[(ih/a)^2 + (jh/b)^2]$ equal to 0·0125, 0·02, 0·05, 0·1 and 0·2, respectively. Thus for these boundary conditions $[(ih/a)^2 + (jh/b)^2] \le 0.02$ is the criterion to be satisfied for thin-plate theory to give the natural frequency of mode i/j with reasonable accuracy. For example, if $h/a = 0.01$ and $b/a = 0.5$, sixty-nine modes satisfy this condition; if $h/a = 0.02$ and $b/a = 0.8$, twenty-three modes satisfy it.

5.3. Transverse Vibrations of Circular Plates

The assumptions made in the previous section are applicable to circular plates. Figure 5.9a shows a circular plate extending from the inner radius R_1 to the outer radius R_2. Taking the origin O at the centre of the plate, a point P on the middle plane is defined by the coordinates (r, θ). The Rayleigh–Ritz method will be used to determine

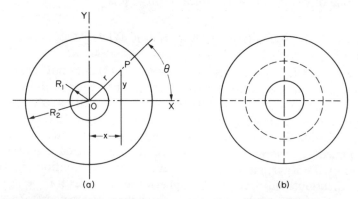

FIG. 5.9. Circular plate. (a) Definition of coordinates. (b) Typical nodal pattern (nodal lines are represented by broken lines).

approximately natural frequencies. It is necessary to convert the energy expressions (5.13) and (5.14) from Cartesian coordinates (x, y) to polar coordinates (r, θ). From Fig. 5.9a the relations between these coordinates are

$$x = r \cos \theta, \qquad y = r \sin \theta$$

or

$$r^2 = x^2 + y^2, \qquad \tan \theta = y/x. \qquad (5.44)$$

Thus

$$2r \cdot \frac{\partial r}{\partial x} = 2x \quad \text{and} \quad \frac{\partial r}{\partial x} = \cos \theta.$$

Similarly

$$\frac{\partial r}{\partial y} = \sin \theta.$$

Also

$$\sec^2 \theta \frac{\partial \theta}{\partial x} = -\frac{y}{x^2},$$

so

$$\frac{\partial \theta}{\partial x} = -\frac{y}{r^2} = -\frac{1}{r} \sin \theta.$$

Similarly

$$\frac{\partial \theta}{\partial y} = \frac{1}{r} \cos \theta.$$

Considering the derivatives of w, the transverse displacement of a point on the plate,

$$\frac{\partial w}{\partial x} = \frac{\partial w}{\partial r} \frac{\partial r}{\partial x} + \frac{\partial w}{\partial \theta} \frac{\partial \theta}{\partial x} = \cos \theta \frac{\partial w}{\partial r} - \frac{\sin \theta}{r} \frac{\partial w}{\partial \theta}, \qquad (5.45)$$

$$\frac{\partial w}{\partial y} = \frac{\partial w}{\partial r} \frac{\partial r}{\partial y} + \frac{\partial w}{\partial \theta} \frac{\partial \theta}{\partial y} = \sin \theta \frac{\partial w}{\partial r} + \frac{\cos \theta}{r} \frac{\partial w}{\partial \theta}. \qquad (5.46)$$

Hence

$$\frac{\partial^2 w}{\partial x^2} = \left[\cos \theta \frac{\partial}{\partial r} - \frac{\sin \theta}{r} \frac{\partial}{\partial \theta}\right]\left[\cos \theta \frac{\partial w}{\partial r} - \frac{\sin \theta}{r} \frac{\partial w}{\partial \theta}\right]$$

$$= \cos^2 \theta \frac{\partial^2 w}{\partial r^2} + \frac{\sin^2 \theta}{r^2} \frac{\partial^2 w}{\partial \theta^2} - \frac{2 \sin \theta \cos \theta}{r} \frac{\partial^2 w}{\partial r \partial \theta}$$

$$+ \frac{2 \sin \theta \cos \theta}{r^2} \frac{\partial w}{\partial \theta} + \frac{\sin^2 \theta}{r} \frac{\partial w}{\partial r}. \tag{5.47}$$

Similarly $\partial^2 w / \partial y^2$ and $\partial^2 w / \partial x\, \partial y$ can be transformed into polar coordinates. Substituting in equation (5.13), the strain energy of a circular plate in polar coordinates reduces (after some algebra) to

$$\mathfrak{S} = \frac{D}{2} \int_{R_1}^{R_2} \int_0^{2\pi} \left[\left(\frac{\partial^2 w}{\partial r^2} + \frac{1}{r} \frac{\partial w}{\partial r} + \frac{1}{r^2} \frac{\partial^2 w}{\partial \theta^2}\right)^2 - 2(1 - v) \right.$$

$$\times \left.\left\{\frac{\partial^2 w}{\partial r^2}\left(\frac{1}{r} \frac{\partial w}{\partial r} + \frac{1}{r^2} \frac{\partial^2 w}{\partial \theta^2}\right) - \left(\frac{1}{r} \frac{\partial^2 w}{\partial r \partial \theta} - \frac{1}{r^2} \frac{\partial w}{\partial \theta}\right)^2\right\}\right] r\, d\theta\, dr. \tag{5.48}$$

The kinetic energy

$$\mathfrak{T} = \frac{1}{2} \rho h \int_{R_1}^{R_2} \int_0^{2\pi} \left(\frac{\partial w}{\partial t}\right)^2 r\, d\theta\, dr. \tag{5.49}$$

For a uniform plate the nodal lines consist of circles and diameters; Fig. 5.9b shows a typical pattern with one nodal circle and two nodal diameters. Thus the variation of the displacement w with θ can be represented by $\cos n\theta$, where n is the number of nodal diameters and $n = 0$ corresponds to axisymmetric vibrations. A solution of the form

$$w(r, \theta, t) = \sum_i \Gamma_{in} F_i(r) \cos n\theta \sin (\omega t + \alpha) \tag{5.50}$$

is assumed; each of the functions $F_i(r)$ must satisfy any displacement and slope conditions at the boundaries $r = R_1$ and $r = R_2$. After substituting from equation (5.50) in equations (5.48) and (5.49) and integrating, natural frequencies are determined from the set of equations

$$\frac{\partial(\omega^2)}{\partial \Gamma_{in}} = 0, \qquad i = 1, 2, 3, \ldots \tag{5.51}$$

as described in Section 3.7.

For example, the deformation in the fundamental mode for a plate, which is clamped at the outer edge $r = R$ and has no inner boundary, is symmetrical about the origin (i.e. independent of θ) and has no nodal circles. Thus a suitable vibration form is

$$w(r, \theta, t) = [\Gamma_{10}(1 - r^2/R^2)^2 + \Gamma_{20}(1 - r^2/R^2)^3 + \cdots] \sin(\omega t + \alpha)$$
$$(5.52)$$

as each function of r satisfies the geometric boundary conditions. The use of equation (5.52) with equations (5.48), (5.49) and (5.51) leads to an accurate value of the fundamental natural frequency[75].

Natural frequencies can be determined from solutions of the plate equation (equation (5.18) transformed into polar coordinates), which satisfy all the boundary conditions. Mode shapes contain Bessel functions with arguments proportional to the radius. Frequency determinants and numerical results are given by Leissa[48] and McLeod and Bishop[59].

5.4. Finite Element Method for In-plane Vibrations of Plates

In the conventional theory of plate vibration (Section 5.2) the stresses in the middle plane of the plate are neglected. This is a valid assumption for small vibrations of uniform plates. However, for a plate reinforced by eccentric stiffeners or for a system of plates, built up to represent approximately a curved or shell structure, bending and membrane, or in-plane, deformations are coupled. In this and the following section finite elements for in-plane and bending vibrations respectively will be developed. The elements for in-plane vibrations are presented first, as they are simpler in concept.

Figure 5.10 shows a rectangular element of sides l_x and l_y. The middle surface of the element lies in the plane OXY. We consider a state of plane stress; i.e. the non-zero components of stress are σ_x, σ_y and σ_{xy}. As the stresses are assumed to be uniform through the thickness h, the strain energy of the element

$$\mathfrak{S}_e = \tfrac{1}{2}h \int_0^{l_y} \int_0^{l_x} (\sigma_x \varepsilon_x + \sigma_y \varepsilon_y + \sigma_{xy} \varepsilon_{xy}) \, dx \, dy \qquad (5.53)$$

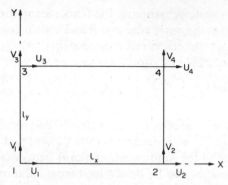

FIG. 5.10. Rectangular element for in-plane deformation of a plate.

from equation (5.6). If u are v are the displacements in the X- and Y-directions, respectively, the strain energy can be expressed in terms of displacements, using the stress–strain relations (5.12) and the strain-displacement relations (5.4) and (5.5),

$$
\begin{aligned}
\mathfrak{S}_e &= \frac{1}{2}\frac{Eh}{1-v^2}\int_0^{l_y}\int_0^{l_x}\left[\left(\frac{\partial u}{\partial x}\right)^2 + \left(\frac{\partial v}{\partial y}\right)^2 + 2v\frac{\partial u}{\partial x}\frac{\partial v}{\partial y}\right. \\
&\quad \left. + \frac{(1-v)}{2}\left(\frac{\partial u}{\partial y} + \frac{\partial v}{\partial x}\right)^2\right] dx\,dy \\
&= \frac{1}{2}\int_0^{l_y}\int_0^{l_x}\left[\frac{\partial u}{\partial x}, \frac{\partial v}{\partial y}, \left(\frac{\partial u}{\partial y} + \frac{\partial v}{\partial x}\right)\right]\mathbf{D}\begin{bmatrix}\partial u/\partial x \\ \partial v/\partial y \\ \partial u/\partial y + \partial v/\partial x\end{bmatrix} dx\,dy \quad (5.54)
\end{aligned}
$$

where

$$
\mathbf{D} = \frac{Eh}{1-v^2}\begin{bmatrix} 1 & v & 0 \\ v & 1 & 0 \\ 0 & 0 & \dfrac{1-v}{2} \end{bmatrix}.
$$

The element of Fig. 5.10 has nodes at the four corners; thus nodal variables are u_j and v_j with $j = 1, 2, 3$ and 4. The assumed displacement functions are

$$
\begin{aligned}
u &= a_1 + a_2 x + a_3 y + a_4 xy, \\
v &= a_5 + a_6 x + a_7 y + a_8 xy
\end{aligned}
\quad (5.55)
$$

or

$$\begin{bmatrix} u \\ v \end{bmatrix} = \begin{bmatrix} \mathbf{g} & 0 \\ 0 & \mathbf{g} \end{bmatrix} \mathbf{a}$$

where

$$\mathbf{g} = \begin{bmatrix} 1 & x & y & xy \end{bmatrix}$$

and $\mathbf{a}$ is a vector containing the coefficients $a_1, a_2, \ldots, a_8$. Substituting the nodal values

$$\mathbf{u}_e = \begin{bmatrix} u_1 \\ u_2 \\ u_3 \\ u_4 \\ v_1 \\ v_2 \\ v_3 \\ v_4 \end{bmatrix} = \mathbf{Na} = \begin{bmatrix} 1 & 0 & 0 & 0 & 0 & 0 & 0 & 0 \\ 1 & l_x & 0 & 0 & 0 & 0 & 0 & 0 \\ 1 & 0 & l_y & 0 & 0 & 0 & 0 & 0 \\ 1 & l_x & l_y & l_x l_y & 0 & 0 & 0 & 0 \\ 0 & 0 & 0 & 0 & 1 & 0 & 0 & 0 \\ 0 & 0 & 0 & 0 & 1 & l_x & 0 & 0 \\ 0 & 0 & 0 & 0 & 1 & 0 & l_y & 0 \\ 0 & 0 & 0 & 0 & 1 & l_x & l_y & l_x l_y \end{bmatrix} \begin{bmatrix} a_1 \\ a_2 \\ a_3 \\ a_4 \\ a_5 \\ a_6 \\ a_7 \\ a_8 \end{bmatrix}.$$

$$(5.56)$$

Thus

$$\mathbf{a} = \mathbf{Bu}_e$$

where

$$\mathbf{B} = \mathbf{N}^{-1}.$$

As

$$\partial u/\partial x = a_2 + a_4 y, \qquad \partial v/\partial y = a_7 + a_8 x$$

and

$$\left(\frac{\partial u}{\partial y} + \frac{\partial v}{\partial x} \right) = a_3 + a_4 x + a_6 + a_8 y$$

$$\begin{bmatrix} \partial u/\partial x \\ \partial v/\partial y \\ \partial u/\partial y + \partial v/\partial x \end{bmatrix} = \mathbf{Ga} \qquad (5.57)$$

where

$$\mathbf{G} = \begin{bmatrix} 0 & 1 & 0 & y & 0 & 0 & 0 & 0 \\ 0 & 0 & 0 & 0 & 0 & 0 & 1 & x \\ 0 & 0 & 1 & x & 0 & 1 & 0 & y \end{bmatrix}.$$

Substituting from equation (5.57) in equation (5.54)

$$
\begin{aligned}
\mathfrak{S}_e &= \frac{1}{2} \int_0^{l_y} \int_0^{l_x} \mathbf{a}^T \mathbf{G}^T \mathbf{D} \mathbf{G} \mathbf{a} \; dx \; dy \\
&= \frac{1}{2} \mathbf{u}_e^T \mathbf{B}^T \left[\int_0^{l_y} \int_0^{l_x} \mathbf{G}^T \mathbf{D} \mathbf{G} \; dx \; dy \right] \mathbf{B} \mathbf{u}_e \\
&= \frac{1}{2} \mathbf{u}_e^T \mathbf{K}_e \, \mathbf{u}_e
\end{aligned}
$$

where the element stiffness matrix

$$
\mathbf{K}_e = \mathbf{B}^T \left[\int_0^{l_y} \int_0^{l_x} \mathbf{G}^T \mathbf{D} \mathbf{G} \; dx \; dy \right] \mathbf{B}. \tag{5.58}
$$

The kinetic energy of the element

$$
\begin{aligned}
\mathfrak{T}_e &= \frac{1}{2} \rho h \int_0^{l_y} \int_0^{l_x} \left[\left(\frac{\partial u}{\partial t} \right)^2 + \left(\frac{\partial v}{\partial t} \right)^2 \right] dx \; dy \\
&= \frac{1}{2} \int_0^{l_y} \int_0^{l_x} \left[\frac{\partial u}{\partial t} \quad \frac{\partial v}{\partial t} \right] \rho h \left[\frac{\partial u / \partial t}{\partial v / \partial t} \right] dx \; dy \\
&= \frac{1}{2} \dot{\mathbf{u}}_e^T \mathbf{M}_e \, \dot{\mathbf{u}}_e
\end{aligned} \tag{5.59}
$$

where the element mass matrix

$$
\mathbf{M}_e = \mathbf{B}^T \rho h \int_0^{l_x} \int_0^{l_y} \left[\begin{array}{c|c} \mathbf{g}^T \mathbf{g} & \mathbf{0} \\ \hline \mathbf{0} & \mathbf{g}^T \mathbf{g} \end{array} \right] dy \; dx \; \mathbf{B}. \tag{5.60}
$$

Before considering the assembly of elements to represent the structure the conditions to be satisfied by the assumed displacement functions (5.55) will be discussed. These will be presented in general terms, so that they are applicable to other types of element.

1. Displacements and their derivatives up to the order one less than that occurring in the strain energy expression should be continuous across element boundaries.
2. The displacement functions should be able to represent appropriate rigid-body motions.
3. The displacement functions should be able to represent states of constant strain.

If these conditions are satisfied, we have conforming or compatible elements; for eigenvalue calculations with conforming elements representing a structure the eigenvalues converge monotonically from above to the correct values with progressive subdivision of the element mesh. For non-conforming elements with condition (3) satisfied eigenvalues converge to the correct values eventually, as the mesh is refined, but convergence is not monotonic. If condition (3) is not satisfied, the elements are too stiff and eigenvalues converge on values higher than the correct ones. Considering the membrane element of Fig. 5.10 and assuming that there is another similar element with a common boundary along the line joining nodes 1 and 2, the nodal values of u_1, v_1, u_2 and v_2 for the two elements are identical by definition. From equations (5.55) u and v at any point along the line AB are defined uniquely in terms of u_1 and u_2 and v_1 and v_2, respectively. As only first derivatives of u and v appear in $\mathfrak{S}_e$, condition (1) is satisfied. The terms a_1 and a_5 in equations (5.55) give the appropriate rigid-body translations; the terms $a_3 y$ in u and $a_6 x$ in v give rigid-body rotations in the plane OXY. The term $a_2 x$ in u, $a_7 y$ in v and $a_3 y$ and $a_6 x$ in u and v, respectively, ensure that the strains ε_x, ε_y and ε_{xy} can each undertake a constant value. Thus all the conditions are satisfied and this element is conforming. (It can be shown that the beam elements of Section 3.9 are conforming.)

Figure 5.11 shows a plate consisting of four membrane elements, designated by the letters A, B, C and D. The nodes are numbered from 1 to 9 and the nodal variables, shown by arrows, from 1 to 18. Thus if

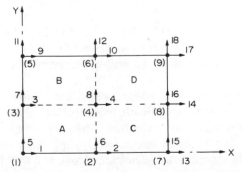

Fig. 5.11. Plate consisting of four in-plane elements.

u_j and v_j are the nodal displacements of node j, the vector of displacements for the structure is

$$\mathbf{u}^T = [u_1, u_2, u_3, u_4, v_1, v_2, v_3, v_4,$$
$$u_5, u_6, v_5, v_6, u_7, u_8, v_7, v_8, u_9, v_9].$$

The stiffness and mass matrices for the structure are of order 18×18. Contributions from element A, using equations (5.58) and (5.60), are located in rows and columns 1, 2, ..., 8; contributions from element B in rows and columns 3, 4, 9, 10, 7, 8, 11 and 12, contributions from element C in rows and columns 2, 13, 4, 14, 6, 15, 8 and 16, contributions from element D in rows and columns 4, 14, 10, 17, 8, 16, 12 and 18. Thus some reordering of the element matrices is necessary when assembling the structural matrices. The matrices are reduced by eliminating rows and columns corresponding to zero boundary displacements; for example, if $u = 0$ at nodes 1, 3 and 5 and $v = 0$ at node 3, rows and columns 1, 3, 7 and 9 are eliminated. The Lagrange equation is applied to these reduced expressions for strain energy and kinetic energy to obtain equations of the form (2.45) or (2.40) for eigenvalue or response problems, respectively, following the analysis in Section 3.9.

When the nodal displacements for a particular element of the plate, $\mathbf{u}_e$, have been found by solving the response equation, the normal and

$$\begin{bmatrix} \sigma_x \\ \sigma_y \\ \sigma_{xy} \end{bmatrix}_e = \frac{1}{h}[\mathbf{DGBu}_e]$$

using equations (5.4), (5.5), (5.12), (5.56) and (5.57). The stresses in the element at a point with specified local coordinates (x, y) are found by substituting these values in $\mathbf{G}$ and performing the matrix multiplications.

There are two possible disadvantages of the element derived in this section: firstly, it may be difficult to represent by rectangular finite elements a practical plate structure which contains holes and an irregular boundary; second, the stresses within an element are linear functions of the coordinates [equations (5.4), (5.5), (5.12) and (5.57)] and thus it will require a large number of small elements to represent accurately dynamic stresses that change rapidly with the coordinates. The first disadvantage can be overcome by using triangular or quadrilateral elements. If elements with additional nodes (i.e. nodes on the edges of the element as well as at the vertices) are used, the assumed polynomials must be of higher order than equation (5.55) and the variation of stresses within an element is a quadratic or cubic function of the coordinates. Elements of various shapes with additional nodes exist. Here isoparametric elements will be described, as both disadvantages can be overcome by using this family of elements. With the exception of the lowest member (the four-node quadrilateral element) these elements can have curved sides; they are transformed into square elements in the (ζ, η)-plane in order to derive stiffness and mass matrices. Figure 5.12a shows the four-node quadrilateral element in the XY-plane with the transformed element in the $\zeta\eta$-plane underneath. Figure 5.12b shows the eight-node element, which can have curved sides; each side contains one node. Again the transformed square element in the $\zeta\eta$-plane is shown underneath the actual element.

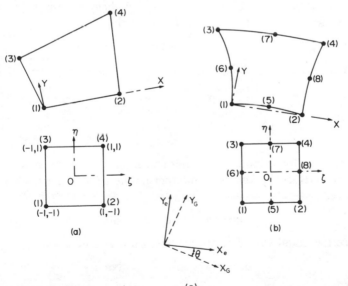

FIG. 5.12. Isoparametric elements. (a) Four-node quadrilateral element. (b) Eight-node quadrilateral element. (c) Element and global axes.

To transform from the isoparametric element to the square of side 2 units we use

$$[x \quad y] = \mathbf{g}[\mathbf{a} \quad \mathbf{a}'] \tag{5.61}$$

where for the four-node element

$$\mathbf{g} = [1 \quad \zeta \quad \eta \quad \zeta\eta],$$
$$\mathbf{a}^T = [a_1 \quad a_2 \quad a_3 \quad a_4]$$

and

$$\mathbf{a}'^T = [a_5 \quad a_6 \quad a_7 \quad a_8]$$

and for the eight-node element

$$\mathbf{g} = [1 \quad \zeta \quad \eta \quad \zeta\eta \quad \zeta^2 \quad \eta^2 \quad \zeta^2\eta \quad \zeta\eta^2]$$

and

$$\mathbf{a}^T = [a_1 \quad a_2, \ldots, a_8], \text{ etc.}$$

The x coordinates $\mathbf{x}_e$ of all the nodes and their corresponding (ζ, η) values are inserted into equation (5.61), giving

$$\mathbf{x}_e = \mathbf{N}\mathbf{a}.$$

Thus

$$\mathbf{a} = \mathbf{N}^{-1}\mathbf{x}_e \tag{5.62}$$

and similarly

$$\mathbf{a}' = \mathbf{N}^{-1}\mathbf{y}_e.$$

The basis of the isoparametric method is that the displacements vary over the element in the same manner as the transformation assumption, i.e.

$$u = \mathbf{g}\mathbf{a}'' \quad \text{and} \quad v = \mathbf{g}\mathbf{a}'''. \tag{5.63}$$

Hence the nodal values of displacement are given by

$$\mathbf{u}_e = \mathbf{N}\mathbf{a}'' \quad \text{and} \quad \mathbf{v}_e = \mathbf{N}\mathbf{a}'''.$$

Thus

$$[u \quad v] = \mathbf{g}\mathbf{N}^{-1}[\mathbf{u}_e \quad \mathbf{v}_e]. \tag{5.64}$$

The strain energy expression (5.54) is in terms of $\partial u/\partial x$, etc., and this has to be rearranged in terms of $\partial u/\partial \zeta$, $\partial u/\partial \eta$, etc. From equation (5.61)

$$\frac{\partial x}{\partial \zeta} = \frac{\partial \mathbf{g}}{\partial \zeta} \mathbf{a}$$

$$= \mathbf{g}_\zeta \mathbf{N}^{-1} \mathbf{x}_e \qquad (5.65)$$

using equation (5.62) and putting $\mathbf{g}_\zeta = \partial \mathbf{g}/\partial \zeta$. From equation (5.65) and similar expressions we form the matrix

$$\mathbf{J} = \begin{bmatrix} \partial x/\partial \zeta & \partial y/\partial \zeta \\ \partial x/\partial \eta & \partial y/\partial \eta \end{bmatrix}. \qquad (5.66)$$

This matrix is inverted and gives

$$\mathbf{J}^{-1} = \begin{bmatrix} \partial \zeta/\partial x & \partial \eta/\partial x \\ \partial \zeta/\partial y & \partial \eta/\partial y \end{bmatrix}. \qquad (5.67)$$

If $\mathbf{g}_x = \partial \mathbf{g}/\partial x$, $\mathbf{g}_x = \mathbf{g}_\zeta(\partial \zeta/\partial x) + \mathbf{g}_\eta(\partial \eta/\partial x)$. Thus

$$\begin{bmatrix} \mathbf{g}_x \\ \mathbf{g}_y \end{bmatrix} = \mathbf{J}^{-1} \begin{bmatrix} \mathbf{g}_\zeta \\ \mathbf{g}_\eta \end{bmatrix}. \qquad (5.68)$$

Thus

$$\begin{bmatrix} \partial u/\partial x \\ \partial v/\partial y \\ \partial u/\partial y + \partial v/\partial x \end{bmatrix} = \begin{bmatrix} \mathbf{g}_x & 0 \\ 0 & \mathbf{g}_y \\ \mathbf{g}_y & \mathbf{g}_x \end{bmatrix} \begin{bmatrix} \mathbf{N}^{-1} & 0 \\ 0 & \mathbf{N}^{-1} \end{bmatrix} \begin{bmatrix} \mathbf{u}_e \\ \mathbf{v}_e \end{bmatrix}$$

$$= \mathbf{GB}\,\boldsymbol{\delta}_e \qquad \text{(say)}. \qquad (5.69)$$

Use of equation (5.68) gives $\mathbf{G}$ in terms of ζ and η. Substituting in equation (5.54) the element strain energy

$$\mathfrak{S}_e = \tfrac{1}{2}\boldsymbol{\delta}_e^T \mathbf{K}_e\,\boldsymbol{\delta}_e$$

where the element stiffness matrix

$$\mathbf{K}_e = \mathbf{B}^T \iint \mathbf{G}^T \mathbf{D} \mathbf{G}\ dx\ dy\ \mathbf{B}$$

$$= \mathbf{B}^T \int_{-1}^{+1} \int_{-1}^{+1} \mathbf{G}^T \mathbf{D} \mathbf{G}\ \det |\mathbf{J}|\ d\zeta\ d\eta\ \mathbf{B} \qquad (5.70)$$

and

$$\delta_e = \begin{bmatrix} \mathbf{u}_e \\ \mathbf{v}_e \end{bmatrix}.$$

It is noted that the elementary area

$$dx\, dy = \det |\mathbf{J}|\, d\zeta\, d\eta \qquad (5.71)$$

where $\det |\mathbf{J}|$ represents the Jacobian of the transformation. Substituting from equation (5.64) in equation (5.59), the element kinetic energy is

$$\mathfrak{T}_e = \frac{1}{2} \iint \begin{bmatrix} \dfrac{\partial u}{\partial t} & \dfrac{\partial v}{\partial t} \end{bmatrix} \rho h \begin{bmatrix} \mathbf{g} & 0 \\ 0 & \mathbf{g} \end{bmatrix} \begin{bmatrix} \mathbf{N}^{-1} & 0 \\ 0 & \mathbf{N}^{-1} \end{bmatrix} \begin{bmatrix} \dot{\mathbf{u}}_e \\ \dot{\mathbf{v}}_e \end{bmatrix} dx\, dy$$

$$= \tfrac{1}{2}\dot{\delta}_e^T \mathbf{M}_e\, \dot{\delta}_e$$

where the element mass matrix

$$\mathbf{M}_e = \mathbf{B}^T \int_{-1}^{+1} \int_{-1}^{+1} \rho h \begin{bmatrix} \mathbf{g}^T \mathbf{g} & 0 \\ 0 & \mathbf{g}^T \mathbf{g} \end{bmatrix} \det |\mathbf{J}|\, d\zeta\, d\eta\, \mathbf{B}. \qquad (5.72)$$

Gauss numerical integration can be used to evaluate the element stiffness and mass matrices. Element matrices are assembled to give structural matrices as described previously, except that transformation from local to global or system coordinates, mentioned previously as a possibility, is now essential. In Fig. 5.12 the local X-axis passes through nodes 1 and 2 for each element and is unlikely to be a common direction for all the elements. The displacements u_j and v_j of node j in the directions of the local coordinates for a particular element can be transformed into displacements u_{jG} and v_{jG} in the global directions by

$$\begin{bmatrix} u_j \\ v_j \end{bmatrix} = \begin{bmatrix} \cos\theta & \sin\theta \\ -\sin\theta & \cos\theta \end{bmatrix} \begin{bmatrix} u_{jG} \\ v_{jG} \end{bmatrix} \qquad (5.73)$$

where the angle θ is defined in Fig. 5.12c. With the aid of this 2×2 transformation matrix strain and kinetic energy expressions for an element can be found in terms of δ_{eG} and $\dot{\delta}_{eG}$, respectively, the vectors of displacements and velocities in global coordinates, and modified element stiffness and mass matrices derived. Further details of two- and three-dimensional isoparametric elements are given by Zienkiewicz[90].

5.5. Finite Element Method for Transverse Vibrations of Plates

The assumptions for small deformations of thin plates and the resulting strain energy expression (5.13) are given in Section 5.2. Figure 5.13 shows a rectangular element with sides of length l_x and l_y and having nodes at the four corners. The nodal variables are the transverse displacement w_j and the rotations ϕ_j $(\equiv -\partial w_j/\partial x)$ and ψ_j $(\equiv \partial w_j/\partial y)$ with $j = 1, 2, 3$ and 4. (For a right-hand set of axes the Z-axis in Fig. 5.13 is outwards from the plane of the diagram. The displacement w is in the Z-direction; ϕ and ψ are positive (i.e. clockwise) rotations about the Y- and X-axes, respectively. These definitions cause the negative sign in the relation between ϕ and $\partial w/\partial x$.) The assumed displacement function is

$$w = a_1 + a_2 x + a_3 y + a_4 x^2 + a_5 xy + a_6 y^2 + a_7 x^3 + a_8 x^2 y$$
$$+ a_9 xy^2 + a_{10} y^3 + a_{11} x^3 y + a_{12} xy^3 = \mathbf{ga} \qquad (5.74)$$

where $\mathbf{g}$ is a row matrix of polynomial terms and the vector $\mathbf{a}$ contains the twelve coefficients a_i.

Considering the conditions listed in Section 5.4, the term in a_1 ensures rigid-body translation; those in a_2 and a_3 ensure rigid-body rotations; those in a_4 and a_6 ensure states of uniform curvature; and that in a_5 ensures a state of uniform twist. Thus the second and third conditions are satisfied. As the strain energy expression (5.13) contains

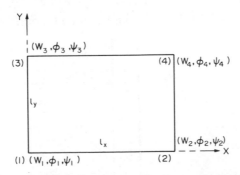

FIG. 5.13. Rectangular element for flexure of a plate.

second derivatives of w, w and its first derivatives should be continuous across element boundaries in order to satisfy the first condition and give conforming elements. Along OX, the line joining nodes 1 and 2 in Fig. 5.13, putting $y = 0$ in equation (5.74),

$$w = a_1 + a_2 x + a_4 x^2 + a_7 x^3$$

and

$$\phi = -[a_2 + 2a_4 x + 3a_7 x^2].$$

Thus the coefficients a_1, a_2, a_4 and a_7 are uniquely defined in terms of the four nodal values w_1, ϕ_1, w_2 and ϕ_2. As the latter are common to the two elements, for which OX is a common boundary, there is continuity of w and ϕ across the inter-element boundary. Also along OX,

$$\psi = a_3 + a_5 x + a_8 x^2 + a_{11} x^3.$$

The values of ψ_1 and ψ_2 at nodes 1 and 2 are insufficient to define uniquely the four coefficients a_3, a_5, a_8 and a_{11}. Thus there will not be continuity of the rotation ψ across element boundaries and this element is not conforming. As the mesh of elements, which represents a structure, is refined, convergence of eigenvalues to the correct values will not be monotonic. Conforming rectangular plate bending elements have been developed by subdividing the basic rectangle into four triangles and using different displacement assumptions in these triangles and also by using higher-order derivatives as additional degrees of freedom at the nodes[14], but these developments will not be described here.

Substituting nodal values of x and y in equation (5.74),

$$\mathbf{w}_e = \begin{bmatrix} w_1 \\ w_2 \\ \vdots \\ \phi_1 \\ \phi_2 \\ \vdots \\ \psi_1 \\ \psi_2 \\ \vdots \end{bmatrix} = \mathbf{N}\mathbf{a}$$

and

$$\mathbf{a} = \mathbf{N}^{-1}\mathbf{w}_e = \mathbf{B}\mathbf{w}_e . \tag{5.75}$$

From equation (5.13) the strain energy expression for the element can be written

$$\mathfrak{S}_e = \frac{1}{2}\int_0^{lx}\int_0^{ly}\begin{bmatrix}\dfrac{\partial^2 w}{\partial x^2} & \dfrac{\partial^2 w}{\partial y^2} & \dfrac{2\partial^2 w}{\partial x\,\partial y}\end{bmatrix}\mathbf{D}\begin{bmatrix}\partial^2 w/\partial x^2 \\ \partial^2 w/\partial y^2 \\ 2\partial^2 w/\partial x\,\partial y\end{bmatrix} dy\ dx \tag{5.76}$$

where

$$\mathbf{D} = D\begin{bmatrix}1 & v & 0 \\ v & 1 & 0 \\ 0 & 0 & \frac{1}{2}(1-v)\end{bmatrix}.$$

From equation (5.74)

$$\begin{bmatrix}\partial^2 w/\partial x^2 \\ \partial^2 w/\partial y^2 \\ 2\partial^2 w/\partial x\,\partial y\end{bmatrix}$$

$$= \begin{bmatrix}0 & 0 & 0 & 2 & 0 & 0 & 6x & 2y & 0 & 0 & 6xy & 0 \\ 0 & 0 & 0 & 0 & 0 & 2 & 0 & 0 & 2x & 6y & 0 & 6xy \\ 0 & 0 & 0 & 0 & 2 & 0 & 0 & 4x & 4y & 0 & 6x^2 & 6y^2\end{bmatrix}\mathbf{a}$$

$$= \mathbf{Ga}. \tag{5.77}$$

Using equations (5.75) and (5.77) in equation (5.76)

$$\mathfrak{S}_e = \frac{1}{2}\mathbf{w}_e^T\mathbf{K}_e\mathbf{w}_e$$

where the element stiffness matrix

$$\mathbf{K}_e = \mathbf{B}^T\left[\int_0^{lx}\int_0^{ly}\mathbf{G}^T\mathbf{D}\mathbf{G}\ dy\ dx\right]\mathbf{B}. \tag{5.78}$$

The kinetic energy of the element

$$\mathfrak{T}_e = \frac{1}{2}\int_0^{lx}\int_0^{ly}\rho h(\partial w/\partial t)^2\ dy\ dx$$

and

$$\frac{\partial w}{\partial t} = \mathbf{gB\dot{w}}_e$$

from equations (5.74) and (5.75). Thus

$$\mathfrak{T}_e = \tfrac{1}{2}\mathbf{\dot{w}}_e^T \mathbf{M}_e\, \mathbf{\dot{w}}_e$$

where the element mass matrix

$$\mathbf{M}_e = \mathbf{B}^T \left[\int_0^{l_x} \int_0^{l_y} \mathbf{g}^T \rho h \mathbf{g}\, dy\, dx \right] \mathbf{B}. \tag{5.79}$$

The structure matrices are assembled in the manner described in Section 5.4, noting that element matrices are now of order 12×12 and there are three degrees of freedom per node. Dawe[28] gives explicit expressions for matrices $\mathbf{K}_e$ and $\mathbf{M}_e$ and demonstrates that accurate natural frequencies of uniform plates with various boundary conditions are predicted with reasonably small numbers of elements. However, as the element is non-conforming, the natural frequencies may be higher or lower than the true values.

For forced vibration the matrix equation

$$\mathbf{M\ddot{w} + Kw = p} \tag{5.80}$$

is obtained from the Lagrange equation, and its solution has been discussed in Chapter 2. In equation (5.80) $\mathbf{K}$ and $\mathbf{M}$ are the stiffness and mass matrices of the structure, obtained by assembling element matrices after converting from local to global coordinates if necessary and after eliminating rows and columns associated with zero nodal values at boundaries. The vector $\mathbf{w}$ comprises all the degrees of freedom of the constrained structure; the vector $\mathbf{p}$ consists of the generalized forces associated with each nodal variable. Considering the contribution to $\mathbf{p}$ from a particular element, $\mathbf{p}_e$, and supposing that this element is subjected to a transverse applied force per unit area of $p(x, y)f(t)$, application of the principle of virtual work gives

$$\delta\mathfrak{W} = \mathbf{p}_e^T\, \delta\mathbf{w}_e = \int_0^{l_x} \int_0^{l_y} p(x, y)f(t)\, \delta w(x, y)\, dy\, dx$$

where δw_e lists the virtual increments in the element nodal values and $\delta w(x, y)$ is the virtual displacement at point (x, y). Using equations (5.74) and (5.75),

$$\mathbf{p}_e^T \, \delta \mathbf{w}_e = f(t) \int_0^{lx} \int_0^{ly} p(x, y) \mathbf{gB} \, \delta \mathbf{w}_e \, dy \, dx$$

and

$$\mathbf{p}_e^T = f(t) \left[\int_0^{lx} \int_0^{ly} p(x, y) \mathbf{g} \, dy \, dx \right] \mathbf{B}. \tag{5.81}$$

With the aid of equation (5.81) the generalized force vector for the structure, $\mathbf{p}$, can be assembled.

After solving equation (5.80), the stress resultants (i.e. the bending moments M_x and M_y and the twisting moment M_{xy}), and hence the stresses, for a particular element can be determined, as the stress resultants can be expressed in terms of the vector of nodal displacements $\mathbf{w}_e$ by the relation [from equations (5.17), (5.75) and (5.77)]

$$\begin{bmatrix} M_x \\ M_y \\ M_{xy} \end{bmatrix}_e = -\mathbf{DGBw}_e.$$

The disadvantages of this element are that plates with holes or of complex planform cannot be represented accurately by an assembly of rectangular elements. Triangular or quadrilateral elements are more versatile. Brebbia and Connor[14] describe several triangular and rectangular plate bending elements, some of which are conforming, and give some comparative static results. A curved quadrilateral bending element, with some properties similar to the isoparametric elements of Section 5.3, has been developed recently[38].

Figure 5.14 shows a structure consisting of four vertical bars, PA, QB, RC and SD, four horizontal stiffening bars AB, BD, AC and CD and a plate $ABDC$. In the general case, if all possible types of deformation are considered, the vertical bars undergo bending in planes XZ and YZ, torsion about the X-axis and extension in the Z-direction; the horizontal bars AB and CD undergo bending in planes XY and XZ, torsion about the X-axis and extension in the X-direction; deformations of bars AC and BD are similar with X and Y interchanged; the

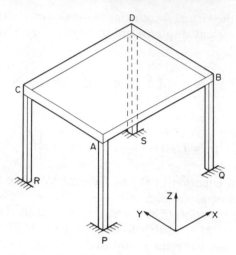

FIG. 5.14. Simple structure.

plate is subjected simultaneously to flexural and in-plane deformations. Each bar is divided into a number of elements; the number of elements for bars AB and CD and for AC and BD define the mesh of rectangular elements for the plate. The element matrices for each beam element are of order 12×12, assembled as diagonal supermatrices from two 4×4 beam bending element matrices (with appropriate properties to account for bending in the two planes) from equations (3.89), a 2×2 beam torsion element matrix and a 2×2 extensional element matrix from equation (3.76), i.e.

$$\mathbf{K}_e = \begin{bmatrix} \mathbf{K}_{eB1} & 0 & 0 & 0 \\ 0 & \mathbf{K}_{eB2} & 0 & 0 \\ 0 & 0 & \mathbf{K}_{eT} & 0 \\ 0 & 0 & 0 & \mathbf{K}_{eE} \end{bmatrix}$$

where the subscripts $B1$, $B2$, T and E refer to bending in the two planes, torsion and extension. Similarly, a plate element stiffness matrix is

$$\mathbf{K}_e = \begin{bmatrix} \mathbf{K}_{eB} & 0 \\ 0 & \mathbf{K}_{eM} \end{bmatrix}$$

where $\mathbf{K}_{eB}$ is the 12×12 bending element matrix of equation (5.78) and $\mathbf{K}_{eM}$ is the 8×8 membrane element matrix of equation (5.58). (As the plate structure is rectangular, the simpler rectangular in-plane elements are used.) Mass matrices for plate and beam elements can be assembled similarly.

A node on one of the horizontal beams is common to two beam elements and two plate elements, except for the corners A, B, C and D where three mutually perpendicular beam elements and one plate element meet. Appropriate continuity conditions must be imposed at all nodes. If the joints are rigid, the continuity conditions at A are:

$$(u)_{AP} = (u)_{AB} = (u)_{AC} = (u)_{\text{plate}}$$

where the subscripts represent the part of the structure, with similar conditions for v and w. Also, if θ represents the rotation of a beam cross-section about its longitudinal axis,

$$\left(\frac{\partial u}{\partial z}\right)_{AP} = -\left(\frac{\partial w}{\partial x}\right)_{AB} = (\theta)_{AC} = -\left(\frac{\partial w}{\partial x}\right)_{\text{plate}},$$

$$\left(\frac{\partial v}{\partial z}\right)_{AP} = -(\theta)_{AB} = -\left(\frac{\partial w}{\partial y}\right)_{AC} = -\left(\frac{\partial w}{\partial y}\right)_{\text{plate}},$$

$$(\theta)_{AP} = \left(\frac{\partial v}{\partial x}\right)_{AB} = -\left(\frac{\partial u}{\partial y}\right)_{AC}.$$

If the natural frequencies are required for the structure when all the feet P, Q, R and S are completely constrained, the overall stiffness and mass matrices $\mathbf{K}''$ and $\mathbf{M}''$ for the whole structure are assembled, the rows and columns in $\mathbf{K}''$ and $\mathbf{M}''$ corresponding to all degrees of freedom at P, Q, R and S deleted to give $\mathbf{K}$ and $\mathbf{M}$ and the eigenvalues determined from

$$\det |\mathbf{K} - \mathbf{M}\omega^2| = 0.$$

The size of the matrices $\mathbf{K}$ and $\mathbf{M}$ may be large. If each horizontal beam and also each vertical member are divided into eight elements with a corresponding 8×8 mesh for the plate, the system has 605 degrees of freedom in the general case after application of the boundary conditions. Similarly, for four elements per beam and a 4×4 mesh for the plate there are 165 degrees of freedom. It may be required to reduce the size of the matrices, and still obtain eigenvalues of acceptable accuracy. If a structure such as that of Fig. 5.14 has two geometrical planes of symmetry, it is possible to analyse only one-quarter of the structure and to treat separately the modes with different classes of symmetry; however, in practice the presence of a concentrated mass or a hole, asymmetrically placed on the surface of plate $ABCD$, or a change of thickness in the plate or a bar destroys this symmetry. In Appendix 5 the "eigenvalue economizer" method of reducing the size of the matrix equation is given. In this method degrees of freedom, associated with minor contributions to the kinetic energy function, are reduced from the equations; for the structure of Fig. 5.14 all degrees of freedom associated with rotations, extension of the bars and in-plane displacements of the plate could be eliminated, leaving for the original 8×8 mesh 173 degrees of freedom associated with bending displacements of the plate and bars. If the lower eigenvalues of this structure are computed, the set based on the 8×8 mesh with reduction to 173 degrees of freedom will be more accurate than the set based on a 4×4 mesh, although the numbers of degrees of freedom are approximately equal.

If the response of the structure of Fig. 5.14 is required for prescribed inputs u, v and w at each of the feet P, Q, R and S (e.g. due to an earthquake), the analysis of Sections 2.6 and 2.8 can be used. The overall stiffness and mass matrices $\mathbf{K}''$ and $\mathbf{M}''$ are reduced to $\mathbf{K}'$ and $\mathbf{M}'$ by eliminating rows and columns, corresponding to the three rotations at each of P, Q, R and S, if these are assumed to remain zero during the prescribed motion. The matrices $\mathbf{K}'$ and $\mathbf{M}'$ are partitioned in terms of $\mathbf{K}$ and $\mathbf{M}$, the matrices of the completely constrained structure, and other sub-matrices. The vector $\mathbf{x}$ in equation (2.83) represents the nodal degrees of freedom of the constrained structure; the additional N displacements, which with $\mathbf{x}$ comprise the vector $\mathbf{x}'$, are u, v and w for each of P, Q, R and S. Then the response of the structure can be determined in terms of the normal modes of the constrained structure from equations (2.87) to (2.92) or by using equations (2.122) and (2.123) together with a numerical integration method.

5.6. Vibrations of Shells: General Comments

Problems concerning the vibration of shells are considerably more complicated than their counterparts for beams or plates. Primarily this is caused by the effects of the curvature on the shell equations and on the dynamic behaviour. For beams and plates it is possible to consider separately the flexural and extensional vibrations and only necessary to combine these effects for complex problems; for shells membrane and flexural deformations are coupled, and any theory must consider these effects simultaneously. (An inextensional theory of shells, i.e. a theory assuming no deformation of the middle surface, exists[54], but has few practical applications.) For flexural vibrations of beams and plates there are well-established theories, which lead to equations (3.33) and (5.18), respectively, in terms of displacement. For shells, because of the coupling, three mutually perpendicular components of displacement must be considered and thus three equilibrium equations in terms of these displacement components can be derived. However, there is no universally accepted set of equations; in fact, many sets of slightly different equations exist. The differences depend upon the assumptions made in the derivation. Leissa[49] gives a comprehensive survey of shell theories, their application to simply supported cylindrical shells and their accuracy in determining natural frequencies; some additional data are given by the author[83]. These results show that complete thin shell theories, e.g. those of Novozhilov[62], Flügge[31] or Sanders[67], provided that no additional simplifications are made, predict natural frequencies that agree very closely.

A derivation of a general shell theory is outside the scope of this book, and solutions exist only for certain special cases. In Sections 5.7 and 5.8 energy expressions for cylindrical and shallow shells, respectively, are obtained so that the Rayleigh–Ritz method can be used to determine natural frequencies. In both cases for certain boundary conditions the solutions are exact within the context of the particular shell theory used, i.e. the same frequencies would be obtained from a solution of the shell equations. For cylindrical shells a method of determining response is outlined. In Section 5.9 the finite element method is extended to shells for which the geometry is symmetric about the axis. Other approximate methods exist; Kraus[45] and the author[83] have given surveys of the available methods.

As for beams and plates, higher-order theories, which include transverse shear deformation and rotatory inertia, exist and some numerical results from three-dimensional elasticity theory for cylindrical shells have been obtained. References 2, 3, 45, 49 and 83 give details and numerical comparisons.

5.7. Cylindrical Shells

Only the modes of vibration of cylindrical shells which are likely to occur in structural components of this form will be considered. If w is the radial displacement of the middle surface of the shell, then in these modes w is proportional to cos $n\theta$, where θ is the angular coordinate (Fig. 5.15) and n is an integer. The variation of the radial displacement (greatly exaggerated as usual) round the circumference at a particular cross-section of the shell is shown in Fig. 5.15 for $n = 1$, 2 and 3. The modes, corresponding to $n = 2$ and 3 and to higher values of n, are characteristic of the vibration of shells. Figure 5.15 illustrates the vibration form at a typical cross-section; however, for all practical end conditions the radial displacement is a function of the axial coordinate. Also the middle surface of the shell has components of displacement in

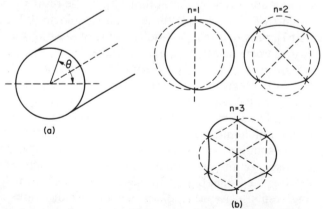

FIG. 5.15. (a) Cylindrical shell. (b) Examples of circumferential modes.

the tangential and axial directions, though in the modes likely to be stimulated in cylindrical structures the maximum values of these components are less than the maximum radial displacement. There are two methods of approach to the vibration of cylindrical shells, similar to those for rectangular plates. By considering the equilibrium of an element and expressing forces and moments in terms of the three components of displacement, three equations of motion are obtained; the assumed vibration forms have to satisfy these equations and also the conditions at the ends of the cylinder. For most practical end conditions solutions are complicated, so this method will not be presented here. The alternative method to be given here is to express the strain and kinetic energies as integrals in terms of the components of displacement, assume reasonable vibration forms for these components and use the Rayleigh–Ritz method to determine the natural frequencies.

An element of a cylindrical shell of mean radius a, uniform thickness h, with h/a assumed to be small, and length l is shown in Fig. 5.16. The coordinate directions are: axial, X; tangential, denoted by the angle θ (Fig. 5.3); and radial, Z, measured positive *outwards* from the middle surface. The components of displacement of a point on the middle surface are u, v and w in the X-, θ- and Z-directions respectively. As for thin plates (see assumptions 2 and 4 of Section 5.2), the direct stress σ_z and the shear strains ε_{xz} and $\varepsilon_{\theta z}$ are zero. At a distance z from the middle surface the strain is equal to the appropriate strain in the middle surface plus the strain due to the change of curvature or the twist. It is to be noted that the middle surface is not assumed to be

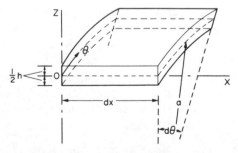

FIG. 5.16. Element of a cylindrical shell.

inextensible; this assumption would lead to a theory inapplicable to shells of finite length.

If ε'_x, ε'_θ and $\varepsilon'_{x\theta}$ are the direct strains in the X- and θ-directions and the shear strain, respectively, at a distance z from the middle surface, and for the middle surface

ε_x ε_θ are the strains in the X- and θ-directions,

κ_x κ_θ are the changes of curvature in the X- and θ-directions,

$\varepsilon_{x\theta}$ $\kappa_{x\theta}$ are the shear strain and twist,

then the strains at a distance z from the middle surface are given approximately by

$$\varepsilon'_x = \varepsilon_x + z\kappa_x \, , \; \varepsilon'_\theta = \varepsilon_\theta + z\kappa_\theta \, , \; \varepsilon'_{x\theta} = \varepsilon_{x\theta} + 2z\kappa_{x\theta}$$

where

$$\varepsilon_x = \frac{\partial u}{\partial x}, \, \varepsilon_\theta = \frac{1}{a}\frac{\partial v}{\partial \theta} + \frac{w}{a}, \, \varepsilon_{x\theta} = \frac{\partial v}{\partial x} + \frac{1}{a}\frac{\partial u}{\partial \theta},$$

$$\kappa_x = -\frac{\partial^2 w}{\partial x^2}, \, \kappa_\theta = -\frac{1}{a^2}\left(\frac{\partial^2 w}{\partial \theta^2} - \frac{\partial v}{\partial \theta}\right), \, \kappa_{x\theta} = -\frac{1}{a}\left(\frac{\partial^2 w}{\partial x \, \partial \theta} - \frac{\partial v}{\partial x}\right).$$

$$(5.82)$$

The strains in the middle surface are obtained from equations (5.4), (5.7) and (5.9), noting that the radius of the middle surface is a. The expression for κ_x and the first terms in κ_θ and $\kappa_{x\theta}$ may be inferred from comparable equations, which occur in plate theory [equations (5.10) and (5.11)]. The second terms in κ_θ and $\kappa_{x\theta}$ arise from the curvature of the element, which also causes the term w/a to occur in ε_θ; these strain–displacement relations are derived by Love[54] and Timoshenko[74].

From equation (5.6), noting that σ_z, ε_{xz} and $\varepsilon_{\theta z}$ have been assumed to be zero and neglecting the trapezoidal form of the faces of the element perpendicular to the X-axis, the strain energy

$$\mathfrak{S} = \int_{-h/2}^{h/2} \int_0^l \int_0^{2\pi} \tfrac{1}{2}[\sigma_x \varepsilon'_x + \sigma_\theta \varepsilon'_\theta + \sigma_{x\theta} \varepsilon'_{x\theta}]a \; d\theta \; dx \; dz. \qquad (5.83)$$

The stress–strain relations, similar to equations (5.12), are

$$\sigma_x = \frac{E}{1 - v^2} (\varepsilon_x' + v\varepsilon_\theta'), \quad \sigma_\theta = \frac{E}{1 - v^2} (\varepsilon_\theta' + v\varepsilon_x'),$$

$$\sigma_{x\theta} = \frac{E}{2(1 + v)} \varepsilon_{x\theta}'. \tag{5.84}$$

Substituting from equations (5.82) and (5.84) in equation (5.83) and integrating with respect to z,

$$
\begin{aligned}
\mathfrak{S} = \int_0^l \int_0^{2\pi} \frac{Eh}{2(1 - v^2)} & \left[\left(\frac{\partial u}{\partial x} \right)^2 + \frac{1}{a^2} \left(\frac{\partial v}{\partial \theta} + w \right)^2 \right. \\
& + \frac{2v}{a} \frac{\partial u}{\partial x} \left(\frac{\partial v}{\partial \theta} + w \right) + \frac{1}{2}(1 - v) \left(\frac{\partial v}{\partial x} + \frac{1}{a} \frac{\partial u}{\partial \theta} \right)^2 \\
& + \frac{h^2}{12} \left\{ \left(\frac{\partial^2 w}{\partial x^2} \right)^2 + \frac{1}{a^4} \left(\frac{\partial^2 w}{\partial \theta^2} - \frac{\partial v}{\partial \theta} \right)^2 + \frac{2v}{a^2} \frac{\partial^2 w}{\partial x^2} \left(\frac{\partial^2 w}{\partial \theta^2} - \frac{\partial v}{\partial \theta} \right) \right. \\
& \left. \left. + \frac{2(1 - v)}{a^2} \left(\frac{\partial^2 w}{\partial x \, \partial \theta} - \frac{\partial v}{\partial x} \right)^2 \right\} \right] a \, d\theta \, dx. \tag{5.85}
\end{aligned}
$$

After integrating with respect to z the kinetic energy is

$$\mathfrak{T} = \int_0^l \int_0^{2\pi} \frac{1}{2} \rho h \left[\left(\frac{\partial u}{\partial t} \right)^2 + \left(\frac{\partial v}{\partial t} \right)^2 + \left(\frac{\partial w}{\partial t} \right)^2 \right] a \, d\theta \, dx. \tag{5.86}$$

Integration with respect to θ and x can be performed only after vibration forms for u, v and w have been assumed. It has been stated that w is proportional to $\cos n\theta$, where the integer n is the number of circumferential waves. Thus it is assumed that

$$u = U(t) f_u(x) \cos n\theta,$$

$$v = V(t) f_v(x) \sin n\theta,$$

$$w = W(t) f_w(x) \cos n\theta. \tag{5.87}$$

The forms of f_u, f_v and f_w depend upon the end conditions; in order to use the Rayleigh–Ritz method all deflection and slope conditions must be satisfied. A suitable form for f_w is the vibration form for a uniform beam with the same end conditions as the cylindrical shell. (More general expressions for u, v and w are given when discussing response, equation (5.98).)

To illustrate the method the natural frequencies of a cylindrical shell, "simply supported" at the ends $x = 0$ and $x = l$, will be investigated. At a "simply supported" end of a cylindrical shell the radial and tangential displacements are zero and the axial force and moment are zero. However, the axial displacement u is not zero, analogous to the axial movement allowed at the end of a simply supported beam. It is assumed that

$$u = U(t) \cos \frac{j\pi x}{l} \cos n\theta,$$

$$v = V(t) \sin \frac{j\pi x}{l} \sin n\theta,$$

$$w = W(t) \sin \frac{j\pi x}{l} \cos n\theta \qquad (5.88)$$

where the integer j is the number of half-waves in the axial direction. Thus the integers j and n define the nodal pattern. From equations (5.82), (5.84) and (5.88) σ_x is zero at the ends of the shell. Thus the axial force and moment are zero also, and *all* boundary conditions are satisfied by equation (5.88). Substituting from equation (5.88) in equations (5.85) and (5.86) and integrating with respect to x and θ,

$$\mathfrak{S} = \frac{\pi Eh}{4a(1 - v^2)} \left[\lambda^2 U^2 + (nV + W)^2 - 2v\lambda U(W + nV) \right.$$

$$+ \frac{1}{2}(1 - v)(\lambda V - nU)^2 + \frac{h^2}{12a^2} \left\{ \lambda^4 W^2 + (nV + n^2 W)^2 \right.$$

$$\left. + 2v\lambda^2 W(n^2 W + nV) + 2(1 - v)(\lambda V + \lambda n W)^2 \right\} \right] \qquad (5.89)$$

and

$$\mathfrak{T} = \tfrac{1}{4}\pi\rho hla(\dot{U}^2 + \dot{V}^2 + \dot{W}^2) \qquad (5.90)$$

where

$$\lambda = j\pi a/l.$$

Either assuming harmonic motion $(U(t) = U \sin \omega t, \quad V(t) = V \sin \omega t$ and $W(t) = W \sin \omega t)$, equating $\mathfrak{S}_{max}$ and $\mathfrak{T}_{max}$ to obtain an equation for ω^2 and applying the Rayleigh–Ritz method in the form

$$\frac{\partial \omega^2}{\partial U} = \frac{\partial \omega^2}{\partial V} = \frac{\partial \omega^2}{\partial W} = 0 \qquad (5.91)$$

or applying the Lagrange equation with respect to $U(t)$, $V(t)$ and $W(t)$ and then assuming harmonic motion, the following three equations are obtained,

$$[\lambda^2 + \tfrac{1}{2}(1 - v)n^2 - \Delta]U - \tfrac{1}{2}(1 + v)\lambda nV - v\lambda W = 0,$$

$$-\tfrac{1}{2}(1 + v)\lambda nU + [\tfrac{1}{2}(1 - v)\lambda^2 + n^2 - \Delta + \beta\{n^2 + 2(1 - v)\lambda^2\}]V$$
$$+ [n + \beta\{n^3 + (2 - v)\lambda^2 n\}]W = 0, \qquad (5.92)$$

$$-v\lambda U + [n + \beta\{n^3 + (2 - v)\lambda^2 n\}]V + [1 - \Delta + \beta(\lambda^2 + n^2)^2]W = 0$$

where

$$\Delta = \rho a^2 (1 - v^2)\omega^2/E \qquad \text{and} \qquad \beta = h^2/12a^2.$$

From equations (5.92) the non-dimensional frequency factor Δ is a function of the four parameters: thickness to radius ratio, h/a; number of circumferential waves, n; axial wavelength parameter, λ (=mean circumference/axial wavelength); and Poisson's ratio, v. Elimination of U, V and W leads to a cubic equation in Δ; thus for a specific set of values of the four parameters there are three values of Δ, each corresponding to different values of the ratio $U : V : W$. Only the lowest value of Δ is of structural interest; for this value $W \simeq nV$ and $V > U$, i.e. the radial component of displacement is the greatest, but the other two components cannot be neglected.

The cubic equation in Δ is

$$\Delta^3 - K_2\Delta^2 + K_1\Delta - K_0 = 0 \qquad (5.93)$$

where

$$K_0 = \tfrac{1}{2}(1 - v)^2(1 + v)\lambda^4$$
$$+ \tfrac{1}{2}(1 - v)\beta[(\lambda^2 + n^2)^4 - 2(4 - v^2)\lambda^4 n^2 - 8\lambda^2 n^4 - 2n^6$$
$$+ 4(1 - v^2)\lambda^4 + 4\lambda^2 n^2 + n^4],$$
$$K_1 = \tfrac{1}{2}(1 - v)(\lambda^2 + n^2)^2 + \tfrac{1}{2}(3 - v - 2v^2)\lambda^2 + \tfrac{1}{2}(1 - v)n^2$$
$$+ \beta[\tfrac{1}{2}(3 - v)(\lambda^2 + n^2)^3 + 2(1 - v)\lambda^4 - (2 - v^2)\lambda^2 n^2$$
$$- \tfrac{1}{2}(3 + v)n^4 + 2(1 - v)\lambda^2 + n^2]$$

and

$$K_2 = \tfrac{1}{2}(3 - v)(\lambda^2 + n^2) + 1 + \beta[(\lambda^2 + n^2)^2 + 2(1 - v)\lambda^2 + n^2].$$

As β ($= h^2/12a^2$) is a small quantity for a thin shell, powers of β in K_i have been neglected. For given shell dimensions and Poisson's ratio the lowest root of equation (5.93) can be found for any mode, described by the integers j and n. It is noted that the natural frequency is proportional to $\Delta^{1/2}$. For most values of the shell and modal parameters it is possible to neglect some small terms within the square brackets in the expressions for K_i and also to obtain the lowest root of equation (5.93) approximately as

$$\Delta = \frac{K_0}{K_1} + \frac{K_2}{K_1}\left(\frac{K_0}{K_1}\right)^2. \tag{5.94}$$

Results from the above theory have been published[5]. For these end conditions there is no error in using the Rayleigh–Ritz method, as the assumed vibration forms (5.88) satisfy Novozhilov's shell equations[62] and all the boundary conditions.

Natural frequencies of cylindrical shells with (a) both ends clamped, (b) both ends free, and (c) one end clamped and the other free have been obtained, using approximate expressions, suggested by beam theory, for $f_u(x)$, $f_v(x)$ and $f_w(x)$ in equations (5.87) and the Rayleigh–Ritz method. By comparison with solutions that satisfy the shell equa-

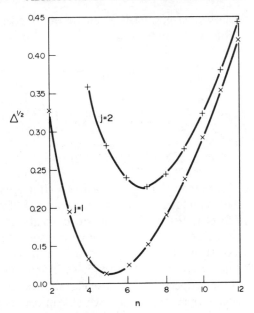

FIG. 5.17. Frequency factors for cylindrical shell with simply supported ends: $h/a = 0·01$, $l/a = 2$, $v - 0·3$.

tions and all the boundary conditions these frequencies have been shown to be reasonably accurate[82, 84].

Figure 5.17 shows the frequency factor $\Delta^{1/2}$ for various values of n and $j = 1$ and 2 for a simply supported cylindrical shell with $l/a = 2$, $h/a = 0·01$ and $v = 0·3$. In order to illustrate the trends the curves have been plotted as $\Delta^{1/2}$ against n for the two values of j, although frequencies for non-integer values of n have no physical significance for a complete shell. For a particular value of n the natural frequency increases as the number of axial half-waves j increases, but the converse is not true; indeed, for this example the lowest natural frequency corresponds to the mode $j/n = 1/5$. Natural frequencies for the flexural modes ($n = 1$) and for the axisymmetric modes ($n = 0$) lie above the range of values of $\Delta^{1/2}$ in the figure.

Response. The response of a simply supported cylindrical shell to a radial force per unit area $p(x, \theta)f(t)$ can be obtained by the normal mode method. Thus we assume that

$$u = \sum_j \sum_n U_{jn} \cos \frac{j\pi x}{l} \cos n\theta \, q_{jn}(t),$$

$$v = \sum_j \sum_n V_{jn} \sin \frac{j\pi x}{l} \sin n\theta \, q_{jn}(t),$$

$$w = \sum_j \sum_n \sin \frac{j\pi x}{l} \cos n\theta \, q_{jn}(t) \tag{5.95}$$

where U_{jn} and V_{jn} are the values of U/W and V/W respectively for the mode j/n with natural frequency ω_{jn}. In equations (5.95) it is assumed that the applied force is distributed symmetrically about the line $\theta = 0$. (For a distribution that is antisymmetric with respect to $\theta = 0$, $\sin n\theta$ and $\cos n\theta$ in equations (5.95) must be interchanged.) Substituting equations (5.95) in equations (5.85) and (5.86), integrating with respect to x and θ, using the properties of the normal modes and applying the Lagrange equation with respect to q_{jn}, we obtain the set of uncoupled equations

$$\tfrac{1}{2}b_n \pi \rho h l a (U_{jn}^2 + V_{jn}^2 + 1)(\ddot{q}_{jn} + \omega_{jn}^2 q_{jn}) = Q_{jn}$$

where Q_{jn} is the generalized force corresponding to q_{jn}; $b_n = 1$ for $n > 0$ and $b_n = 2$ for $n = 0$. Applying the principle of virtual work for an increment δq_{jn}, it can be shown that

$$Q_{jn} = f(t) \int_0^l \int_0^{2\pi} p(x, \theta) \sin \frac{j\pi x}{l} \cos n\theta \, a \, d\theta \, dx.$$

Thus

$$\ddot{q}_{jn} + c\omega_{jn}^2 \dot{q}_{jn} + \omega_{jn}^2 q_{jn} = K_{jn} f(t) \tag{5.96}$$

where viscous internal damping has been introduced by replacing E and G by $E[1 + c(\partial/\partial t)]$ and $G[1 + c(\partial/\partial t)]$, respectively, as in Section 3.4, and

$$K_{jn} = \frac{\displaystyle\int_0^{2\pi} \int_0^l p(x, \theta) \sin \frac{j\pi x}{l} \cos n\theta \, dx \, d\theta}{\tfrac{1}{2}b_n \pi \rho h l (U_{jn}^2 + V_{jn}^2 + 1)}. \tag{5.97}$$

The solution of equation (5.96) can be determined from Duhamel's integral and then the response is obtained from equations (5.95). If the excitation is harmonic with frequency ω, hysteretic damping can be included by replacing c in equation (5.96) by μ/ω. If the excitation is a radial force $Pf(t)$ applied at the point $x = x_1$, $\theta = 0$, the numerator of K_{jn} in equation (5.97) is replaced by

$$\frac{P}{a} \sin \frac{j\pi x_1}{l}.$$

Figure 5.18 shows the response for the radial component of displacement, expressed non-dimensionally as $|wEa/P|$, of a simply supported cylindrical shell plotted against the frequency factor $\Delta^{1/2}$ $[=\omega a\{(1 - v^2)\rho/E\}^{1/2}]$, when the excitation is a radial harmonic force $P \sin \omega t$. The shell dimensions are the same as in Fig. 5.17, namely $l/a = 2$, $h/a = 0\cdot01$; also $v = 0\cdot3$. The hysteretic damping factor $\mu = 0\cdot01$. The curves show the response at the point of excitation, which is $(l/4, 0)$ for Fig. 5.18a and $(l/2, 0)$ for Fig. 5.18b. The curves illustrate the relative complexity of the response of a shell. At the top of Fig. 5.18 the values of $\Delta_{jn}^{1/2} = \omega_{jn} a[(1 - v^2)\rho/E]^{1/2}$, where ω_{jn} is the natural frequency of mode j/n, are indicated. (These factors, corresponding to natural frequencies, are identical to those shown in Fig. 5.17.) If the excitation is at $x_1 = l/2$, modes with $j = 2$ do not respond as $\sin (j\pi x_1/l) = 0$; thus resonances associated with $j = 2$ occur only on Fig. 5.18a.

For boundary conditions, other than simply supported, the response can be determined approximately by the Rayleigh–Ritz method. (The method is similar to that given for rectangular plates in Section 5.2.) It is assumed that

$$u = \sum_j \sum_n U_{jn} f_{ju}(x) \cos n\theta \; \Gamma_{jn}(t),$$

$$v = \sum_j \sum_n V_{jn} f_{jv}(x) \sin n\theta \; \Gamma_{jn}(t),$$

$$w = \sum_j \sum_n f_{jw}(x) \cos n\theta \; \Gamma_{jn}(t) \tag{5.98}$$

where U_{jn} and V_{jn} are the values of U/W and V/W, respectively, for the mode j/n with natural frequency ω_{jn}; U_{jn}, V_{jn} and ω_{jn} are found by the Rayleigh–Ritz method for natural frequencies. Substituting from equa-

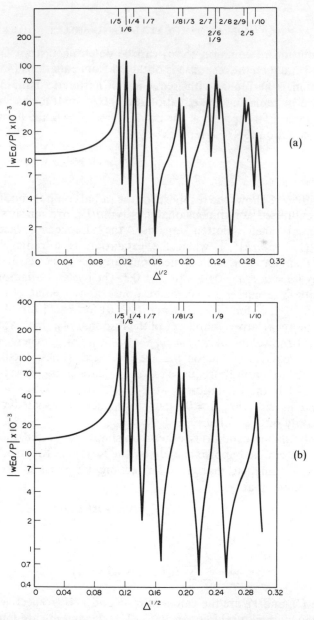

FIG. 5.18. Response of cylindrical shell with simply supported ends to radial harmonic force; $h/a = 0.01$, $l/a = 2$, $\nu = 0.3$, $\mu = 0.01$. (a) Force and response at $x = l/4$, $\theta = 0$. (b) Force and response at $x = l/2$, $\theta = 0$.

tions (5.98) in the strain and kinetic energy expressions (5.85) and (5.86), integrating with respect to x and θ and applying the Lagrange equation with respect to each of Γ_{jn}, j, $n = 1, 2, 3, \ldots$, we obtain the matrix equation

$$M\ddot{g} + Kg = p \tag{5.99}$$

where the vector g is a list of the coordinates Γ_{jn}. Appropriate beam modal functions can be used for each of the sets of functions f_{ju}, f_{jv} and f_{jw}. The elements of the generalized force vector p can be found from the applied forces, using the principle of virtual work. Damping can be included in equation (5.99) by the methods of Sections 3.4 and 3.5.

5.8. Shallow Shells

A shallow shell may be regarded as a slightly curved plate, whose smallest radius of curvature at any point is large compared with the largest length measured along the middle surface of the shell. Alternatively, Vlasov[79] defines a shallow shell as a thin-walled structure with a comparatively small rise above the base plane covered by the structure. For example, for a structure with a rectangular floor plan the rise, or maximum height of any point on the shell from the base plane on which the shell is supported, should not be greater than one-fifth of the smaller side of the rectangle. For a spherical cap with a circular base Reissner states that the shell is shallow if the rise is less than one-eighth of the base diameter[64].

Figure 5.19 shows a shallow shell whose projection on the base plane is the rectangle $ABCD$. A point on the middle surface of the shell is defined by the curvilinear coordinates x and y. The arc lengths of the middle surface are l_x and l_y. These lengths can be only a few per cent greater than the lengths of the corresponding sides of the base rectangle, if the shell is truly shallow. At any point on the middle surface the radii of curvature are R_x and R_y, which are assumed here to be constants. The components of displacement of the middle surface, u, v and w (u and v are in the X- and Y-directions, respectively, and w is in the direction normal to the surface), are functions of the coordinates x and y and of time t. As with cylindrical shells in Section 5.7, strain and kinetic energy expressions will be established, rather than equilibrium

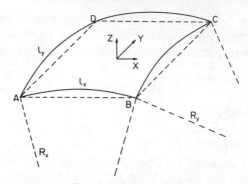

FIG. 5.19. Shallow shell.

equations. The strains at a distance z from the middle surface are related to the middle surface strains and changes of curvature by equation (5.82), with the following definitions. The middle surface strains, allowing for the double curvature, are:

$$\varepsilon_x = \frac{\partial u}{\partial x} + \frac{w}{R_x}, \, \varepsilon_y = \frac{\partial v}{\partial y} + \frac{w}{R_y}, \, \varepsilon_{xy} = \frac{\partial v}{\partial x} + \frac{\partial u}{\partial y} \qquad (5.100)$$

and the changes of curvature and twist, analogous to equations (5.10) and (5.11) from plate theory, are:

$$\kappa_x = -\frac{\partial^2 w}{\partial x^2}, \, \kappa_y = -\frac{\partial^2 w}{\partial y^2}, \, \kappa_{xy} = -\frac{\partial^2 w}{\partial x \, \partial y}. \qquad (5.101)$$

Substituting from the stress–strain relations (5.84) and equations (5.100) and (5.101) in the strain energy expression (5.83) and integrating with respect to z, the strain energy

$$\mathfrak{S} = \frac{Eh}{2(1-v^2)} \int_0^{l_x} \int_0^{l_y} \left[\left(\frac{\partial u}{\partial x} + \frac{w}{R_x} \right)^2 + \left(\frac{\partial v}{\partial y} + \frac{w}{R_y} \right)^2 + 2v \left(\frac{\partial u}{\partial x} + \frac{w}{R_x} \right) \right.$$

$$\times \left(\frac{\partial v}{\partial y} + \frac{w}{R_y} \right) + \left(\frac{1-v}{2} \right) \left(\frac{\partial v}{\partial x} + \frac{\partial u}{\partial y} \right)^2 + \frac{h^2}{12} \left\{ \left(\frac{\partial^2 w}{\partial x^2} \right)^2 \right.$$

$$\left. + \left(\frac{\partial^2 w}{\partial y^2} \right)^2 + 2v \frac{\partial^2 w}{\partial x^2} \frac{\partial^2 w}{\partial y^2} + 2(1-v) \left(\frac{\partial^2 w}{\partial x \, \partial y} \right)^2 \right\} \right] \, dy \, dx.$$

$$(5.102)$$

When deriving the kinetic energy only the transverse component of displacement is considered; this is equivalent to neglecting the in-plane components of inertia in the appropriate equilibrium equations. Thus

$$\mathfrak{T} = \frac{1}{2}\rho h \int_0^{l_x} \int_0^{l_y} \left(\frac{\partial w}{\partial t}\right)^2 dy \, dx. \tag{5.103}$$

The Rayleigh–Ritz method can be used, with equations (5.102) and (5.103), to determine natural frequencies following the procedure given for rectangular plates. The functions chosen to approximate u, v and w must satisfy all geometric boundary conditions. Beam modal functions of Section 3.3 are appropriate for the transverse displacement w; for the in-plane components u and v the alternative simple boundary conditions are completely constrained and zero constraint and thus harmonic functions are suitable, similar to those established for comparable boundary conditions for extensional and torsional vibrations of bars.

If all four edges of the shallow shell are simply supported, u and w are zero on the edges parallel to OX and v and w are zero on the edges parallel to OY. Suitable expressions for u, v and w are

$$u = U(t) \cos \frac{j\pi x}{l_x} \sin \frac{n\pi y}{l_y},$$

$$v = V(t) \sin \frac{j\pi x}{l_x} \cos \frac{n\pi y}{l_y},$$

$$w = W(t) \sin \frac{j\pi x}{l_x} \sin \frac{n\pi y}{l_y} \tag{5.104}$$

where j and n are integers, defining the mode shape. Substituting equation (5.104) in equations (5.102) and (5.103), integrating, applying the Lagrange equation (A4.17) with respect to U, V and W, putting $\ddot{W} = -\omega^2 W$ for free vibrations and eliminating U, V and W from the resulting three equations, the natural frequency of mode j/n, ω_{jn}, is given by

$$\rho \frac{\omega_{jn}^2}{E} (\lambda_x^2 + \lambda_y^2)^2 = \left(\frac{\lambda_x^2}{R_y} + \frac{\lambda_y^2}{R_x}\right)^2 + \frac{h^2}{12(1-v^2)} (\lambda_x^2 + \lambda_y^2)^4 \tag{5.105}$$

where $\lambda_x = j\pi/l_x$ and $\lambda_y = n\pi/l_y$. Equations (5.104) satisfy the equilibrium equations for a shallow shell and all the boundary conditions[83]; thus the frequency equation (5.105) is the exact solution of shallow-shell theory for the specified boundary conditions. However, simple solutions of the form of equation (5.104) cannot be obtained for other boundary conditions.

Equation (5.105) can be used to demonstrate the appreciable increase in the fundamental frequency of a plate caused by slight curvature. Considering the fundamental mode of a shallow spherical shell of square planform,

$$R_x = R_y = R; \; l_x = l_y = l \quad \text{and} \quad j = n = 1.$$

Thus

$$\rho \frac{\omega_{11}^2}{E} = \frac{1}{R^2} + \frac{h^2}{12(1 - v^2)} \left(\frac{2\pi^2}{l^2} \right)^2 \tag{5.106}$$

from equation (5.105). If z_c is the central (maximum) rise of the middle surface above the base plane, then approximately for z_c/l small (i.e. ≤ 0.05)

$$2Rz_c = (l/2)^2. \tag{5.107}$$

Using equation (5.107) to eliminate R from equation (5.106) and rearranging,

$$\omega_{11}^2 = \omega_0^2 \left[1 + \frac{192(1 - v^2)}{\pi^4} \left(\frac{z_c}{h} \right)^2 \right] \tag{5.108}$$

where ω_0 is the fundamental frequency in flexure of a simply supported square flat plate of side l, obtained from equation (5.106) by making $R \to \infty$ or from equation (5.23). Table 5.1 gives values of ω_{11}/ω_0 for a few values of z_c/h and $v = 0.3$. The assumptions inherent in equation (5.107), namely omission of the term $-z_c^2$ from the left-hand side and use of the arc length, rather than the chord length, in the right-hand side, cause small errors, ≤ 1.2 per cent for $z_c/l \leq 0.05$, in the frequency ratios of Table 5.1. Also included in the table are the minimum values of l/h required to satisfy $z_c/l \leq 0.05$. It is seen that slight spherical curvature increases appreciably the fundamental frequency of a simply supported square plate. For all modes the difference between ω_{jn}^2 and the value of ω^2 for a comparable flat plate in the mode j/n is $E/\rho R^2$; thus for higher modes the ratio of ω_{jn} for a shallow shell to ω_{jn} for the corresponding flat plate is less than that given in Table 5.1 for a specified value of z_c/h.

If shallow shell theory is applied to a cylindrically curved panel $(R_x \to \infty$ and $R_y = a)$ and frequencies from equation (5.105) compared with values from the thin-shell theory of Section 5.7, there is reasonable agreement for panels that are geometrically shallow. The additional assumption and simplification of using the lengths of the sides of the base rectangle, instead of l_x and l_y, in equation (5.105) should be used only when the ratio of the rise to the length of the smaller side is very small.

TABLE 5.1. Effect of rise to thickness ratio on fundamental frequency of shallow spherical shell with square planform and simply supported edges. $v = 0.3$

z_c/h	0.5	1	2	5
ω_{11}/ω_0	1·203	1·671	2·859	6·771
Minimum value of l/h	10	20	40	100

5.9. Finite Element Method for Shell Vibrations

One method of solving problems for shells uses an idealization consisting of flat-plate elements, either triangular or quadrilateral in form. This can be achieved with any suitable combination of a bending and an in-plane element (e.g. the rectangular elements of Sections 5.4 and 5.5), and thus allows for the simultaneous existence of bending and membrane stresses in a shell. Accurate natural frequencies can be obtained provided that a large number of elements is used, so that the artificial geometric discontinuities of the idealization are reduced.

Elements, which are singly or doubly curved and are of triangular or rectangular planform, have been developed. Some of these elements are based on the shallow shell theory of Section 5.8; others are based on general shell theory. Considerable research effort has been expended in the development of general doubly curved shell elements, which can be used to predict with reasonable accuracy and economy of computation the dynamic behaviour of complex structures which may be built up from shells, plates and beams. Details are beyond the scope of this book; Brebbia and Connor[14] describe various elements and give comparative results for some static problems.

In many problems the geometry of the shell is symmetric about an axis, e.g. cooling-towers, chimneys, pressure vessels, containers, etc. For these shells of revolution relatively simple concepts and elements can be used and these will be described. It should be noted that only the geometry of the shell must be axisymmetric, as asymmetric loads and vibrations are treated by Fourier analysis. If θ is the circumferential angular coordinate, components of displacement, stress and applied force can be expanded as Fourier series with respect to this coordinate. For example, the normal component of displacement w and the applied force per unit area in the normal direction can be written

$$w = \sum_n (w_n \cos n\theta + w'_n \sin n\theta)$$

and

$$p = \sum_n (p_n \cos n\theta + p'_n \sin n\theta) \tag{5.109}$$

where $_n$, w'_n, p_n and p'_n are functions of the meridional coordinate and time. The response w_n and w'_n for prescribed excitation p_n and p'_n is determined and the total response obtained by summation with respect to n, i.e. the calculations for each value of n are "uncoupled".

Figure 5.20 shows the meridian of an axisymmetric element; the

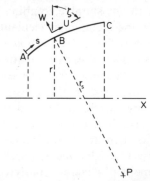

FIG. 5.20. Axisymmetric shell element.

complete element is obtained by rotating the line ABC about the X-axis. The simplest element of this type is the conical element, occurring when ABC is a straight line. In general, ABC will be curved; the meridional radius of curvature r_s ($= PB$) may vary over the length of the element and thus enable the element to be used to represent the exact geometry of the actual shell; or r_s may be treated as a constant for a particular element, giving an approximate representation of the shell. If u and w are the meridional and normal components of displacement (Fig. 5.20) and v is the tangential component, positive into the paper, then

$$u = \sum_n u_n \cos n\theta,$$

$$v = \sum_n v_n \sin n\theta,$$

$$w = \sum_n w_n \cos n\theta \qquad (5.110)$$

where the positive directions of v and θ coincide. Equations (5.110) can be expanded to include components such as $w'_n \sin n\theta$, if necessary. The subsequent analysis will consider only a single harmonic component, i.e. $u_n \cos n\theta$, $v_n \sin n\theta$ and $w_n \cos n\theta$, recalling that for free vibrations the harmonic components are uncoupled and for response problems the applied forces must be expressed in terms of their Fourier components by equation (5.109). The displacement assumptions are

$$\mathbf{u} = \begin{bmatrix} u_n \cos n\theta \\ v_n \sin n\theta \\ w_n \cos n\theta \end{bmatrix} = \begin{bmatrix} \mathbf{g} \cos n\theta & 0 & 0 \\ 0 & \mathbf{g} \sin n\theta & 0 \\ 0 & 0 & \mathbf{g}^+ \cos n\theta \end{bmatrix} \mathbf{a} \qquad (5.111)$$

where for the element shown in Fig. 5.20, which has nodes at A and C only, $\mathbf{g} = \begin{bmatrix} 1 & s \end{bmatrix}$ and $\mathbf{g}^+ = \begin{bmatrix} 1 & s & s^2 & s^3 \end{bmatrix}$, and for an element with one mid-side node $\mathbf{g} = \begin{bmatrix} 1 & s & s^2 \end{bmatrix}$ and $\mathbf{g}^+ = \begin{bmatrix} 1 & s & s^2 & s^3 & s^4 & s^5 \end{bmatrix}$. The similarity of $\mathbf{g}$ and $\mathbf{g}^+$ to expressions used in extension and flexure of beams, respectively, is noted. The vector $\mathbf{a}$ is obtained in terms of the vector of nodal variables $\mathbf{u}_e$, defined as the values of u_n, v_n, w_n and the rotation β_n

$$\left(= \frac{\partial w_n}{\partial s} - \frac{u_n}{r_s} \right)$$

at each node. Flügge[31] gives a strain energy expression for a shell of revolution. Substituting in such an expression, integration through the thickness h and with respect to the coordinate θ is straightforward (the direct strains are proportional to cos $n\theta$ and the shear strain to sin $n\theta$); this leaves the element stiffness matrix in terms of integrals, with respect to the coordinate s, and the vector $\mathbf{u}_e$. These integrals can be evaluated, numerically if necessary.

The kinetic energy

$$\mathfrak{T}_e = \tfrac{1}{2} \int_0^l \int_0^{2\pi} \dot{\mathbf{u}}^T \rho h \dot{\mathbf{u}} r \, d\theta \, ds$$

after integrating through the thickness. Using equation (5.111), and proceeding similarly, $\mathfrak{T}_e$ is expressed in terms of integrals with respect to s and the vector $\dot{\mathbf{u}}_e$. Finally, in order to ensure appropriate continuity conditions between adjacent shell elements and to be able to use this element in conjunction with other types of element for which the nodal displacements are defined in terms of Cartesian coordinates, the vector $\mathbf{u}_e$ is transformed geometrically to $\boldsymbol{\delta}_e$, which lists for each node the displacements in the axial, tangential and radial directions and the rotation (i.e. a transformation from the s and r_s directions in Fig. 5.20 to the directions of x and r). For node j

$$[\boldsymbol{\delta}_e]_j = \begin{bmatrix} \sin \zeta_j & 0 & -\cos \zeta_j & 0 \\ 0 & 1 & 0 & 0 \\ \cos \zeta_j & 0 & \sin \zeta_j & 0 \\ 0 & 0 & 0 & 1 \end{bmatrix} \begin{bmatrix} u_n \\ v_n \\ w_n \\ \beta_n \end{bmatrix}_j$$

where ζ_j is the value of ζ, defined in Fig. 5.20, at node j.

From results for test problems natural frequencies of shells of revolution can be obtained with adequate accuracy using the simplest element of this family, i.e. that with nodal rings only at the ends. However, higher-order elements, i.e. those with one or more intermediate nodal rings, are required to determine dynamic stresses, particularly if these vary rapidly with the axial coordinate[86].

After assembling the stiffness and mass matrices for the structure, the response of the shell to prescribed excitation can be obtained, using the Lagrange equation to establish the matrix equation (2.40) and an appropriate method from Chapter 2 to solve this equation. This gives the response for a single Fourier component of the excitation, e.g. for an applied force per unit area in the normal direction of $p_n(s, t) \cos n\theta$. For a general applied force the calculations must be repeated for those values of n which contribute significantly to the response, and the total response obtained by appropriate summation.

The approximations, which were introduced in Section 2.9 in order to obtain comparatively simple expressions for the response to random excitation, require the natural frequencies of the structure to be well separated, although the necessary degree of separation is less for a lightly damped structure. As shown in Figs. 5.17 and 5.18, two or more natural frequencies may lie close together for a shell. For a large shell, or other, structure many natural frequencies lie within a narrow frequency band. For example, if the radius $a = 10$ m for the shell of Fig. 5.17, the lowest natural frequency is 9·6 Hz, there are 126 modes with natural frequencies below 85 Hz and 32 natural frequencies within the band 75 to 85 Hz. The statistical energy method can be used to determine the response to random excitation for structures with closely spaced natural frequencies[76]. In this method the individual normal modes are not considered. Analysis is in terms of modal density, i.e. the average number of modes with natural frequencies in a band of width 1 Hz; simple expressions exist for the modal densities of plates and shells in the higher frequency ranges[12, 30].

Problems

1. A uniform square plate of thickness h has sides of length a and is simply supported on all edges. Using the Rayleigh method and assuming that $W(x, y) = f_1(x)f_2(y)$, where $f_1(x)$ and $f_2(y)$ are appropriate polynomials, obtain an approximate expression for the fundamental natural frequency of transverse vibration. Determine the percentage error in this expression by comparison with the solution from the plate equation.

2. A uniform square plate has sides of length a; it is clamped on two opposite edges and simply supported on the other pair. Using the Rayleigh method, obtain an approximate expression for the fundamental natural frequency of transverse vibration. For these boundary conditions the equation of motion for the plate can be solved to give $\omega_1 = 8·06[Eh^2/\rho a^4(1 - v^2)]^{1/2}$; what is the error in the approximate natural frequency found above?

3. A uniform rectangular plate is simply supported at its edges, $x = 0$, $x = a$, $y = 0$ and $y = b$. A small concentrated mass, equal to 2 per cent of the mass of the plate, is attached to the plate at the point $(a/3, b/3)$. If $a/b = 3$, find approximately the percentage change in the natural frequencies of the first three modes of vibration caused by the addition of the mass.

4. A constant transverse force P is suddenly applied at time $t = 0$ at the centre of a square plate with simply supported edges. Find the resulting maximum displacement of the plate. Neglect damping. ($v = 0.3$.)

5. A uniform square plate with sides of length a is clamped on all its edges; it is subjected to a transverse force $P \sin \omega t$ at its centre. The hysteretic damping constant of the plate material is μ. Assuming that $w(x, y, t) = \Gamma \phi_1(x) \phi_1(y) \exp(i\omega t)$, where ϕ_1 is the normalized first mode shape of a uniform clamped–clamped beam, show that the steady-state amplitude at the centre of the plate, when the excitation frequency is equal to the fundamental natural frequency of the plate, is approximately $0.00620 Pa^2/\mu D$. (Note:

$$a^3 \int_0^a \left(\frac{d^2\phi_1}{dx^2} \right)^2 dx = (\lambda_1 a)^4 = (4.730)^4;$$

$$a \int_0^a \frac{d^2\phi_1}{dx^2} \phi_1 \, dx = a \int_0^a \left(\frac{d\phi_1}{dx} \right)^2 dx = 12.303;$$

and

$$\phi_1(a/2) = 1.588.)$$

6. Determine approximately the fundamental natural frequency of transverse vibration of a circular plate, clamped at the outer boundary $r = R$.

7. A plate of uniform thickness h is in the form of a sector of a circle, being bounded by two arcs of radii $\frac{1}{2}R$ and R and subtending an angle of $60°$ at the centre. If all the edges are simply supported, find an approximate expression for the fundamental natural frequency of transverse vibration. ($v = 0.3$.)

8. Determine the first three natural frequencies of an aluminium cylindrical shell, which is 2 m diameter, 10 mm wall thickness, 4 m length and is simply supported at the ends. (For aluminium $E = 69$ GN/m^2, $v = 0.3$ and $\rho = 2700$ kg/m^3.)

9. Determine the first three natural frequencies of the shell of Problem 8, using shallow-shell theory. For which of these modes is the shallow-shell criterion of Vlasov satisfied? What is the lowest value of the number of circumferential waves, n, for which this criterion is satisfied?

10. A cylindrically curved panel has a square planform with sides of length l and is simply supported on its edges. The thickness $h = 0.01l$. If the maximum rise above the base plane z_c is given by $z_c/h = $ (i) 0, (ii) 2, (iii) 5, find the frequency ratio ω_{jn}/ω_0 for the modes $j/n = 1/1, 2/1, 1/2$ and $1/3$; ω_0 is the fundamental natural frequency of a simply supported square flat plate with the same dimensions as the panel; j and n are the numbers of half-waves in the axial and circumferential directions, respectively, for the panel. Poisson's ratio $v = 0.3$.

CHAPTER 6

Dynamic Interaction Problems

IN PREVIOUS chapters the response of structures has been determined for prescribed excitation, due either to forces applied to the structure or to motion of the supports. In this chapter we consider structural vibrations, for which the excitation mechanism is affected by the properties of the surrounding or underlying medium. Three important types of dynamic interaction will be introduced: ground–structure interaction; fluid–structure interaction (e.g. the dam–reservoir problem); and wind-induced vibrations of structures. The last of these is also a fluid–structure problem, but the interaction effect arises from the dependence of the wind forces on the relative velocity of structure and air. In the dam–reservoir type of problem inertia and elasticity effects in the water lead to the interaction effect. In each case allowance for the properties of the underlying or surrounding medium is necessary in order to predict the dynamic response with reasonable accuracy. Each type of problem is complex; there is considerable current research in these areas and only a brief introduction can be given here.

6.1. Ground–Structure Dynamic Interaction

The response of a structure to prescribed motion at its base has been considered in Sections 2.6 and 2.8. Provided that the motion of the base is known, these methods are satisfactory. In practice, the response of a structure to an earthquake or other transient phenomenon may be required and from experience of previous records the free field acceleration, i.e. the surface acceleration in the absence of the structure, can be assessed. If dynamic interaction is neglected, the acceleration of the base of the structure is equated to the appropriate components of the free field acceleration and the methods of Section 2.6 or 2.8 used to determine the response. If dynamic interaction is considered, the acceleration of the base is unknown and this and the response of the structure have to be determined by considering the whole system of flexible structure, rigid base and underlying soil. In this section the dynamics of the interaction problem will be outlined, treating the soil as an elastic medium. The use of elastic models to investigate problems in soil dynamics is discussed by Richart *et al.*[65]. Ranges of values of structural and soil parameters, for which dynamic interaction effects are significant, cannot be easily summarized; neglecting dynamic interaction may lead to an underestimate or an overestimate of response.

Initially, we consider the vibrations of a light, rigid circular disc, which is attached to the surface of an elastic half-space, i.e. an elastic medium bounded by the plane $z = 0$ and occupying the volume $0 \leq z \leq \infty$ (Fig. 6.1), and is subjected to a vertical harmonic force

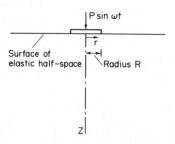

FIG. 6.1. Disc on an elastic half-space.

$P \sin \omega t$ acting along the axis of symmetry. It is to be noted that the solution for this idealized problem has been used extensively in the determination of the response of structures to earthquakes. The problem is symmetrical about the axis OZ, so displacements and stresses are functions of r and z and are harmonic functions of time t. Mathematically, we have a mixed boundary value problem with conditions: the amplitude $W(r, 0) = $ a constant for $r \leq R$, and all stresses zero on the surface $z = 0$ for $r > R$. Solutions have been obtained from elasticity theory, e.g. refs. 4 and 65, and are of the form

$$w = \text{Im}\left\{\frac{P \exp (i\omega t)}{GR} [f_1 + if_2]\right\} \tag{6.1}$$

where w is the displacement of the disc in direction OZ and f_1 and f_2 are functions of the non-dimensional frequency factor $a_0 [=\omega R(\rho/G)^{1/2}]$ and Poisson's ratio v. [G = shear modulus.] Equation (6.1) can be inverted to give a complex dynamic stiffness

$$\frac{GR(f_1 - if_2)}{f_1^2 + f_2^2}. \tag{6.2}$$

Equations (6.1) and (6.2) apply to a light, rigid circular disc. If this is replaced by a rigid foundation of mass m and base radius R subjected to an applied vertical force $P \exp (i\omega t)$ and the reaction between the foundation and the half-space is $Q \exp (i\omega t)$, equation (6.1) is replaced by

$$w = Q \exp (i\omega t)(f_1 + if_2)/GR. \tag{6.3}$$

Equilibrium of the foundation yields:

$$P \exp (i\omega t) - Q \exp (i\omega t) = -m\omega^2 w. \tag{6.4}$$

From equations (6.3) and (6.4), after taking the imaginary part of the solution for w, the amplitude of the foundation W is given by

$$\frac{GRW}{P} = \left[\frac{f_1^2 + f_2^2}{(1 - ba_0^2 f_1)^2 + (ba_0^2 f_2)^2}\right]^{1/2} \tag{6.5}$$

where $b = m/\rho R^3$. Figure 6.2 shows response curves, GRW/P, plotted against frequency factor a_0 for $v = 0$ and various values of the mass ratio b. Although internal damping for the elastic medium has not been included in the analysis, the response curves have finite peaks, because of the radiation damping, i.e. the energy radiated outwards and downwards towards the boundaries at infinity. The response curves of Fig. 6.2 are similar to those for the single-degree-of-freedom system of Fig. 1.6. Thus various authors have considered replacing the elastic half-space by an equivalent spring and dashpot; as the latter give a complex dynamic stiffness $k + i\omega c$, the functions $f_1(f_1^2 + f_2^2)^{-1}$ and $f_2/[a_0(f_1^2 + f_2^2)]$ should be approximately independent of a_0 for this assumption to be valid. (Richart et al.[65] give extensive results for

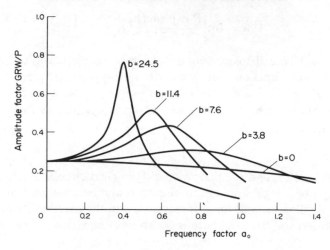

FIG. 6.2. Response curves for the vertical translational mode for a mass on an elastic half-space; $v = 0$ (ref. 4).

these equivalent parameters for all the modes of vibration of a rigid circular foundation on an elastic half-space.) The effect of this assumption on the assessment of dynamic interaction will be considered later.

The disc of Fig. 6.1 has four modes of vibration, vertical translation (considered above), torsion about the Z-axis, horizontal translation and rocking about a diameter; approximate analysis has treated the last two modes as uncoupled for a thin disc[4], but more accurate solutions show that they are coupled, i.e. a horizontal force produces a small rocking motion of the disc in addition to horizontal translation[78]. Complex dynamic stiffnesses, similar to equation (6.2), can be derived for all these modes and assembled in matrix form as

$$\mathbf{p}_F = [\mathbf{K}_G + i\mathbf{C}_G]\mathbf{x}_F \qquad (6.6)$$

where $\mathbf{p}_F$ is the vector of forces and moments on the ground due to the foundation and $\mathbf{x}_F$ is the corresponding vector of displacements of the foundation. Although the above outline has considered a foundation with a circular base on an elastic half-space, an equation similar to (6.6) can be obtained in principle for foundations with rectangular and other bases, on media whose properties vary continuously or discontinuously with depth and for foundations embedded in the medium. The matrices $\mathbf{K}_G$ and $\mathbf{C}_G$ can be obtained from elasticity theory, using various approximations, or by using the finite element method to model the medium or from experimental tests.

We consider now an elastic structure mounted on a rigid foundation, which is supported by soil; we require the structural response when the free field acceleration at the ground surface is prescribed. Let $\mathbf{x}_s$ be the vector of displacements of the constrained structure, i.e. the structure with the foundation clamped. For example, for the shear building of Fig. 2.5 $\mathbf{x}_s$ consists of the horizontal displacements x_1, x_2, ..., x_n of the masses m_1, m_2, ..., m_n, respectively; for a finite element idealization of a complex structure such as an arch dam $\mathbf{x}_s$ consists of the displacements, linear and angular, associated with all the unconstrained degrees of freedom. Let $\mathbf{M}_s$, $\mathbf{C}_s$ and $\mathbf{K}_s$ be the mass, damping and stiffness matrices of the constrained structure; each matrix is symmetric. Vector $\mathbf{x}_F$ lists the displacements of the foundation; vector $\mathbf{x}$ lists the displacements of the structure–foundation system, i.e.

$$\mathbf{x} = \begin{bmatrix} \mathbf{x}_s \\ \hline \mathbf{x}_F \end{bmatrix}. \tag{6.7}$$

(The partitioning to be introduced here is identical to that of Section 2.6, but note that there are changes in the definitions of subscripts.) Using matrices with subscripts F for the foundation and with no subscripts for the complete system, the equation

$$\mathbf{M}\ddot{\mathbf{x}} + \mathbf{C}\dot{\mathbf{x}} + \mathbf{K}\mathbf{x} = \mathbf{p} \tag{6.8}$$

for the structure–foundation system can be partitioned as

$$\begin{bmatrix} \mathbf{M}_s & \vdots & \mathbf{M}_{sF} \\ \hline \mathbf{M}_{sF}^T & \vdots & \mathbf{M}_F \end{bmatrix} \begin{bmatrix} \ddot{\mathbf{x}}_s \\ \ddot{\mathbf{x}}_F \end{bmatrix} + \begin{bmatrix} \mathbf{C}_s & \vdots & \mathbf{C}_{sF} \\ \hline \mathbf{C}_{sF}^T & \vdots & \mathbf{C}_F \end{bmatrix} \begin{bmatrix} \dot{\mathbf{x}}_s \\ \dot{\mathbf{x}}_F \end{bmatrix}$$

$$+ \begin{bmatrix} \mathbf{K}_s & \vdots & \mathbf{K}_{sF} \\ \hline \mathbf{K}_{sF}^T & \vdots & \mathbf{K}_F \end{bmatrix} \begin{bmatrix} \mathbf{x}_s \\ \mathbf{x}_F \end{bmatrix} = \begin{bmatrix} \mathbf{0} \\ \hline -\mathbf{p}_F \end{bmatrix} \tag{6.9}$$

assuming that there are no applied forces on the structure and excitation is due to the forces $-\mathbf{p}_F$ on the foundation due to the ground. If the free field displacement of the ground is represented by the vector $\mathbf{y}$, which is of the same order as the displacement vector for the foundation $\mathbf{x}_F$, the interaction (or dynamic magnification) vector $\mathbf{x}_F - \mathbf{y}$ is related to the interaction forces $\mathbf{p}_F$ by modifying equation (6.6) to give[63]

$$\mathbf{p}_F = [\mathbf{K}_G + i\mathbf{C}_G][\mathbf{x}_F - \mathbf{y}] \tag{6.10}$$

or

$$\mathbf{p}_F = \mathbf{K}_G[\mathbf{x}_F - \mathbf{y}] + \mathbf{C}_G[\dot{\mathbf{x}}_F - \dot{\mathbf{y}}]/\omega. \tag{6.11}$$

Substituting from equation (6.11) in equation (6.9) and rearranging terms,

$$\mathbf{M}\ddot{\mathbf{x}} + \mathbf{C}'\dot{\mathbf{x}} + \mathbf{K}'\mathbf{x} = \mathbf{p}' \tag{6.12}$$

where

$$\mathbf{C}' = \left[\begin{array}{c|c} \mathbf{C}_s & \mathbf{C}_{sF} \\ \hline \mathbf{C}_{sF}^T & \mathbf{C}_F + \mathbf{C}_G/\omega \end{array}\right], \quad \mathbf{K}' = \left[\begin{array}{c|c} \mathbf{K}_s & \mathbf{K}_{sF} \\ \hline \mathbf{K}_{sF}^T & \mathbf{K}_F + \mathbf{K}_G \end{array}\right],$$

$$\mathbf{p}' = \left[\begin{array}{c} 0 \\ \hline \mathbf{K}_G\mathbf{y} + \mathbf{C}_g\dot{\mathbf{y}}/\omega \end{array}\right].$$

The interaction equations, (6.10) and (6.11), are based on steady-state solutions for excitation frequency ω. Thus equation (6.12) appears to be restricted to harmonic excitation only. However, if the elements of the matrices $\mathbf{K}_G$ and $\mathbf{C}_G/\omega$ can be treated as independent of frequency, as discussed above and demonstrated in the literature for some cases, the square matrices in equation (6.12) are constants, the vectors $\mathbf{y}$ and $\dot{\mathbf{y}}$ in $\mathbf{p}'$ may be any functions of time and equation (6.12) is of the same form as equation (2.40). Alternatively, if these elements are frequency dependent, the Fourier transform of equation (6.12) is taken, using equation (A1.6) in the form

$$\mathbf{X}(\omega) = \int_{-\infty}^{\infty} \mathbf{x}(t) \exp\left(-i\omega t\right) dt \tag{6.13}$$

with a similar definition for $\mathbf{P}'(\omega)$ in terms of $\mathbf{p}'(t)$. (It is noted that to conform with the convention for Fourier transforms (Appendix 1) capital letters are used for $\mathbf{X}(\omega)$ and $\mathbf{P}'(\omega)$, but they are nevertheless column vectors.) This gives

$$\mathbf{M}\ddot{\mathbf{X}}(\omega) + \mathbf{C}'(\omega)\dot{\mathbf{X}}(\omega) + \mathbf{K}'(\omega)\mathbf{X}(\omega) = \mathbf{P}'(\omega).$$ (6.14)

Equation (6.14) is solved for $\mathbf{X}(\omega)$ and the inverse transform used to determine $\mathbf{x}(t)$, i.e.

$$\mathbf{x}(t) = \frac{1}{2\pi}\int_{-\infty}^{\infty}\mathbf{X}(\omega)\exp(i\omega t)\,d\omega.$$ (6.15)

Some comments on methods of determining the response from equation (6.12) follow. If the elements of $\mathbf{K}_G$ and $\mathbf{C}_G/\omega$ are treated as constants, (a) the numerical integration methods of Section 2.8 can be used; (b) the normal mode method of Section 2.6 is unsuitable, as the damping restriction ($\mathbf{C}' = \lambda_k\mathbf{K}' + \lambda_m\mathbf{M}$ in this case from equation (2.86)) is unlikely to be satisfied; (c) the complex eigenvalue method of Section 2.7 can be used, but only in conjunction with Fourier transforms (also for the complex stiffness matrix of equation (2.104) to be constant, it is necessary to assume that $\mathbf{C}_G$, rather than $\mathbf{C}_G/\omega$, is independent of frequency and this is not supported by the available evidence); and (d) the complex eigenvalue method, described by Hurty and Rubinstein[40] and applied to interaction problems by Jennings and Bielak[43], can be used.

If $\mathbf{K}_G$ and $\mathbf{C}_G/\omega$ are frequency dependent, $\mathbf{X}(\omega)$ can be determined from equation (6.14) by the frequency response method of Section 2.7. In practice, the response vector $\mathbf{x}(t)$ is obtained from $\mathbf{X}(\omega)$ using the fast Fourier transform (FFT) algorithm, which as outlined in Appendix 1 depends upon the discrete Fourier transform and requires evaluation of $\mathbf{X}(\omega)$ from equation (6.14) for a large number of values of ω in the chosen range. This procedure requires considerable computation for a complex structure for which the order of the matrices in equation (6.14) is high; the subsequent inversion process to obtain $\mathbf{x}(t)$ from $\mathbf{X}(\omega)$ is highly efficient. For earthquakes and other practical excitations the response is usually confined to the first few modes. Methods of economizing in computation by taking advantage of the low-frequency nature of this response have been developed[21, 43].

The vector $\mathbf{x}_s$ in equation (6.7) consists of the absolute displacements of the masses in a multi-degree-of-freedom system, e.g. Fig. 2.5, or the absolute displacements in the coordinate directions of the nodes in a finite element idealization. However, the equations are frequently formulated in terms of relative displacements; this is convenient because the excitation vector, $\mathbf{p}'$ in equation (6.12), can then be expressed in terms of the prescribed free field acceleration vector $\ddot{\mathbf{y}}$, instead of in terms of $\mathbf{y}$ and $\dot{\mathbf{y}}$. Reformulation for the general problem, when the vectors $\mathbf{x}$ and $\mathbf{y}$ may contain components in the three coordinate directions, will not be attempted. Vaish and Chopra[77] give the analysis when $\ddot{\mathbf{y}}$ contains components in two perpendicular directions. Here the shear building of Fig. 2.5 will be considered when the free field acceleration has a single component.

Figure 6.3 shows the n-storey shear building in its deformed position. The free field displacement of the surface of the ground in the X-direction is y_0; due to dynamic interaction the foundation has a displacement x_0 and a rotation ϕ. With the conventional assumptions for a shear building (Section 2.4) the mass m_j $(j = 1, 2, \ldots, n)$ has a displacement x_j and a rotation ϕ. If h_j is the height of mass m_j above the foundation, the relative displacement z_j is defined by

$$x_j = x_0 + h_j\phi + z_j. \tag{6.16}$$

(ϕ is assumed to be small.) Thus the deformation in the flexible mem-

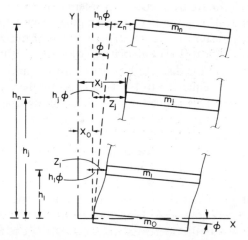

Fig. 6.3. Shear building with n storeys in its deformed position.

bers, which connect masses m_j and m_{j-1} and have a combined stiffness of k_j, is $(z_j - z_{j-1})$. The strain energy of the structure

$$\mathfrak{S} = \tfrac{1}{2} \sum_{j=1}^{n} k_j (z_j - z_{j-1})^2 \tag{6.17}$$

$$= \tfrac{1}{2} \sum_{j=1}^{n} k_j [x_j - x_{j-1} - (h_j - h_{j-1})\phi]^2 \tag{6.18}$$

where $z_0 = 0$ and $h_0 = 0$. Similarly, the dissipation function

$$\mathfrak{F} = \tfrac{1}{2} \sum_{j=1}^{n} c_j (\dot{z}_j - \dot{z}_{j-1})^2 \tag{6.19}$$

$$= \tfrac{1}{2} \sum_{j=1}^{n} c_j [\dot{x}_j - \dot{x}_{j-1} - (h_j - h_{j-1})\dot{\phi}]^2. \tag{6.20}$$

The kinetic energy

$$\mathfrak{T} = \tfrac{1}{2} \sum_{j=0}^{n} m_j \dot{x}_j^2 + \tfrac{1}{2} \sum_{j=0}^{n} I_j \dot{\phi}^2 \tag{6.21}$$

$$= \tfrac{1}{2} \sum_{j=1}^{n} m_j (\dot{x}_0 + h_j \dot{\phi} + \dot{z}_j)^2 + \tfrac{1}{2} m_0 \dot{x}_0^2 + \tfrac{1}{2} \sum_{j=0}^{n} I_j \dot{\phi}^2 \tag{6.22}$$

where I_j is the moment of inertia of mass m_j about an axis through its centre of mass and perpendicular to the plane of Fig. 6.3.

Defining

$$\mathbf{x} = \begin{bmatrix} \mathbf{x}_s \\ \cdots \\ \mathbf{x}_F \end{bmatrix} \quad \text{with} \quad \mathbf{x}_s^I = [z_1, z_2, z_3, \ldots, z_n] \tag{6.23}$$

and $\mathbf{x}_F^T = [\zeta_0, \phi]$, where the dynamic interaction $\zeta_0 = x_0 - y_0$, and applying the Lagrange equation (A4.17) with respect to each term in $\mathbf{x}$ [using equations (6.17), (6.19) and (6.22)],

$$m_j(\ddot{\zeta}_0 + \ddot{y}_0 + h_j\ddot{\phi} + \ddot{z}_j) + c_j(\dot{z}_j - \dot{z}_{j-1}) - c_{j+1}(\dot{z}_{j+1} - \dot{z}_j)$$
$$+ k_j(z_j - z_{j-1}) - k_{j+1}(z_{j+1} - z_j) = 0, \qquad j = 1, 2, \ldots, n,$$

$$\sum_{j=1}^{n} m_j(\ddot{\zeta}_0 + \ddot{y}_0 + h_j\ddot{\phi} + \ddot{z}_j) + m_0(\ddot{\zeta}_0 + \ddot{y}_0) = -Q_x \tag{6.24}$$

and

$$\sum_{j=0}^{n} I_j \ddot{\phi} + \sum_{j=1}^{n} m_j h_j (\ddot{\zeta}_0 + \ddot{y}_0 + h_j \ddot{\phi} + \ddot{z}_j) = -Q_\phi$$

where Q_x and Q_ϕ are the reaction force and moment respectively *on* the ground due to the foundation. Considering the interaction between the foundation and the ground, equation (6.11) becomes

$$\begin{bmatrix} Q_x \\ Q_\phi \end{bmatrix} = \begin{bmatrix} k_{xx} & k_{x\phi} \\ k_{\phi x} & k_{\phi\phi} \end{bmatrix} \begin{bmatrix} \zeta_0 \\ \phi \end{bmatrix} + \begin{bmatrix} c_{xx} & c_{x\phi} \\ c_{\phi x} & c_{\phi\phi} \end{bmatrix} \begin{bmatrix} \dot{\zeta}_0 \\ \dot{\phi} \end{bmatrix} \qquad (6.25)$$

where k_{xx} and $k_{\phi\phi}$ are the real parts of the complex stiffnesses in the horizontal translation and rocking modes respectively, $k_{x\phi}$ ($=k_{\phi x}$) allows for coupling between the horizontal translation and rocking modes, c_{xx}, $c_{\phi\phi}$ and $c_{x\phi}$ ($=c_{\phi x}$) are elements of the matrix $\mathbf{C}_G/\omega$ and are similarly defined. Either k_{xx}, $k_{x\phi}$, $k_{\phi\phi}$, c_{xx}, etc., are treated as constants in subsequent analysis or equation (6.25) applies to a particular excitation frequency ω and the Fourier transform of equations (6.24) has to be taken before proceeding with the analysis. Using equation (6.25) to eliminate Q_x and Q_ϕ from equations (6.24) and rearranging terms, the resulting matrix equation is formulated in terms of the relative displacement vector $\mathbf{x}$, defined by equation (6.23), and the elements of the excitation vector $\mathbf{p}'$ are in terms of the free field acceleration $\ddot{y}_0$, as

$$\mathbf{p}'^T = - \left[m_1 \ddot{y}_0, m_2 \ddot{y}_0, \ldots, m_n \ddot{y}_0, \sum_{j=0}^{n} m_j \ddot{y}_0, \sum_{j=1}^{n} m_j h_j \ddot{y}_0 \right]. \qquad (6.26)$$

Alternatively, if the Lagrange equation is applied with respect to a vector $\mathbf{x}$, consisting of the absolute displacements $x_1, x_2, \ldots, x_n, x_0$ and ϕ, to equations (6.18), (6.20) and (6.21), a matrix equation in terms of the absolute displacements is eventually obtained and the excitation vector is defined in terms of y_0 and $\dot{y}_0$. Numerical results for the response of multi-storey structures to harmonic and earthquake inputs are given by Jennings and Bielak[43]. For harmonic excitation the interaction gives values of $|x_0/y_0|$ and $|R\phi/y_0|$ up to 4 and 10, respectively, for particular values of the system parameters.

The above method of treating the structure and the ground

separately, i.e. the sub-structure approach, and matching their dynamic behaviour through the ground–foundation interaction equations (6.11) or (6.25) has the advantage that different techniques can be used for the analysis of the structure and the ground. For example, a finite element model of a complex structure can be used in conjunction with an analytical solution for the ground. Even if finite element models are generated for both structure and ground, the sub-structure approach reduces the size of matrices to be handled. A three-dimensional finite element model of the ground for a problem that is not symmetrical about the vertical axis through the centroid of the contact area may lead to large matrices. The model must have finite boundaries and still allow radiation damping to occur, i.e. reflected waves from these boundaries must be prevented from returning to the source. Lysmer and Kuhlemeyer[55] have given a method of overcoming this difficulty. Weaver *et al.*[85] have given a general analysis for the response of multi-storey frame structures to earthquakes, idealizing the soil by a finite element mesh of three-dimensional rectangular prisms, using the non-reflecting boundaries of Lysmer and Kuhlemeyer, allowing for embedment, pile foundations, etc., and subjecting the bedrock underlying the soil to an earthquake accelerogram. They obtain the response of a ten-storey building to an earthquake with and without a soil layer between the bedrock and the foundation and show that the maximum shear force at each storey is increased appreciably when allowance is made for soil–structure interaction.

6.2. Fluid–Structure Dynamic Interaction

We consider a long dam, or similar structure, with a vertical face in contact with a reservoir of water of depth H (Fig. 6.4). The base of the dam and reservoir is subjected to a horizontal acceleration in the direction OX, e.g. due to an earthquake. If the dam is sufficiently long, we have a plane strain problem and all displacements, stresses, etc., are independent of the Z-coordinate. It is assumed that the water stretches to infinity in the X-direction and surface waves are neglected; these two assumptions have been discussed and shown to be reasonable[18]. (The length of the reservoir should be greater than $3H$ for the former assumption to be valid.)

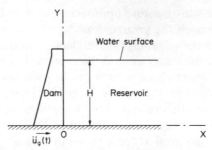

FIG. 6.4. Cross-section of dam and reservoir.

We are interested in the dynamic displacements and stresses caused by the imposed base acceleration; these are superposed on their static counterparts, which are caused by the static pressure distribution of the water, gravity forces, etc., and will not be considered. The dam is subjected to the base acceleration together with the dynamic pressures on its vertical face, resulting from this acceleration. The classical method of solution was to determine the dynamic pressures associated with a rigid-body acceleration of the dam, thus neglecting interaction effects, and obtain the corresponding response. The analysis, without interaction effects, will be outlined using concepts and equations from earlier sections. The cross-section of the dam is represented by a finite element model for in-plane deformations, using isoparametric elements. The isoparametric elements of Section 5.4, which were applied to the in-plane vibrations of thin plates, are based on plane stress theory, i.e. all stress components in the Z-direction are zero. Here we have a plane strain problem, in which all strain components in the Z-direction are zero. The only modifications required to use the analysis of Section 5.4 for plane strain problems are to consider unit length of the dam by putting $h = 1$ and to redefine the matrix $\mathbf{D}$ of equation (5.54) as

$$\mathbf{D} = \frac{E}{(1 + v)(1 - 2v)} \begin{bmatrix} 1 - v & v & 0 \\ v & 1 - v & 0 \\ 0 & 0 & \dfrac{1 - 2v}{2} \end{bmatrix}.$$

As support excitation is prescribed at all nodes along the base of the dam, equation (2.84) is obtained with the N additional displacements in the vector $\mathbf{x}'$ [equation (2.83)] corresponding to the displacements in the X-direction of each node on the base; the force vector $\mathbf{p}'$ is given by

$$\mathbf{p}'^T = [\mathbf{p}^T \mathbin{\vdots} p_{n+1}, p_{n+2}, \ldots, p_{n+N}].$$

The vector $\mathbf{p}$ has non-zero components at degrees of freedom associated with nodes for elements bounding the vertical face of the dam; these components are derived from the consistent force vector for each element along the vertical face and these vectors can be obtained from the dynamic pressure distribution, associated with a rigid-body acceleration of the dam. This pressure distribution will be derived before considering the effects of dynamic interaction.

The governing equation for irrotational, compressible motion of a fluid in the plane OXY is[22]:

$$\frac{\partial^2 \phi}{\partial x^2} + \frac{\partial^2 \phi}{\partial y^2} = \frac{\rho}{K} \frac{\partial^2 \phi}{\partial t^2} \tag{6.27}$$

where the velocity potential $\phi(x, y, t)$ is related to the displacements u and v in the X- and Y-directions by

$$\frac{\partial u}{\partial t} = -\frac{\partial \phi}{\partial x}, \qquad \frac{\partial v}{\partial t} = -\frac{\partial \phi}{\partial y}. \tag{6.28}$$

K is the bulk modulus of the fluid. The dynamic pressure

$$p = \rho \frac{\partial \phi}{\partial t} \tag{6.29}$$

where ρ is the density of the fluid; the wave velocity $C = (K/\rho)^{1/2}$. The boundary conditions are

$$p(x, H, t) = 0, \qquad v(x, 0, t) = 0. \tag{6.30}$$

$$\phi(x, y, t) \to 0 \quad \text{as} \quad x \to \infty \tag{6.31}$$

and

$$\frac{\partial^2 u}{\partial t^2}(0, y, t) = \ddot{u}_g(t) \tag{6.32}$$

where $\ddot{u}_g(t)$ is the imposed acceleration of the base of the dam and the dynamic fluid pressures are calculated assuming that the vertical wall of the dam has the rigid body acceleration $\ddot{u}_g(t)$. Putting $\ddot{u}_g(t) = \ddot{U}_g \exp(i\omega t)$, the governing equation (6.27) and the boundary conditions (6.30) to (6.32) are satisfied, using equations (6.28) and (6.29), if

$$\phi(x, y, t) = \frac{4\ddot{U}_g \exp(i\omega t)}{i\omega\pi} \times$$

$$\sum_j \frac{(-1)^{j-1} \exp\left[-x(\beta_j^2 - \omega^2/C^2)^{1/2}\right] \cos \beta_j y}{(2j-1)(\beta_j^2 - \omega^2/C^2)^{1/2}} \quad (6.33)$$

where $\beta_j = (2j-1)\pi/2H$. The pressure on the dam face is

$$p(0, y, t) = \frac{4\ddot{U}_g\rho \exp(i\omega t)}{\pi} \sum_j \frac{(-1)^{j-1} \cos \beta_j y}{(2j-1)(\beta_j^2 - \omega^2/C^2)^{1/2}}. \quad (6.34)$$

Using equation (6.34) for the pressure, the consistent load vectors for the elements with OY as a boundary can be formulated [following the method of equation (5.81)]. Assembling the element matrices, we obtain the matrix equation (6.8), where in equation (6.7) the vector x_s lists the displacements of the constrained structure and x_F the displacement components of the base of the dam, where motion is imposed, i.e. x_F lists the displacements in the X-direction of nodes on the base of the dam. Equation (6.8) can be written in the partitioned form (6.9), except that

$$p = \left[\begin{array}{c} p_s \\ \hline -p_F \end{array}\right]$$

where the non-zero components of p_s are the nodal forces on the dam due to the fluid pressure and $-p_F$ lists the reaction forces on the foundation due to the ground. The nodal and reaction forces are for unit length of the dam and foundation respectively in the Z-direction. The equations above the partition in equation (6.9) can be written

$$M_s \ddot{x}_s + C_s \dot{x}_s + K_s x_s = -M_{sF} \ddot{x}_F - C_{sF} \dot{x}_F - K_{sF} x_F + p_s \quad (6.35)$$

where M_s, C_s and K_s are the mass, damping and stiffness matrices for unit length of the constrained structure. As $\ddot{x}_F = 1\ddot{U}_g \exp(i\omega t)$, where 1 is a vector with each component equal to unity, and $\ddot{x}_F =$

$i\omega\dot{\mathbf{x}}_F = -\omega^2\mathbf{x}_F$ and $\ddot{\mathbf{x}}_s = i\omega\dot{\mathbf{x}}_s = -\omega^2\mathbf{x}_s$, equation (6.35) can be solved by the frequency response method of Section 2.7 or in terms of the normal modes of the constrained structure to give $\mathbf{x}_s$. From the form of equation (6.34) the vector $\mathbf{p}_s$ is a complicated function of frequency. The right-hand side of equation (6.35) can be written $\mathbf{a}\exp(i\omega t)$; if $\beta_1 > \omega/C$, i.e. the fundamental frequency of the reservoir is greater than the excitation frequency, the components of $\mathbf{a}$ are real; otherwise they will be complex. For a general excitation $\ddot{u}_g(t)$ applied to the base, as in an earthquake problem, Fourier transforms are used to determine the response. Using equation (A1.6),

$$\ddot{U}_g(\omega) = \int_{-\infty}^{\infty} \ddot{u}_g(t) \exp(-i\omega t)\, dt. \tag{6.36}$$

Replacing $\ddot{U}_g$ by $\ddot{U}_g(\omega)$ in equation (6.35) and solving yields $\mathbf{X}_s(\omega)$ and from the inverse transform

$$\mathbf{x}_s(t) = \frac{1}{2\pi} \int_{-\infty}^{\infty} \mathbf{X}_s(\omega) \exp(i\omega t)\, d\omega. \tag{6.37}$$

When interaction effects are included, boundary condition (6.32) is rejected. A solution for $\phi(x, y, t)$, satisfying the governing equation (6.27) and boundary conditions (6.30) and (6.31), is

$$\phi(x, y, t) = \frac{\exp(i\omega t)}{i\omega} \sum_j \frac{a_j \exp\left[-x(\beta_j^2 - \omega^2/C^2)^{1/2}\right] \cos\beta_j y}{(\beta_j^2 - \omega^2/C^2)^{1/2}} \tag{6.38}$$

where excitation of frequency ω has been assumed. If $y_1, y_2, \ldots, y_s$ are the heights of the nodes of the finite element idealization on the vertical face of the dam and $u_1, u_2, \ldots, u_s$ are the displacements of the dam in the X-direction at these nodes, the continuity conditions for this face are

$$\ddot{u}_q = -\frac{\partial^2\phi(0, y_q, t)}{\partial x\, \partial t}$$

$$= \exp(i\omega t) \sum_{j=1}^{s} a_j \cos\beta_j y_q, \qquad q = 1, 2, \ldots, s. \tag{6.39}$$

Also the dynamic fluid pressure at node q

$$p_q(t) = \rho \exp(i\omega t) \sum_{j=1}^{s} \frac{a_j \cos\beta_j y_q}{(\beta_j^2 - \omega^2/C^2)^{1/2}}, \qquad q = 1, 2, \ldots, s. \tag{6.40}$$

Equation (6.39) can be written in matrix form as

$$\ddot{\mathbf{x}}_D = -\omega^2 \mathbf{x}_D = \mathbf{A}\mathbf{x}_R \tag{6.41}$$

where $\ddot{\mathbf{x}}_D$ is a vector of the s nodal accelerations $\ddot{u}_q$ and $\mathbf{x}_R$ a vector of the s generalized coordinates for the reservoir, a_j. From the pressures of equation (6.40) the consistent force vectors for elements with the vertical face as a boundary are formulated as

$$\mathbf{p}_D = -\mathbf{B}\mathbf{x}_R \tag{6.42}$$

where $\mathbf{p}_D$ is the vector of nodal forces on unit length of the dam due to the water. Due to the method of forming consistent force vectors, $\mathbf{p}_D$ is of order $s' \times 1$ where $s' > s$. From equations (6.41) and (6.42)

$$\mathbf{p}_D = \omega^2 \mathbf{B}\mathbf{A}^{-1}\mathbf{x}_D . \tag{6.43}$$

Vectors $\mathbf{x}_D$ and $\mathbf{p}_D$ form parts of vectors $\mathbf{x}_s$ and $\mathbf{p}_s$, respectively, in equation (6.35); substituting for $\mathbf{p}_D$ from equation (6.43), rearranging equation (6.35) and putting $\ddot{\mathbf{x}}_F = \mathbf{1}\ddot{U}_g \exp(i\omega t)$,

$$[\mathbf{K}_s - \omega^2 \mathbf{M}'_s + i\omega \mathbf{C}_s]\mathbf{x}_s = \left[\frac{1}{\omega^2}\mathbf{K}_{sF} - \mathbf{M}_{sF} + \frac{i}{\omega}\mathbf{C}_{sF}\right]\mathbf{1}\ddot{U}_g \exp(i\omega t) \tag{6.44}$$

where $\mathbf{K}_s$, $\mathbf{M}_s$ and $\mathbf{C}_s$ are the stiffness, mass and damping matrices of the constrained structure, $\mathbf{K}_{sF}$, $\mathbf{M}_{sF}$ and $\mathbf{C}_{sF}$ are the stiffness, mass and damping matrices for coupled structure–foundation motion and the modified mass matrix

$$\mathbf{M}'_s = \mathbf{M}_s + \begin{bmatrix} \mathbf{B}\mathbf{A}^{-1} & \vdots & \mathbf{0} \\ \hdashline \mathbf{0} & \vdots & \mathbf{0} \end{bmatrix}. \tag{6.45}$$

All matrices relate to unit length of the structure.

The rectangular matrix $\mathbf{B}\mathbf{A}^{-1}$ is of order $s' \times s$ and its components are frequency dependent [compare equations (6.40) and (6.42)]. The response $\mathbf{x}_s$ can be found from equation (6.44) by the frequency response method of Section 2.7. The response to a general excitation $\ddot{u}_g(t)$ can be obtained using Fourier transforms, as outlined above. The standard substitution $\mathbf{x}_s = \mathbf{Z}\mathbf{q}$ does not give a set of uncoupled equations, because of the form of $\mathbf{M}'_s$. However, Chopra[19] uses a normal mode approach, which is restricted to the fundamental mode as the major response of dams to earthquakes is associated with that mode; thus the analysis is simplified.

It is seen that allowing for dynamic interaction complicates the analysis as it leads to a frequency-dependent modified mass matrix, but whether or not dynamic interaction is included the response to a general acceleration of the base $\ddot{u}_g(t)$ requires the use of Fourier transforms. A simplification occurs if the water is assumed to be incompressible. Then the bulk modulus $K \to \infty$ and the wave velocity $C \to \infty$. The terms within the summation in the expressions for the pressure on the dam face [equations (6.34) and (6.40)] are independent of frequency. Then if interaction effects are included, the rectangular matrix $\mathbf{BA}^{-1}$ of equation (6.45) is independent of frequency; if interaction effects are neglected, the time-dependence of the vector $\mathbf{p}_s$ in equation (6.35) is $\ddot{u}_g(t)$, equation (6.35) is of the same form as equation (2.40) and can be solved directly by the methods of Sections 2.6 and 2.8 to give $\mathbf{x}_s(t)$. For a dam–reservoir system subjected to a specific earthquake accelerogram Chopra[19, 20] showed that significant errors in the maximum dynamic response can be caused by neglecting dynamic interaction or by assuming water to be incompressible; these errors depend upon the fundamental periods of the dam and reservoir, T_{D1} and T_{R1}, respectively, and are small for $T_{D1}/T_{R1} \geq 1\cdot4$.

In the method outlined the dam and the fluid are treated separately as sub-structures and appropriate continuity conditions at the interface imposed to obtain the response of the complete system. Due to the assumptions it was possible to treat the problem as two-dimensional and use a finite element model of the dam together with an analytical solution of the equation for the fluid (6.27). The sub-structure approach, which was used also in Section 6.1, can be applied to more general problems. If the dam, or another structure in contact with a fluid medium, has a cross-section which varies along its length or the structure–fluid interface is a curved surface (instead of the vertical plane assumed in Fig. 6.4), three-dimensional finite element models of the dam and of the fluid can be developed as sub-structures and appropriate continuity conditions at the interface applied to determine the response. This will require less computer store than assembling a finite element model of the complete system.

The response of towers, which are submerged or partially submerged in water, when accelerations are imposed at their bases, is another example of possible fluid–structure dynamic interaction. If the geometry of the tower is symmetric about its vertical axis, but its cross-section varies with height, finite element models of the tower and the surrounding fluid can be developed, using axisymmetric elements and treating the tower and fluid as sub-structures. For a cylindrical tower an analytical solution for the dynamic pressure distribution in the fluid can be derived[51, 52].

In this section it has been assumed that the acceleration of the base of the dam or other structure is equal to the free field acceleration of the ground. Combining the methods of this and the preceding section, simultaneous interaction between the ground and the structure and between the structure and the fluid can be investigated.

6.3. Wind-induced Vibrations of Structures

When a structure is exposed to a steady wind of velocity V_0 relative to the ground, the wind force on the structure is a function of the relative velocity between the air and the structure. Thus the excitation force depends upon the motion of the structure and we have an interaction effect. There are several types of practical problem associated with wind-induced vibrations of structures and some will be discussed briefly. An analytical treatment requires knowledge of aerodynamics, as well as structural vibrations, and is outside the scope of this book. Scanlan and Wardlaw[68] have given a survey of the subject and include an extensive bibliography. In general, the excitation mechanisms cause self-excited vibrations of the structure; thus first we consider this type of vibration for a simple structure.

If the mass BC of the single-storey shear building of Fig. 1.2 is subjected to a wind force in the X-direction, the applied force on the mass

$$P(t) = f(V) \tag{6.46}$$

where the relative velocity

$$V = V_0 - \dot{x} \tag{6.47}$$

and V_0 is the constant wind velocity. If the function $f(V)$ is expanded as a power series, i.e.

$$f(V) = a_0 + a_1 V + a_2 V^2 + \cdots \qquad (6.48)$$

and we consider initially only the first two terms, then using equations (6.47) and (6.48) in (6.46)

$$P(t) = a + b\dot{x} \qquad (6.49)$$

where a and b are constants, dependent upon V_0 and a_j. Thus the equation of motion of the mass m is

$$m\ddot{x} + c\dot{x} + kx = a + b\dot{x},$$

i.e.

$$m\ddot{x} + (c - b)\dot{x} + kx = a. \qquad (6.50)$$

The complete solution of equation (6.50) is

$$x = \frac{a}{k} + A \exp \left[\frac{(b - c)t}{2m} \right] \sin (\omega' t + \alpha) \qquad (6.51)$$

where the constants A and α depend upon the initial conditions. The frequency $\omega' \simeq \omega_n \left[= (k/m)^{1/2} \right]$, provided that

$$\frac{(b - c)^2}{4km} \ll 1.$$

If $b > c$, the amplitude of vibration increases exponentially with time, as shown in Fig. 6.5a. This is the theoretical curve for self-excited vibrations. When the amplitude increases, the approximation involved

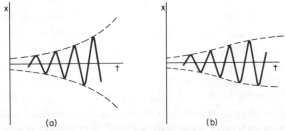

FIG. 6.5. Self-excited vibration. (a) Theoretical. (b) Practical, tending to a limit amplitude.

in considering only two terms in equation (6.48) is not justified and the form of equation (6.49) is modified. In practice, self-excited vibrations often build up to a limit amplitude, as shown in Fig. 6.5b. If $b < c$, the vibrations decay with time and the system is stable; when b is negative, the effective damping of the system is increased.

Galloping. Electrical transmission lines have been observed to vibrate with large amplitudes and low frequency (about 1 Hz, which is of the order of the fundamental natural frequency of the span) when there is a strong transverse wind and ice has formed on the wire. The stability conditions for this phenomenon of galloping are given by Den Hartog[29]. A circular cross-section is stable in a wind. An elongated symmetric cross-section, e.g. an ellipse, is stable if the wind is in the direction of the longer axis, but may be unstable if it is in the direction of the shorter axis. This latter condition applies to the ice-coated trans-mission line. Also towers and pylons of polygonal cross-section, e.g. squares, octagons, etc., may be unstable and develop galloping oscilla-tions for certain wind directions[68]. Wake galloping can occur when the disturbed or wake flow behind one bluff object strikes a second bluff object, e.g. the subconductor oscillations of the parallel conduc-tors used in modern high-voltage transmission lines.

Vortex shedding. When a fluid flows past a rigid body, the flow consists of streamlines at very low speeds, but at higher speeds vortices are formed in the wake. The flow pattern depends on the value of Reynolds number $R_e = Vd/v_0$, where V is the fluid velocity, d is a typical projected dimension, e.g. the diameter for a circular cross-section, of the structure and v_0 is the kinematic viscosity of the fluid. For a wide range of values of R_e the vortices are shed from the body in a regular pattern, alternately clockwise and anti-clockwise from either side, and at a definite frequency f (one cycle corresponds to the shed-ding downstream of two vortices, one in each rotational direction). The Strouhal number $S, = fd/V$, has values from 0·1 to 0·25 depending on the shape of the body and Reynolds number; for circular cross-sections $S \simeq 0·2$. The vortex shedding on alternate sides of a cylinder causes a harmonic force of frequency f to act on the cylinder in the direction transverse to the flow. If the cylinder is not rigid, and thus has natural frequencies, vibration will occur. The phenomenon is complex. As the velocity V increases, vortices are shed and vibration commences only when the Strouhal frequency coincides with the fundamental natural frequency of the structure. However, for some further appreciable velocity range vibration and vortex shedding at this natural frequency persist, although the Strouhal relation is no longer satisfied exactly. Practical examples of vibrations of structures due to vortex shedding include: electrical transmission lines, and other cables, vibrating in higher modes (i.e. at considerably higher frequencies than the galloping vibrations described earlier), chimneys, towers, antennae, bridge decks of bluff cross-section, etc.

Suspension-bridge flutter. When the deck section is not streamlined, the aerodynamic forces may contribute a negative damping term [similar to the term $b\dot{x}$ in equation (6.50)] to the equation governing the torsional mode of vibration. This type of flutter eventually destroyed the Tacoma Narrows suspension bridge in 1940. Streamlining of the deck section gives stability[68].

APPENDIX 1

Fourier Transforms

DEFINITIONS and brief descriptions of some properties of Fourier transforms are given in order to understand their use in Chapters 1, 2 and 6. The reader should consult Brigham[15] for further details of Fourier transforms and the related discrete Fourier transforms and the fast Fourier transform algorithm.

We consider a transient function $f(t)$, which exists only for a limited range of values of t, e.g. the function $f(t)$ shown in Fig. A1.1a. The actual function $f(t)$ is replaced by a periodic function f_T, which is identical with $f(t)$ during the time interval $-T/2 < t < T/2$ and has a period T [Fig. A1.1b]. The period T must be large compared with the duration of $f(t)$. The generated function f_T can be expanded into a complex Fourier series as

$$f_T = \sum_n a_n \exp\left(in\omega_0 t\right) \tag{A1.1}$$

where $\omega_0 T = 2\pi$. Multiplying equation (A1.1) by $\exp\left(-in\omega_0 t\right)$ and integrating from $-T/2$ to $T/2$,

$$a_n = \frac{1}{T} \int_{-T/2}^{T/2} f_T \exp\left(-in\omega_0 t\right) dt. \tag{A1.2}$$

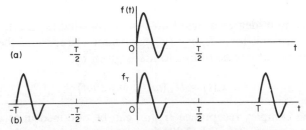

FIG. A1.1. (a) Actual function $f(t)$. (b) Periodic function f_T.

305

To make the generated function a good approximation we let $T \to \infty$, put $n\omega_0 = \omega$ and $\omega_0 = \Delta\omega$. Then as $T = 2\pi/\Delta\omega$,

$$f_T = \sum_n (Ta_n) \exp{(i\omega t)} \frac{\Delta\omega}{2\pi}. \qquad (A1.3)$$

Applying the limiting procedure $T \to \infty$ to equation (A1.2)

$$\lim_{T \to \infty} (Ta_n) = \int_{-\infty}^{\infty} f(t) \exp{(-i\omega t)} \, dt$$

$$= F(\omega) \qquad (A1.4)$$

and also to equation (A1.3), where the infinite series is replaced by an infinite integral,

$$f(t) = \frac{1}{2\pi} \int_{-\infty}^{\infty} F(\omega) \exp{(i\omega t)} \, d\omega. \qquad (A1.5)$$

The function $F(\omega)$ is the *Fourier transform* of $f(t)$ and equations (A1.4) and (A1.5) define a transform pair. The notation of lower-case letters for the time function and capitals for the frequency function is adopted. For example, the Fourier transform of a typical displacement $x_s(t)$ is $X_s(\omega)$ and they are related by

$$x_s(t) = \frac{1}{2\pi} \int_{-\infty}^{\infty} X_s(\omega) \exp{(i\omega t)} \, d\omega \qquad (A1.6)$$

with

$$X_s(\omega) = \int_{-\infty}^{\infty} x_s(t) \exp{(-i\omega t)} \, dt.$$

For the multi-degree-of-freedom systems of Chapter 2, if the excitation is a harmonic force $P_j \exp{(i\omega t)}$ at degree of freedom j, the steady-state response at degree of freedom s is given by

$$x_s(t) = H_{sj}(\omega)P_j \exp{(i\omega t)} \qquad (A1.7)$$

where the complex receptance $H_{sj}(\omega)$ can be obtained in series form, from equations (2.94) and (2.99) for viscous and hysteretic damping respectively, from the normal mode method of analysis, and in closed form from equation (2.103) from the frequency response method. Considering a general transient excitation $f_j(t)$ at degree of freedom j, assuming that it can be expanded as a sum of harmonics as in equation (A1.1), the response

$$x_s(t) = \sum_n a_n \exp(in\omega_0 t)H_{sj}(n\omega_0).$$

Following the limiting procedures used in equations (A1.2) to (A1.5), this leads to

$$x_s(t) = \frac{1}{2\pi} \int_{-\infty}^{\infty} F_j(\omega)H_{sj}(\omega) \exp(i\omega t) \, d\omega \qquad (A1.8)$$

where

$$F_j(\omega) = \int_{-\infty}^{\infty} f_j(t) \exp(-i\omega t) \, dt. \qquad (A1.9)$$

Equation (A1.8) is used when deriving the response to random excitation of (a) single-degree-of-freedom systems in Section 1.9 and (b) multi-degree-of-freedom systems in Section 2.9 with $H_{sj}(\omega)$ expressed as a series.

If $h_{sj}(t - \tau)$ is the response at time t at degree of freedom s when a unit impulse is applied at time τ at degree of freedom j, the response at s for a force $f_j(t)$ applied at j is given by

$$x_s(t) = \int_0^t f_j(\tau)h_{sj}(t - \tau) \, d\tau. \qquad (A1.10)$$

For single-degree-of-freedom systems, where $s = j$, an explicit expression for $h_{sj}(t - \tau)$ is given by the Duhamel integral of equation (1.36); for multi-degree-of-freedom systems a series expression for $h_{sj}(t - \tau)$ can be inferred from equation (2.75). Equation (A1.10) assumes that the excitation $f_j(t)$ commences at $t = 0$; if the excitation exists for $t > -\infty$, the lower limit of the integral is changed to $-\infty$. Also changing the variable from τ to τ', where $\tau' = t - \tau$,

$$x_s(t) = \int_0^{\infty} f_j(t - \tau')h_{sj}(\tau') \, d\tau'. \qquad (A1.11)$$

If the excitation $f_j(t)$ is the unit impulse $\delta_j(t)$ applied at j at time $t = 0$, the response at s is $h_{sj}(t)$. From the definition (A1.9) of $F_j(\omega)$ and noting that $\delta_j(t) = 0$ for $t \neq 0$ and $\delta_j(0) \, \Delta t \rightarrow 1$, $F_j(\omega) = 1$. In general, comparing equations (A1.6) and (A1.8), the Fourier transform of the response

$$X_s(\omega) = H_{sj}(\omega)F_j(\omega). \tag{A1.12}$$

Thus for excitation by the unit impulse $X_s(\omega) = H_{sj}(\omega)$. From equation (A1.8)

$$h_{sj}(t) = \frac{1}{2\pi} \int_{-\infty}^{\infty} H_{sj}(\omega) \exp{(i\omega t)} \, d\omega. \tag{A1.13}$$

Thus the complex frequency response $H_{sj}(\omega)$ is the Fourier transform of the unit impulse response $h_{sj}(t)$; hence

$$H_{sj}(\omega) = \int_{-\infty}^{\infty} h_{sj}(t) \exp{(-i\omega t)} \, dt. \tag{A1.14}$$

Equation (A1.14) is used in Section 1.9 when deriving the relation (1.51) between spectral densities.

The majority of the equations given in this appendix are expressed in terms of infinite integrals, which are unsuitable for rapid computation. Also in practice response records extend over finite times and frequency spectra over finite ranges of frequency. Thus discrete Fourier transforms where quantities are summed at a finite number of sampling times or frequencies are advantageous. The discrete Fourier transform pair, analogous to equations (A1.6), can be inferred from equations (A1.1) and (A1.2) and is

$$x_s(k \, \Delta t) = \sum_{j=0}^{N-1} X_s(j \, \Delta f) \exp{(i2\pi jk/N)},$$

$$X_s(j \, \Delta f) = \frac{1}{N} \sum_{k=0}^{N-1} x_s(k \, \Delta t) \exp{(-i2\pi jk/N)}. \tag{A1.15}$$

Δt and Δf (in Hertz) are the time and frequency intervals at which x_s and X_s, respectively, are sampled. The frequency range $F = N \, \Delta f$, and $T = N \, \Delta t$, where N is the number of samples. In obtaining equations (A1.15) Δt has been equated to $1/F$. When using equations (A1.15), Δt is chosen so that $X_s(f) \simeq 0$ for $|f| > \frac{1}{2}F$ or $|f| > 1/(2 \, \Delta t)$. As

$\Delta f = 1/(N \, \Delta t)$, N is chosen to be large enough to give a sufficiently fine spacing of the frequency estimates.

The fast Fourier transform algorithm is a highly efficient procedure for computing discrete Fourier transforms from equations (A1.15). It takes advantage of the fact that the coefficients can be calculated iteratively. For the large values of N normally employed its use, compared with conventional evaluation of the series, reduces the number of arithmetical operations by factors of over 100.

APPENDIX 2

Some Properties of Matrices

IN THIS appendix the properties of matrices which are used in this book are defined.

A set of linear simultaneous algebraic equations can be written

$$
\begin{aligned}
k_{11}x_1 + k_{12}x_2 + k_{13}x_3 + \cdots + k_{1n}x_n &= b_1, \\
k_{21}x_1 + k_{22}x_2 + k_{23}x_3 + \cdots + k_{2n}x_n &= b_2, \\
&\vdots \\
k_{n1}x_1 + k_{n2}x_2 + k_{n3}x_3 + \cdots + k_{nn}x_n &= b_n
\end{aligned}
\tag{A2.1}
$$

where the coefficients k_{rs} and the constants b_r are known and the values of x_r are to be determined. The coefficients k_{rs} form a *square matrix*

$$
\mathbf{K} = \begin{bmatrix}
k_{11} & k_{12} & k_{13} & \cdots & k_{1n} \\
k_{21} & k_{22} & k_{23} & \cdots & k_{2n} \\
\vdots & \vdots & \vdots & & \vdots \\
k_{n1} & k_{n2} & k_{n3} & \cdots & k_{nn}
\end{bmatrix}.
\tag{A2.2}
$$

The column of constants b_r forms a *column matrix* or *vector* $\mathbf{b}$; the unknowns x_r form a *vector* $\mathbf{x}$. Then equation (A2.1) may be written in matrix form

$$
\mathbf{Kx} = \mathbf{b}.
\tag{A2.3}
$$

The rule for pre-multiplication of a vector by a matrix is given by the left-hand sides of equations (A2.1).

The solution of equations (A2.1) can be written

$$
\begin{aligned}
x_1 &= \kappa_{11}b_1 + \kappa_{12}b_2 + \kappa_{13}b_3 + \cdots + \kappa_{1n}b_n, \\
x_2 &= \kappa_{21}b_1 + \kappa_{22}b_2 + \kappa_{23}b_3 + \cdots + \kappa_{2n}b_n, \\
&\vdots \\
x_n &= \kappa_{n1}b_1 + \kappa_{n2}b_2 + \kappa_{n3}b_3 + \cdots + \kappa_{nn}b_n
\end{aligned}
\tag{A2.4}
$$

or in matrix notation

$$
\mathbf{x} = \mathbf{K}^{-1}\mathbf{b}
\tag{A2.5}
$$

where the matrix K^{-1} has the same form as equation (A2.2) with κ_{rs} replacing k_{rs}. The matrix K^{-1} is called the *inverse* or *reciprocal* of matrix K. The coefficients κ_{rs} depend only on the coefficients k_{rs}, but their evaluation is not trivial. Putting $b_1 = 1, b_2 = b_3 = \cdots = b_n = 0$ in equations (A2.1) and solving the resulting set of equations, $\kappa_{11}, \kappa_{21}, \kappa_{31}, \ldots, \kappa_{n1}$ are obtained as the values of $x_1, x_2, x_3, \ldots, x_n$, respectively, from equations (A2.4). Similarly, putting $b_1 = b_3 = b_4 = \cdots = b_n = 0$, $b_2 = 1$ in equations (A2.1) leads to the values of the second column of K^{-1}. Thus the columns of K^{-1} can be determined by solving successively equations (A2.1) with their right-hand sides replaced by successive columns of the matrix

$$I = \begin{bmatrix} 1 & 0 & 0 & \cdots & 0 \\ 0 & 1 & 0 & \cdots & 0 \\ 0 & 0 & 1 & \cdots & 0 \\ \cdot & \cdot & \cdot & \cdot & \cdot \\ 0 & 0 & 0 & \cdots & 1 \end{bmatrix}. \tag{A2.6}$$

This is the *unit matrix* or *identity matrix* of order n, from the property

$$Ix = x. \tag{A2.7}$$

Pre-multiplying equation (A2.5) by K,

$$Kx = KK^{-1}b$$

$$= b$$

from equation (A2.3). Thus

$$KK^{-1} = I. \tag{A2.8}$$

Before considering further the solution of equation (A2.1) some additional definitions are necessary.

A square matrix K in which only the diagonal elements k_{rr} are nonzero is called a *diagonal* matrix. The matrix 0 in which all elements are zero is called the *null matrix*.

Matrices can be added or subtracted only when of the same order, defined as $m \times n$ where m and n are the numbers of rows and columns, respectively. If a_{rs} is a term in A and b_{rs} in B, then $a_{rs} + b_{rs}$ is the corresponding term in $A + B$. If A is multiplied by a scalar c, the resulting matrix has elements ca_{rs}.

If A is of order $m \times n$ and B is of order $n \times p$, then the product matrix $C = AB$ is of order $m \times p$ with terms given by:

$$c_{rs} = \sum_{j=1}^{n} a_{rj} b_{js}, \qquad (A2.9)$$

i.e.

$$AB = \begin{bmatrix} (a_{11}b_{11} + a_{12}b_{21} + \cdots) & (a_{11}b_{12} + a_{12}b_{22} + \cdots) & \cdots \\ (a_{21}b_{11} + a_{22}b_{21} + \cdots) & (a_{21}b_{12} + a_{22}b_{22} + \cdots) & \cdots \\ \cdot \quad \cdot \quad \cdot \quad \cdot \quad \cdot \quad \cdot \quad \cdot \quad \cdot \quad \cdot \quad \cdot \quad \cdot \quad \cdot \quad \cdot \quad \cdot \quad \cdot \quad \cdot \\ (a_{m1}b_{11} + a_{m2}b_{21} + \cdots) & (a_{m1}b_{12} + a_{m2}b_{22} + \cdots) & \cdots \end{bmatrix}.$$

It will be seen from equation (A2.9) that in general $AB \neq BA$; in the above example the product BA cannot be formed unless $m = p$. For clarity the product AB is termed B *pre-multiplied* by A or A *post-multiplied* by B.

The *transposed matrix* of A, called A^T, is derived from A by interchanging the rows and columns. For example, if

$$A = \begin{bmatrix} a_{11} & a_{12} \\ a_{21} & a_{22} \end{bmatrix}$$

then

$$A^T = \begin{bmatrix} a_{11} & a_{21} \\ a_{12} & a_{22} \end{bmatrix}.$$

If a matrix is *symmetric*, i.e. $a_{rs} = a_{sr}$, then $A^T = A$ and hence $A^T A = A A^T$. Other special cases in which the order of multiplication is immaterial are: $AI = IA$ and $AA^{-1} = A^{-1}A = I$. The transpose of a product is given by

$$(AB)^T = B^T A^T \qquad (A2.10)$$

and the inverse of a product by

$$(AB)^{-1} = B^{-1}A^{-1}. \qquad (A2.11)$$

Associated with a square matrix K is its *determinant*, written det $|\mathbf{K}|$; whereas the matrix is an array of numbers, the determinant is a pure number. If

$$\mathbf{K} = \begin{bmatrix} k_{11} & k_{12} & k_{13} \\ k_{21} & k_{22} & k_{23} \\ k_{31} & k_{32} & k_{33} \end{bmatrix}$$

then

$$
\begin{aligned}
\det |\mathbf{K}| &= \begin{vmatrix} k_{11} & k_{12} & k_{13} \\ k_{21} & k_{22} & k_{23} \\ k_{31} & k_{32} & k_{33} \end{vmatrix} \\
&= k_{11} \begin{vmatrix} k_{22} & k_{23} \\ k_{32} & k_{33} \end{vmatrix} - k_{12} \begin{vmatrix} k_{21} & k_{23} \\ k_{31} & k_{33} \end{vmatrix} + k_{13} \begin{vmatrix} k_{21} & k_{22} \\ k_{31} & k_{32} \end{vmatrix} \\
&= k_{11}(k_{22}k_{33} - k_{23}k_{32}) - k_{12}(k_{21}k_{33} - k_{23}k_{31}) \\
&\quad + k_{13}(k_{21}k_{32} - k_{22}k_{31}).
\end{aligned}
\tag{A2.12}
$$

The *minor* of k_{rs} is the determinant formed by removing the rth row and sth column from the original determinant $\det |\mathbf{K}|$ (i.e. removing the row and column containing the particular element). Thus for the above example the minor of k_{11} is

$$\begin{vmatrix} k_{22} & k_{23} \\ k_{32} & k_{33} \end{vmatrix}$$

and the minor of k_{12} is

$$\begin{vmatrix} k_{21} & k_{23} \\ k_{31} & k_{33} \end{vmatrix}.$$

The *co-factor* or signed minor, K_{rs}, of k_{rs} is equal to $(-1)^{r+s} \times$ (minor of k_{rs}). For the above example,

$$K_{11} = \begin{vmatrix} k_{22} & k_{23} \\ k_{32} & k_{33} \end{vmatrix} \quad \text{and} \quad K_{12} = - \begin{vmatrix} k_{21} & k_{23} \\ k_{31} & k_{33} \end{vmatrix}.$$

Also

$$\det |\mathbf{K}| = k_{11}K_{11} + k_{12}K_{12} + k_{13}K_{13}. \tag{A2.13}$$

Equation (A2.13) illustrates a method of expanding determinants; its extension to matrices of higher order is apparent. If the co-factors are assembled into a matrix and this matrix is transposed, we obtain the *adjoint matrix* of $\mathbf{K}$, adj $[\mathbf{K}]$. If $\mathbf{A}$ is defined as adj $[\mathbf{K}]$, then an element a_{rs} of $\mathbf{A}$ is given by

$$a_{rs} = K_{sr}.\tag{A2.14}$$

Although the formal solution of equation (A2.3) is $\mathbf{x} = \mathbf{K}^{-1}\mathbf{b}$ [equation (A2.5)], an expression for the inverse matrix $\mathbf{K}^{-1}$ is required and can be given in terms of the adjoint matrix of $\mathbf{K}$, $\mathbf{A}$, as

$$\mathbf{K}^{-1} = \frac{\mathbf{A}}{\det |\mathbf{K}|}.\tag{A2.15}$$

Hence

$$\mathbf{x} = \frac{\mathbf{Ab}}{\det |\mathbf{K}|}$$

and a particular element of $\mathbf{x}$

$$\begin{aligned}
x_r &= \sum_s \frac{(a_{rs}b_s)}{\det |\mathbf{K}|} \\
&= \sum_s \frac{(K_{sr}b_s)}{\det |\mathbf{K}|}
\end{aligned}\tag{A2.16}$$

from equation (A2.14).

An alternative form for the solution of equation (A2.3) is obtained by Cramer's rule as

$$x_r = \frac{\det |\mathbf{K}_r|}{\det |\mathbf{K}|}\tag{A2.17}$$

where $\det |\mathbf{K}_r|$ is the determinant formed by replacing the rth column in $\det |\mathbf{K}|$ by the vector $\mathbf{b}$. The equivalence of equations (A2.16) and (A2.17) can be shown, but will be illustrated only for a simple example.

Considering the three equations

$$\begin{aligned}
k_{11}x_1 + k_{12}x_2 + k_{13}x_3 &= b_1, \\
k_{21}x_1 + k_{22}x_2 + k_{23}x_3 &= b_2, \\
k_{31}x_1 + k_{32}x_2 + k_{33}x_3 &= b_3,
\end{aligned}$$

standard algebraic manipulation yields a solution

$$x_1 = [b_1(k_{22}k_{33} - k_{23}k_{32}) - b_2(k_{12}k_{33} - k_{13}k_{32})$$
$$+ b_3(k_{12}k_{23} - k_{13}k_{22})]/\det |\mathbf{K}| \qquad (A2.18)$$

where $\det |\mathbf{K}|$ is given by equation (A2.12). From its definition the determinant

$$\det |\mathbf{K}_1| = \begin{vmatrix} b_1 & k_{12} & k_{13} \\ b_2 & k_{22} & k_{23} \\ b_3 & k_{32} & k_{33} \end{vmatrix}.$$

Expansion of this determinant shows that solution (A2.18) is a particular case of equation (A2.17). Using equation (A2.14) and the definition of co-factors, the adjoint matrix

$$\mathbf{A} = \begin{bmatrix} K_{11} & K_{21} & K_{31} \\ K_{12} & K_{22} & K_{32} \\ K_{13} & K_{23} & K_{33} \end{bmatrix}$$
$$= \begin{bmatrix} (k_{22}k_{33} - k_{23}k_{32}) & -(k_{12}k_{33} - k_{13}k_{32}) & (k_{12}k_{23} - k_{13}k_{22}) \\ -(k_{21}k_{33} - k_{23}k_{31}) & (k_{11}k_{33} - k_{13}k_{31}) & -(k_{11}k_{23} - k_{13}k_{21}) \\ (k_{21}k_{32} - k_{22}k_{31}) & -(k_{11}k_{32} - k_{12}k_{31}) & (k_{11}k_{22} - k_{12}k_{21}) \end{bmatrix}.$$

$$(A2.19)$$

From equation (A2.16)

$$x_1 = (K_{11}b_1 + K_{21}b_2 + K_{31}b_3)/\det |\mathbf{K}|. \qquad (A2.20)$$

Comparison of equations (A2.18), (A2.19) and (A2.20) shows that the two solutions for x_1 are equivalent.

If in equations (A2.1) all the components of the vector $\mathbf{b}$ are zero, non-zero solutions for x_r exist only if $\det |\mathbf{K}| = 0$ [from equation (A2.16) or (A2.17)]. This is used in Section 2.5 when determining the natural frequencies for systems with several degrees of freedom. If the determinant $\det |\mathbf{K}|$ is zero, the matrix $\mathbf{K}$ is *singular*.

It is sometimes convenient to divide matrix equations into sub-matrices using the method of *partitioning*. The set of equations

$$a_{11}x_1 + a_{12}x_2 + a_{13}x_3 + c_{11}y_1 + c_{12}y_2 = b_1,$$
$$a_{21}x_1 + a_{22}x_2 + a_{23}x_3 + c_{21}y_1 + c_{22}y_2 = b_2,$$
$$a_{31}x_1 + a_{32}x_2 + a_{33}x_3 + c_{31}y_1 + c_{32}y_2 = b_3,$$

$$d_{11}x_1 + d_{12}x_2 + d_{13}x_3 + e_{11}y_1 + e_{12}y_2 = f_1,$$
$$d_{21}x_1 + d_{22}x_2 + d_{23}x_3 + e_{21}y_1 + e_{22}y_2 = f_2$$

can be written in partitioned matrix form as

$$\left[\begin{array}{c:c} \mathbf{A} & \mathbf{C} \\ \hdashline \mathbf{D} & \mathbf{E} \end{array}\right] \left[\begin{array}{c} \mathbf{x} \\ \hdashline \mathbf{y} \end{array}\right] = \left[\begin{array}{c} \mathbf{b} \\ \hdashline \mathbf{f} \end{array}\right] \tag{A2.21}$$

where $\mathbf{A}$, $\mathbf{C}$, $\mathbf{D}$ and $\mathbf{E}$ are 3×3, 3×2, 2×3 and 2×2 matrices in terms of a_{rs}, c_{rs}, d_{rs} and e_{rs}, respectively, and $\mathbf{x}$, $\mathbf{y}$, $\mathbf{b}$ and $\mathbf{f}$ are vectors. The partitioned form may be used to eliminate the vector $\mathbf{y}$. Expanding equation (A2.21)

$$\mathbf{Ax} + \mathbf{Cy} = \mathbf{b}$$

and

$$\mathbf{Dx} + \mathbf{Ey} = \mathbf{f}.$$

From the second equation

$$\mathbf{y} = \mathbf{E}^{-1}\mathbf{f} - \mathbf{E}^{-1}\mathbf{Dx}.$$

Substituting for $\mathbf{y}$ in the first equation,

$$[\mathbf{A} - \mathbf{CE}^{-1}\mathbf{D}]\mathbf{x} = \mathbf{b} - \mathbf{CE}^{-1}\mathbf{f}. \tag{A2.22}$$

In this way the original set of five equations is reduced to three.
    The matrix

$$\mathbf{K} = \begin{bmatrix} \mathbf{A} & \mathbf{0} & \mathbf{0} \\ \mathbf{0} & \mathbf{B} & \mathbf{0} \\ \mathbf{0} & \mathbf{0} & \mathbf{C} \end{bmatrix} \tag{A2.23}$$

where $\mathbf{A}$, $\mathbf{B}$ and $\mathbf{C}$ are square, symmetric matrices of specified order and $\mathbf{0}$ is the null matrix, is referred to as a diagonal supermatrix and occurs frequently in assembly processes with the finite element method (Section 5.5).

APPENDIX 3

Orthogonality Conditions for Beams

For a non-uniform beam the governing equation for free vibrations [equations (3.14)] is:

$$\frac{\partial^2}{\partial x^2}\left[-EI\frac{\partial^2 v}{\partial x^2}\right] = \rho A\frac{\partial^2 v}{\partial t^2}.$$

If $\phi_r(x)$ is a normal mode, so that

$$v = \phi_r(x)\sin(\omega_r t + \alpha_r),$$

then the equation governing the mode shape is:

$$\frac{d^2}{dx^2}\left[EI\frac{d^2\phi_r}{dx^2}\right] = \rho A\omega_r^2\phi_r. \tag{A3.1}$$

Multiplying equation (A3.1) by ϕ_s, where ϕ_s is another mode shape, and integrating over the length of the beam,

$$\int_0^l \phi_s\frac{d^2}{dx^2}\left[EI\frac{d^2\phi_r}{dx^2}\right]dx = \omega_r^2\int_0^l \rho A\phi_r\phi_s\,dx. \tag{A3.2}$$

Integrating the left-hand side of equation (A3.2) by parts twice,

$$\omega_r^2\int_0^l \rho A\phi_r\phi_s\,dx = \left|\phi_s\frac{d}{dx}\left[EI\frac{d^2\phi_r}{dx^2}\right]\right|_0^l - \int_0^l \frac{d\phi_s}{dx}\frac{d}{dx}\left[EI\frac{d^2\phi_r}{dx^2}\right]dx$$

$$= \left|\phi_s\frac{d}{dx}\left[EI\frac{d^2\phi_r}{dx^2}\right] - \frac{d\phi_s}{dx}\left[EI\frac{d^2\phi_r}{dx^2}\right]\right|_0^l$$

$$+ \int_0^l EI\frac{d^2\phi_r}{dx^2}\cdot\frac{d^2\phi_s}{dx^2}\,dx. \tag{A3.3}$$

For simple end conditions, i.e. any combination of free, simply supported and clamped ends at $x = 0$ and $x = l$, the terms in the brackets | | are zero and equation (A3.3) reduces to

$$\omega_r^2 \int_0^l \rho A \phi_r \phi_s \, dx = \int_0^l EI \frac{d^2\phi_r}{dx^2} \frac{d^2\phi_s}{dx^2} \, dx. \tag{A3.4}$$

Writing down the equation governing the shape of the sth mode, and multiplying it by ϕ_r and integrating with respect to x over the length of the beam,

$$\int_0^l \phi_r \frac{d^2}{dx^2} \left[EI \frac{d^2\phi_s}{dx^2} \right] \, dx = \omega_s^2 \int_0^l \rho A \phi_s \phi_r \, dx. \tag{A3.5}$$

If the left-hand side of equation (A3.5) is integrated by parts twice, then for any combination of the above end conditions

$$\omega_s^2 \int_0^l \rho A \phi_s \phi_r \, dx = \int_0^l EI \frac{d^2\phi_r}{dx^2} \frac{d^2\phi_s}{dx^2} \, dx. \tag{A3.6}$$

From equations (A3.4) and (A3.6) for $r \neq s$, it follows that

$$\int_0^l \rho A \phi_r \phi_s \, dx = 0 \tag{A3.7}$$

and

$$\int_0^l EI \frac{d^2\phi_r}{dx^2} \frac{d^2\phi_s}{dx^2} \, dx = 0 \tag{A3.8}$$

and from equation (A3.2)

$$\int_0^l \phi_s \frac{d^2}{dx^2} \left[EI \frac{d^2\phi_r}{dx^2} \right] \, dx = 0. \tag{A3.9}$$

Equations (A3.7), (A3.8) and (A3.9) are true for $r \neq s$ for uniform and non-uniform beams for any combination of free, simply supported and clamped ends.

For the general end conditions of Fig. 3.5, where linear and rotational elastic constraints, k and K, and concentrated masses m are introduced, with subscripts L and R denoting the left-hand and right-hand ends of the beam, respectively, substitution of the appropriate boundary conditions (3.27) to (3.30), written in terms of the mode shape ϕ_r, for

$$\left| \frac{d}{dx} \left[EI \frac{d^2\phi_r}{dx^2} \right] \right|_0^l \quad \text{and} \quad \left| EI \frac{d^2\phi_r}{dx^2} \right|_0^l$$

in equation (A3.3) yields

$$
\begin{aligned}
\omega_r^2 \int_0^l \rho A \phi_r \phi_s \, dx &= \phi_s(l)[k_R \phi_r(l) - m_r \omega_r^2 \phi_r(l)] \\
&\quad + \phi_s(0)[k_L \phi_r(0) - m_L \omega_r^2 \phi_r(0)] \\
&\quad + \frac{d\phi_s(l)}{dx} K_R \frac{d\phi_r(l)}{dx} + \frac{d\phi_s(0)}{dx} K_L \frac{d\phi_r(0)}{dx} \\
&\quad + \int_0^l EI \frac{d^2\phi_r}{dx^2} \frac{d^2\phi_s}{dx^2} \, dx.
\end{aligned}
\tag{A3.10}
$$

Writing down the equation governing the shape of the sth mode, multiplying by ϕ_r, integrating with respect to x over the length of the beam and integrating by parts twice, an equation, similar to equation (A3.10) but with the subscripts r and s interchanged, is obtained. Subtracting these two equations and taking $\omega_r \neq \omega_s$, the orthogonality condition for general end conditions, replacing equation (A3.7), is

$$\int_0^l \rho A \phi_r \phi_s \, dx + m_R \phi_s(l)\phi_r(l) + m_L \phi_s(0)\phi_r(0) = 0. \tag{A3.11}$$

From equations (A3.10) and (A3.11) the general form of equation (A3.8) is

$$
\begin{aligned}
\int_0^l EI \frac{d^2\phi_r}{dx^2} \frac{d^2\phi_s}{dx^2} \, dx &+ k_R \phi_r(l)\phi_s(l) + k_L \phi_r(0)\phi_s(0) \\
&+ K_R \frac{d\phi_r(l)}{dx} \frac{d\phi_s(l)}{dx} + K_L \frac{d\phi_r(0)}{dx} \frac{d\phi_s(0)}{dx} = 0. \tag{A3.12}
\end{aligned}
$$

From equations (A3.2) and (A3.11), equation (A3.9) is replaced by

$$\int_0^l \phi_s \frac{d^2}{dx^2} \left[EI \frac{d^2\phi_r}{dx^2} \right] dx + m_R \omega_r^2 \phi_r(l)\phi_s(l) + m_L \omega_r^2 \phi_r(0)\phi_s(0) = 0.$$
$$\tag{A3.13}$$

For the special case of $r = s$,

$$\omega_r^2 \int_0^l \rho A \phi_r^2 \, dx = \int_0^l \phi_r \frac{d^2}{dx^2} \left[EI \frac{d^2 \phi_r}{dx^2} \right] dx \qquad (A3.14)$$

from equation (A3.2); this equation is used to derive the general expression for the dynamic response of beams (Section 3.4).

APPENDIX 4

The Lagrange Equation

THE Lagrange equation has been used to derive matrix equations in conjunction with the Rayleigh–Ritz and finite element methods in Chapters 3 and 5. It could have been used as an alternative method of deriving the equations of motion for the structures considered in this book. However, for these structures derivation of the equations by Lagrange, instead of directly from Newton's laws, shows no advantages, although for systems where the accelerations are complex functions of the coordinates, e.g. pendulum problems when the angles are not small, the Lagrangian derivation is advantageous. In this appendix the Lagrange equation is derived.

The principle of virtual work states that for any system in equilibrium, and for a dynamical system the inertia forces must be included, the work done in any permissible virtual displacement is zero. A permissible virtual displacement is a small hypothetical displacement of the system under the forces acting on the system at time t and this displacement must not violate any constraint conditions. Considering a system of masses, we assume that the ith mass m_i has a virtual displacement $\delta \bar{r}_i$. (The notation is: the symbol δ signifies a virtual quantity; the "bar" denotes a vector.)

If $\bar{P}_i^{(T)}$ is the total force on m_i, including inertia, constraint and applied forces, the principle of virtual work yields

$$\sum_i \bar{P}_i^{(T)} \cdot \delta \bar{r}_i = 0$$

where the "dot" indicates the scalar product of two vectors. (The scalar product of two vectors $\bar{x}$ and $\bar{y}$ is given by: $\bar{x} \cdot \bar{y} = xy \cos \theta$, where x and y are the magnitudes of the vectors $\bar{x}$ and $\bar{y}$ and θ is the angle between the vectors.) The force $\bar{P}_i^{(T)}$ can be divided into applied,

321

inertia and constraint parts; for most systems the constraint forces do no work. The inertia component of $\bar{P}_i^{(T)}$, $\bar{P}_i^{(I)}$, is related to the acceleration of m_i by

$$\bar{P}_i^{(I)} = -m_i\ddot{\bar{r}}_i$$

by d'Alembert's principle. Thus, if $\bar{P}_i$ is the contribution to $\bar{P}_i^{(T)}$ from the applied forces,

$$\sum_i \bar{P}_i \cdot \delta\bar{r}_i = \sum_i m_i\ddot{\bar{r}}_i \cdot \delta\bar{r}_i. \qquad (A4.1)$$

In the following analysis the subscript i refers to a particular mass m_i and its displacement $\bar{r}_i$, velocity $\dot{\bar{r}}_i$, etc.; summation with respect to i implies summation with respect to all the masses comprising the system.

Considering a system with n degrees of freedom defined by the generalized (and independent) coordinates $q_1, q_2, q_3, \ldots, q_n$, an increment δq_j can be given to coordinate q_j without changing any other coordinate. For this increment the system displaces and the applied forces acting on the system do work

$$\delta\mathfrak{W} = Q_j\,\delta q_j$$
$$= \sum_i \bar{P}_i \cdot \delta\bar{r}_i \qquad (A4.2)$$

where Q_j is the generalized force corresponding to coordinate q_j. The position vectors of the masses $\bar{r}_i$ can be defined in terms of the generalized coordinates by

$$\bar{r}_i = f_i(q_1, q_2, q_3, \ldots, q_n). \qquad (A4.3)$$

If the applied forces have potential (e.g. spring or gravity forces),

$$\delta\mathfrak{W} = -\delta\mathfrak{U}$$

where $\mathfrak{U}$ is the potential energy of the system. Thus

$$\delta\mathfrak{U} = -Q_j\,\delta q_j.$$

However, $\mathfrak{U}$ is a function of the generalized coordinates q_j; for the increment δq_j the corresponding increment in $\mathfrak{U}$ is

$$\delta\mathfrak{U} = \frac{\partial\mathfrak{U}}{\partial q_j}\delta q_j.$$

Thus for potential forces

$$Q_j = -\frac{\partial \mathcal{U}}{\partial q_j}. \tag{A4.4}$$

The virtual displacement $\delta \bar{r}_i$, corresponding to the increment δq_j, is from equation (A4.3)

$$\delta \bar{r}_i = \frac{\partial \bar{r}_i}{\partial q_j} \delta q_j. \tag{A4.5}$$

From equations (A4.1) and (A4.2)

$$Q_j \, \delta q_j = \sum_i m_i \ddot{\bar{r}}_i \cdot \delta \bar{r}_i$$

$$= \sum_i \left(m_i \ddot{\bar{r}}_i \cdot \frac{\partial \bar{r}_i}{\partial q_j} \right) \delta q_j$$

from equation (A4.5). Hence

$$Q_j = \sum_i m_i \ddot{\bar{r}}_i \cdot \frac{\partial \bar{r}_i}{\partial q_j}. \tag{A4.6}$$

The kinetic energy

$$\mathfrak{T} = \tfrac{1}{2} \sum_i m_i \dot{\bar{r}}_i^2$$

$$= \tfrac{1}{2} \sum_i m_i (\dot{\bar{r}}_i \cdot \dot{\bar{r}}_i).$$

Thus

$$\frac{\partial T}{\partial \dot{q}_j} = \sum_i m_i \left(\dot{\bar{r}}_i \cdot \frac{\partial \dot{\bar{r}}_i}{\partial \dot{q}_j} \right) \tag{A4.7}$$

and

$$\frac{\partial T}{\partial q_j} = \sum_i m_i \left(\dot{\bar{r}}_i \cdot \frac{\partial \dot{\bar{r}}_i}{\partial q_j} \right). \tag{A4.8}$$

From equation (A4.3)

$$\dot{\bar{r}}_i = \frac{d}{dt} (\bar{r}_i)$$

$$= \sum_{s=1}^{n} \frac{\partial \bar{r}_i}{\partial q_s} \dot{q}_s. \tag{A4.9}$$

Thus

$$\frac{\partial \bar{r}_i}{\partial q_j} = \sum_{s=1}^{n} \frac{\partial^2 \bar{r}_i}{\partial q_j \, \partial q_s} \dot{q}_s \qquad \text{(A4.10)}$$

and

$$\frac{d}{dt}\left(\frac{\partial \bar{r}_i}{\partial q_j}\right) = \sum_{s=1}^{n} \frac{\partial^2 \bar{r}_i}{\partial q_j \, \partial q_s} \dot{q}_s \, . \qquad \text{(A4.11)}$$

From equation (A4.9)

$$\frac{\partial \dot{\bar{r}}_i}{\partial \dot{q}_j} = \frac{\partial \bar{r}_i}{\partial q_j} \, . \qquad \text{(A4.12)}$$

From equations (A4.10) and (A4.11)

$$\frac{\partial \dot{\bar{r}}_i}{\partial q_j} = \frac{d}{dt}\left(\frac{\partial \bar{r}_i}{\partial q_j}\right) . \qquad \text{(A4.13)}$$

From equation (A4.6)

$$\begin{aligned}
Q_j &= \sum_i m_i \ddot{\bar{r}}_i \cdot \frac{\partial \bar{r}_i}{\partial q_j} \\
&= \sum_i m_i \frac{d}{dt}\left(\dot{\bar{r}}_i \cdot \frac{\partial \bar{r}_i}{\partial q_j}\right) - \sum_i m_i \dot{\bar{r}}_i \cdot \left[\frac{d}{dt}\left(\frac{\partial \bar{r}_i}{\partial q_j}\right)\right] \\
&= \sum_i m_i \frac{d}{dt}\left(\dot{\bar{r}}_i \cdot \frac{\partial \dot{\bar{r}}_i}{\partial \dot{q}_j}\right) - \sum_i m_i \dot{\bar{r}}_i \cdot \frac{\partial \dot{\bar{r}}_i}{\partial q_j}
\end{aligned}$$

using equations (A4.12) and (A4.13). Thus

$$Q_j = \frac{d}{dt}\left(\frac{\partial \mathfrak{T}}{\partial \dot{q}_j}\right) - \frac{\partial \mathfrak{T}}{\partial q_j} \qquad \text{(A4.14)}$$

from equations (A4.7) and (A4.8). In equation (A4.14) the generalized force Q_j is the total force, with contributions from all the forces on the system, and may be divided into parts, associated with the forces with potential, the damping (or dissipation) forces and the applied forces. Equation (A4.4) gives the contribution to Q_j from the potential forces; if $Q_j^{(D)}$ is the contribution from the damping forces and Q_j is redefined as the generalized force associated with the applied forces only, equation (A4.14) may be rewritten

$$\frac{d}{dt}\left(\frac{\partial T}{\partial \dot{q}_j}\right) - \frac{\partial T}{\partial q_j} + \frac{\partial \mathfrak{U}}{\partial q_j} - Q_j^{(D)} = Q_j. \qquad (A4.15)$$

If the mass m_i is subjected to a viscous damping force

$$\bar{P}_i = -c_i \bar{r}_i$$

the principle of virtual work leads to

$$Q_j^{(D)} \delta q_j = \delta \mathfrak{W} = -\sum_i c_i \bar{r}_i \cdot \delta \bar{r}_i$$

$$= -\sum_i c_i \bar{r}_i \cdot \frac{\partial \bar{r}_i}{\partial \dot{q}_j} \delta q_j$$

using equations (A4.5) and (A4.12). Hence

$$Q_j^{(D)} = -\sum_i \frac{\partial}{\partial \dot{q}_j}(\tfrac{1}{2}c_i \bar{r}_i^2)$$

$$= -\frac{\partial \mathfrak{F}}{\partial \dot{q}_j} \qquad (A4.16)$$

where the dissipation function $\mathfrak{F} = \sum_i \tfrac{1}{2}c_i \bar{r}_i^2$. For damping forces which depend upon relative velocity, a typical force between masses m_i and m_s is $c_{is}(\bar{r}_i - \bar{r}_s)$. Following similar analysis, equation (A4.16) for $Q_j^{(D)}$ applies, if

$$\mathfrak{F} = \sum \tfrac{1}{2}c_{is}(\bar{r}_i - \bar{r}_s)^2$$

where the summation includes all the damping mechanisms c_{is} in the system. Substituting from equation (A4.16) in equation (A4.15), we obtain the Lagrange equation:

$$\frac{d}{dt}\left(\frac{\partial \mathfrak{T}}{\partial \dot{q}_j}\right) - \frac{\partial \mathfrak{T}}{\partial q_j} + \frac{\partial \mathfrak{F}}{\partial \dot{q}_j} + \frac{\partial \mathfrak{U}}{\partial q_j} = Q_j, \qquad j = 1, 2, 3, \ldots, n. \quad (A4.17)$$

In this general form of the Lagrange equation the potential energy $\mathfrak{U}$

may contain contributions from gravity forces and from the strain energy stored in elastic components. In the applications of the equation in this book there are no contributions to $\mathfrak{U}$ from gravity forces, as all displacements are measured relative to the position of static equilibrium, and $\mathfrak{U}$ can be replaced by the strain energy $\mathfrak{S}$. Also in these applications the kinetic energy $\mathfrak{T}$ is a function of the velocities $\dot{q}_j$ and not of the coordinates q_j, so that the term $\partial \mathfrak{T}/\partial q_j$ in equation (A4.17) is zero.

An alternative derivation of the Lagrange equation from Hamilton's principle is given by Hurty and Rubinstein[40].

APPENDIX 5

Eigenvalue Economizers

THE analysis of a complex structure by the finite element method leads to a large number of degrees of freedom and the corresponding matrix equations may be too large to be handled with the available computational facilities. Thus the method of eigenvalue economization (also known as mass condensation or reduction) which reduces the size of the final equation is important in practice. Considering the matrix equation for free vibrations

$$\mathbf{M}\ddot{\mathbf{x}} + \mathbf{K}\mathbf{x} = 0 \qquad (A5.1)$$

we partition $\mathbf{x}$ (and similarly $\ddot{\mathbf{x}}$) into master degrees of freedom $\mathbf{x}_m$ and slave degrees of freedom $\mathbf{x}_s$ with the intention of reducing the vector $\mathbf{x}_s$ from the equations, i.e.

$$\mathbf{x} = \begin{bmatrix} \mathbf{x}_m \\ \hline \mathbf{x}_s \end{bmatrix}.$$

The corresponding partitioned form of equation (A5.1) is

$$\begin{bmatrix} \mathbf{M}_{mm} & \mathbf{M}_{ms} \\ \hline \mathbf{M}_{sm} & \mathbf{M}_{ss} \end{bmatrix} \begin{bmatrix} \ddot{\mathbf{x}}_m \\ \hline \ddot{\mathbf{x}}_s \end{bmatrix} + \begin{bmatrix} \mathbf{K}_{mm} & \mathbf{K}_{ms} \\ \hline \mathbf{K}_{sm} & \mathbf{K}_{ss} \end{bmatrix} \begin{bmatrix} \mathbf{x}_m \\ \hline \mathbf{x}_s \end{bmatrix} = \begin{bmatrix} \mathbf{0} \\ \hline \mathbf{0} \end{bmatrix}. \quad (A5.2)$$

The basis of the method is to use the lower set of equations to express $\mathbf{x}_s$ in terms of $\mathbf{x}_m$, assuming that the relation between $\mathbf{x}_s$ and $\mathbf{x}_m$ is not affected by inertia effects; thus expanding the lower partition of equation (A5.2) and neglecting the inertia terms,

$$\mathbf{K}_{sm}\, \mathbf{x}_m + \mathbf{K}_{ss}\, \mathbf{x}_s = 0$$

or

$$\mathbf{x}_s = -\mathbf{K}_{ss}^{-1}\mathbf{K}_{sm}\, \mathbf{x}_m. \qquad (A5.3)$$

The partitioned forms of the strain and kinetic energy expressions, leading to equation (A5.2), are

$$\mathfrak{S} = \tfrac{1}{2}[\mathbf{x}_m^T \mathrel{\vdots} \mathbf{x}_s^T]\left[\begin{array}{c|c}\mathbf{K}_{mm} & \mathbf{K}_{ms} \\ \hline \mathbf{K}_{sm} & \mathbf{K}_{ss}\end{array}\right]\left[\begin{array}{c}\mathbf{x}_m \\ \mathbf{x}_s\end{array}\right] \tag{A5.4}$$

and

$$\mathfrak{T} = \tfrac{1}{2}[\dot{\mathbf{x}}_m^T \mathrel{\vdots} \dot{\mathbf{x}}_s^T]\left[\begin{array}{c|c}\mathbf{M}_{mm} & \mathbf{M}_{ms} \\ \hline \mathbf{M}_{sm} & \mathbf{M}_{ss}\end{array}\right]\left[\begin{array}{c}\dot{\mathbf{x}}_m \\ \dot{\mathbf{x}}_s\end{array}\right]. \tag{A5.5}$$

Expanding expression (A5.4), applying equation (A5.3) and noting that

$$\mathbf{x}_s^T = -\mathbf{x}_m^T \mathbf{K}_{ms} \mathbf{K}_{ss}^{-1} \quad \text{as} \quad \mathbf{K}_{sm}^T = \mathbf{K}_{ms} \quad \text{and} \quad \mathbf{K}_{ss} = \mathbf{K}_{ss}^T,$$
$$\mathfrak{S} = \tfrac{1}{2}\mathbf{x}_m^T \mathbf{K}' \mathbf{x}_m$$

where

$$\mathbf{K}' = \mathbf{K}_{mm} - \mathbf{K}_{ms} \mathbf{K}_{ss}^{-1} \mathbf{K}_{sm}. \tag{A5.6}$$

Expanding expression (A5.5) for $\mathfrak{T}$ and using equation (A5.3), a similar simplification does not occur and

$$\mathfrak{T} = \tfrac{1}{2}\dot{\mathbf{x}}_m^T \mathbf{M}' \dot{\mathbf{x}}_m$$

where

$$\begin{aligned}\mathbf{M}' = {} & \mathbf{M}_{mm} - \mathbf{M}_{ms} \mathbf{K}_{ss}^{-1}\mathbf{K}_{sm} - \mathbf{K}_{ms} \mathbf{K}_{ss}^{-1}\mathbf{M}_{sm} \\ & + \mathbf{K}_{ms} \mathbf{K}_{ss}^{-1}\mathbf{M}_{ss} \mathbf{K}_{ss}^{-1}\mathbf{K}_{sm}.\end{aligned} \tag{A5.7}$$

Applying the Lagrange equation to equations (A5.6) and (A5.7) with the vector $\mathbf{x}_m$ as the independent coordinates, the reduced equation of motion is

$$\mathbf{M}'\ddot{\mathbf{x}}_m + \mathbf{K}'\mathbf{x}_m = 0. \tag{A5.8}$$

For free vibrations:

$$[\mathbf{K}' - \mathbf{M}'\omega^2]\mathbf{x}_m = 0. \tag{A5.9}$$

Thus we have reduced the frequency determinant by the number of degrees of freedom associated with $\mathbf{x}_s$.

The distribution of the original degrees of freedom between the master and slave vectors, $\mathbf{x}_m$ and $\mathbf{x}_s$, is of vital importance. If there is no kinetic energy associated with the degrees of freedom contained in x_s, M_{sm} and M_{ss} are zero, equation (A5.3) is valid without any assumptions, the reduced equation (A5.8) contains no approximations [other than those used to establish the original equation (A5.1)], K' is given by equation (A5.6) and $M' = M_{mm}$. Alternatively, equation (A5.3) is valid without any assumptions if $M_{sm} = M_{ss}K_{ss}^{-1}K_{sm}$; in this case the reduced equation (A5.8) contains no approximations, K' is given by equation (A5.6) and $M' = M_{mm} - M_{ms}K_{ss}^{-1}K_{sm}$. In general, neither of the above restrictions on the mass matrix will be satisfied, but if we select for x_s those degrees of freedom associated with *small* contributions to the kinetic energy, assumption (A5.3) will cause a small error in the natural frequencies. Suppose that it is possible to calculate the lower eigenvalues of a structure by the following three methods: (a) from a fine mesh with N degrees of freedom; (b) using the same mesh, but reducing the number of degrees of freedom to N_1 before solving the eigenvalue problem; and (c) using a coarser mesh with N_1 degrees of freedom and no reduction procedure. In general, frequencies from (b) will be less accurate than those from (a), but better than those from (c).

Although judgement is required in selecting the degrees of freedom for the vector x_s, those associated with in-plane displacements and with rotations, as opposed to normal displacements, for plate and shell elements are usually associated with small contributions to the kinetic energy for low-frequency vibrations and can be included in x_s. The slave degrees of freedom can be eliminated during the assembly process so that the large matrices, associated with the complete vector x, need never be completely stored in the computer. A method for the automatic selection of the master degrees of freedom by the computer has been given by Henshell and Ong[37].

Although this method of reducing the order of matrix equations is normally associated with the finite element method, where it is frequently used, it can be applied to any problem governed by equation (A5.1). This equation is obtained also when free vibrations of complex structures are analysed by the Rayleigh–Ritz or finite difference methods.

It should be noted that although degrees of freedom associated with small contributions to the kinetic energy are reduced from the matrix equation, the procedure outlined above is *not* equivalent to neglecting these small contributions to the kinetic energy; the latter procedure would lead to $\mathbf{M}' = \mathbf{M}_{mm}$, which is not true in general.

The numerical effects of these approximations are illustrated in the following simple example. Zienkiewicz[90] shows the small increases in the first four natural frequencies of a rectangular cantilever plate, as the number of master degrees of freedom in the finite element idealization is reduced progressively from ninety to six.

Example. Using beam finite elements to represent a uniform beam with both ends clamped, determine the first two natural frequencies using (a) two, three and four elements and no reduction; (b) four elements with reduction of rotational degrees of freedom; (c) four elements and neglecting rotational kinetic energy.

Without reduction the frequency determinant is of order 2×2, 4×4 and 6×6 for two, three and four elements, respectively. With four elements and either reduction of rotational degrees of freedom or neglect of rotational kinetic energy the determinant is of order 3×3. The matrix equation for the three-element model has been given [equation (3.93)] and will not be repeated.

For the four-element model it is simpler to take advantage of the symmetry and consider separately the symmetric and antisymmetric modes of one half of the beam. Figure A5.1 shows the beam divided

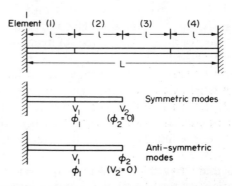

FIG. A5.1. Beam with clamped ends divided into four elements.

into four equal elements; considering the left-hand half of the beam the non-zero degrees of freedom are v_1, ϕ_1 and v_2 for symmetric modes and v_1, ϕ_1 and ϕ_2 for anti-symmetric modes.

Symmetric Modes

Using equations (3.89) and (3.91) for the element stiffness and mass matrices, assembling for the two elements and imposing the boundary conditions (v and ϕ are zero at the left-hand end of element 1 and ϕ is zero at the right-hand end of element 2), the matrix equation for free vibrations is:

$$\left\{ \begin{bmatrix} 24 & 0 & -12 \\ 0 & 8l^2 & -6l \\ -12 & -6l & 12 \end{bmatrix} \frac{EI}{l^3} - \begin{bmatrix} 312 & 0 & 54 \\ 0 & 8l^2 & 13l \\ 54 & 13l & 156 \end{bmatrix} \frac{\rho A l \omega^2}{420} \right\} \begin{bmatrix} v_1 \\ \phi_1 \\ v_2 \end{bmatrix} = 0.$$

(A5.10)

The lowest natural frequency ω_1, obtained from the determinant of equation (A5.10), is given by $(\rho A/EI)^{1/2}L^2\omega_1 = 22\cdot403$. (The length of the beam $L = 4l$.) This is $0\cdot13$ per cent higher than the value obtained from the beam equation (Section 3.3). Comparable errors with three- and two-element idealizations are $0\cdot41$ and $1\cdot62$ per cent, respectively. Equation (A5.10) is rearranged in partitioned form so that ϕ_1 can be eliminated as

$$\left\{ \left[\begin{array}{cc:c} 24 & -12 & 0 \\ -12 & 12 & -6l \\ \hdashline 0 & -6l & 8l^2 \end{array} \right] \frac{EI}{l^3} - \left[\begin{array}{cc:c} 312 & 54 & 0 \\ 54 & 156 & 13l \\ \hdashline 0 & 13l & 8l^2 \end{array} \right] \frac{\rho A l w^2}{420} \right\} \begin{bmatrix} v_1 \\ v_2 \\ \hdashline \phi_1 \end{bmatrix} = 0.$$

Using equation (A5.6) the reduced stiffness matrix

$$\mathbf{K}' = \frac{EI}{l^3} \left\{ \begin{bmatrix} 24 & -12 \\ -12 & 12 \end{bmatrix} - \begin{bmatrix} 0 \\ -6l \end{bmatrix} \frac{1}{8l^2} [0 \quad -6l] \right\}$$

$$= \begin{bmatrix} 24 & -12 \\ -12 & 15/2 \end{bmatrix} \frac{EI}{l^3}.$$

(A5.11)

From equation (A5.7) the reduced mass matrix

$$\mathbf{M}' = \frac{\rho A l}{420}\left\{\begin{bmatrix} 312 & 54 \\ 54 & 156 \end{bmatrix} - \begin{bmatrix} 0 \\ 13l \end{bmatrix}\frac{1}{8l^2}\begin{bmatrix} 0 & -6l \end{bmatrix}\right\} = \begin{bmatrix} 0 \\ -6l \end{bmatrix}\frac{1}{8l^2}\begin{bmatrix} 0 & 13l \end{bmatrix}$$

$$+ \begin{bmatrix} 0 \\ -6l \end{bmatrix}\left(\frac{1}{8l^2}\right)(8l^2)\left(\frac{1}{8l^2}\right)\begin{bmatrix} 0 & -6l \end{bmatrix}\right\} = \begin{bmatrix} 312 & 54 \\ 54 & 180 \end{bmatrix}\frac{\rho A l}{420}. \quad (A5.12)$$

From det $|\mathbf{K}' - \mathbf{M}'\omega^2| = 0$ the lowest natural frequency is given by $(\rho A/EI)^{1/2}L^2\omega_1 = 22\cdot410$, which is $0\cdot17$ per cent higher than the value from the beam equation.

When the kinetic energy associated with the rotational coordinates is neglected, two methods are possible: (a) neglecting all terms in $\mathbf{x}_s$ in the kinetic energy expression, so that $\mathfrak{T} = \frac{1}{2}\dot{\mathbf{x}}_m^T\mathbf{M}_{mm}\dot{\mathbf{x}}_m$, using the notation of equation (A5.5); (b) using lumped masses. Procedure (a) gives

$$\mathbf{M}' = \mathbf{M}_{mm} = \frac{\rho A l}{420}\begin{bmatrix} 312 & 54 \\ 54 & 156 \end{bmatrix}.$$

$\mathbf{K}'$ is given by equation (5.11). From det $|\mathbf{K}' - \mathbf{M}'\omega^2| = 0$ the lower natural frequency is given by $(\rho A/EI)^{1/2}L^2\omega_1 = 23\cdot278$, which is $4\cdot04$ per cent higher than the value from the beam equation.

For procedure (b)

$$\mathfrak{T} = \frac{1}{2}m_1\dot{v}_1^2 + \frac{1}{2}m_2\dot{v}_2^2$$

where m_1 and m_2 are the lumped masses at nodes 1 and 2, respectively.

(In the finite element method presented here consistent mass matrices have been derived, but the lumped mass approach, which leads to diagonal mass matrices, has been mentioned in Section 3.9.) Replacing the distributed mass of $\rho A l$ per element by $\frac{1}{2}\rho A l$ at the end of each element

$$\mathfrak{T} = \frac{1}{2}\rho A l\dot{v}_1^2 + \frac{1}{2}(\frac{1}{2}\rho A l)\dot{v}_2^2.$$

As there is no change in the strain energy expression, equation (A5.10) is replaced by

$$\left\{\begin{bmatrix} 24 & 0 & -12 \\ 0 & 8l^2 & -6l \\ -12 & -6l & 12 \end{bmatrix}\frac{EI}{l^3} - \begin{bmatrix} 1 & 0 & 0 \\ 0 & 0 & 0 \\ 0 & 0 & \frac{1}{2} \end{bmatrix}\rho A l\omega^2\right\}\begin{bmatrix} v_1 \\ \phi_1 \\ v_2 \end{bmatrix} = 0. \quad (A5.13)$$

Elimination of ϕ_1 leads to the frequency determinant

$$\det |\mathbf{K}' - \mathbf{M}''\omega^2| = 0$$

with $\mathbf{K}'$ given by equation (A5.11) and

$$\mathbf{M}'' = \begin{bmatrix} 1 & 0 \\ 0 & \frac{1}{2} \end{bmatrix} \rho A l.$$

The lowest natural frequency is given by $(\rho A/EI)^{1/2}L^2\omega_1 = 22\cdot302$, which is 0·32 per cent lower than the value from the beam equation.

Anti-symmetric Modes

Noting that $v_2 = 0$, but $\phi_2 \neq 0$, the matrix equation for free vibrations is

$$\left\{ \begin{bmatrix} 24 & 0 & 6l \\ 0 & 8l^2 & 2l^2 \\ 6l & 2l^2 & 4l^2 \end{bmatrix} \frac{EI}{l^3} \right.$$

$$\left. - \begin{bmatrix} 312 & 0 & -13l \\ 0 & 8l^2 & -3l^2 \\ -13l & -3l^2 & 4l^2 \end{bmatrix} \frac{\rho A l\omega^2}{420} \right\} \begin{bmatrix} v_1 \\ \phi_1 \\ \phi_2 \end{bmatrix} = 0. \quad (A5.14)$$

The lowest natural frequency for anti-symmetric modes, ω_2, is obtained from equation (A5.14) as $(\rho A/EI)^{1/2}L^2\omega_2 = 62\cdot243$, which is 0·92 per cent higher than the value from the beam equation. With two and three elements the comparable errors are 32·9 and 2·0 per cent, respectively.

Using the partitions, shown in equation (A5.14), and equation (A5.6) to eliminate ϕ_1 and ϕ_2, the reduced stiffness matrix

$$\mathbf{K}' = \frac{EI}{l^3} \left\{ [24] - [0 \quad 6l] \begin{bmatrix} 1/7l^2 & -1/14l^2 \\ -1/14l^2 & 2/7l^2 \end{bmatrix} \begin{bmatrix} 0 \\ 6l \end{bmatrix} \right\}$$

as

$$\begin{bmatrix} 8l^2 & 2l^2 \\ 2l^2 & 4l^2 \end{bmatrix}^{-1} = \begin{bmatrix} 1/7l^2 & -1/14l^2 \\ -1/14l^2 & 2/7l^2 \end{bmatrix}.$$

Thus

$$\mathbf{K}' = \frac{96}{7} \frac{EI}{l^3}. \quad (A5.15)$$

From equation (A5.7)

$$\mathbf{M}' = \frac{\rho Al}{420}\Bigg\{[312] - [0 \quad -13l]\begin{bmatrix} 1/7l^2 & -1/14l^2 \\ -1/14l^2 & 2/7l^2 \end{bmatrix}\begin{bmatrix} 0 \\ 6l \end{bmatrix}$$

$$- [0 \quad 6l]\begin{bmatrix} 1/7l^2 & -1/14l^2 \\ -1/14l^2 & 2/7l^2 \end{bmatrix}\begin{bmatrix} 0 \\ -13l \end{bmatrix}$$

$$+ [0 \quad 6l]\begin{bmatrix} 1/7l^2 & -1/14l^2 \\ -1/14l^2 & 2/7l^2 \end{bmatrix}\begin{bmatrix} 8l^2 & -3l^2 \\ -3l^2 & 4l^2 \end{bmatrix}$$

$$\times \begin{bmatrix} 1/7l^2 & -1/14l^2 \\ -1/14l^2 & 2/7l^2 \end{bmatrix}\begin{bmatrix} 0 \\ 6l \end{bmatrix}\Bigg\}$$

$$= \frac{374 \cdot 20 \rho Al}{420}.$$

Thus the natural frequency ω_2 is given by $(\rho A/EI)^{1/2}L^2\omega_2 = 62 \cdot 77$, which is 1·8 per cent higher than the value from the beam equation.

Neglecting the kinetic energy associated with $\dot{\phi}_1$ and $\dot{\phi}_2$, and following procedure (a) $\mathbf{M}' = 312\rho Al/420$ and $\mathbf{K}'$ is given by equation (A5.15). Thus $(\rho A/EI)^{1/2}L^2\omega_2 = 68 \cdot 75$, which is 11·5 per cent higher than the value from the beam equation. Following procedure (b), $\mathfrak{T} = \frac{1}{2}(\rho Al)\dot{v}_1^2$, as $\dot{v}_2 = 0$; using equation (A5.15) for $\mathbf{K}'$, $(\rho A/EI)^{1/2}L^2\omega_2 = 59 \cdot 252$, which is 3·9 per cent low.

The numerical results, which are collected together in Table A5.1, illustrate the general points made earlier, as the natural frequencies for four elements using eigenvalue economization or mass reduction are less accurate than the corresponding frequencies without reduction, but more accurate than those obtained from a two- or three-element idealization without reduction. Eigenvalue economization applied to the rotational coordinates for which the associated kinetic energy is expected to be small is clearly better than complete neglect of this contribution to the kinetic energy [procedure (a)]. With the lumped mass approach [procedure (b)] the errors are less than for procedure (a), but the natural frequencies obtained are lower than the true values. This illustrates the point that consistent mass matrices must be used to achieve monotonic convergence of natural frequencies to the true values, as the mesh of conforming finite elements is progressively subdivided. However, the lumped mass method, unlike procedure (a), is used in practice when a diagonal mass matrix leads to computational advantages in response calculations (Section 2.8).

TABLE A5.1. Percentage errors in natural frequencies of clamped–clamped beam using simple finite element idealizations

No. of elements	Degrees of freedom after reduction	Procedure	Percentage error in	
(whole beam)			ω_1	ω_2
2	2	No reduction	1·62	32·9
3	4	No reduction	0·41	2·0
4	6	No reduction	0·13	0·92
4	3	Reduction of ϕ_j	0·17	1·8
4	3	Neglect of K.E. associated with ϕ_j [procedure (a)]	4·0	11·5
4	3	Lumped mass approximation	−0·32	−3·9

References

1. ABRAMSON, H. N., *Journal of the Acoustical Society of America*, **29**, 42 (1957).
2. ARMENAKAS, A. E., *Journal of the Engineering Mechanics Division, ASCE*, **93**, no. EM5, 95 (1967).
3. ARMENAKAS, A. E., GAZIS, D. C. and HERRMANN, G., *Free Vibration of Circular Cylindrical Shells*, Pergamon Press (1969).
4. ARNOLD, R. N., BYCROFT, G. N. and WARBURTON, G. B., *Journal of Applied Mechanics, Trans. ASME*, **77**, 391 (1955).
5. ARNOLD, R. N. and WARBURTON, G. B., *Proceedings of the Royal Society*, Series A, **197**, 238 (1949).
6. BATHE, K.-J. and WILSON, E. L., *International Journal for Numerical Methods in Engineering*, **6**, 213 (1973).
7. BATHE, K.-J. and WILSON, E. L., *International Journal of Earthquake Engineering and Structural Dynamics*, **1**, 283 (1973).
8. BISHOP, R. E. D. and GLADWELL, G. M. L., *Philosophical Transactions of the Royal Society*, Series A, **255**, 241 (1963).
9. BISHOP, R. E. D. and JOHNSON, D. C., *Vibration Analysis Tables*, Cambridge University Press, Cambridge (1956).
10. BISHOP, R. E. D. and JOHNSON, D. C., *The Mechanics of Vibration*, Cambridge University Press, Cambridge (1960).
11. BISHOP, R. E. D. and PENDERED, J. W., *Journal of Mechanical Engineering Science*, **5**, 343 (1963).
12. BOLOTIN, V. V., *Proceedings of Vibration Problems, Warsaw*, **6**, 361 (1965).
13. BRADSHAW, A. and WARBURTON, G. B., *Journal of Mechanical Engineering Science*, **9**, 290 (1967).
14. BREBBIA, C. A. and CONNOR, J. J., *Fundamentals of Finite Element Techniques*, Butterworths, London (1973).
15. BRIGHAM, E. O., *The Fast Fourier Transform*, Prentice-Hall, Englewood Cliffs, N.J. (1974).
16. CARR, J. B., *Aeronautical Quarterly*, **21**, 79 (1970).
17. CASE, J. and CHILVER, A. H., *Strength of Materials*, Arnold, London (1959).
18. CHOPRA, A. K., *Journal of the Engineering Mechanics Division, ASCE*, **93**, no. EM6, 205 (1967).
19. CHOPRA, A. K., *Journal of the Engineering Mechanics Division, ASCE*, **94**, 1475 (1968).
20. CHOPRA, A. K., *Journal of the Engineering Mechanics Division, ASCE*, **96**, 443 (1970).

21. CHOPRA, A. K. and GUTIERREZ, J. A., *International Journal of Earthquake Engineering and Structural Dynamics*, **3**, 65 (1974).
22. COULSON, C. A., *Waves*, Oliver & Boyd, Edinburgh (1949).
23. COWPER, G. R., *Journal of Applied Mechanics, Trans. ASME*, **88**, 335 (1966).
24. CRANDALL, S. H. and MARK, W. D., *Random Vibration in Mechanical Systems*, Academic Press, New York (1963).
25. DAVIS, R., HENSHELL, R. D. and WARBURTON, G. B., *International Journal of Earthquake Engineering and Structural Dynamics*, **1**, 165 (1972).
26. DAVIS, R., HENSHELL, R. D. and WARBURTON, G. B., *Journal of Sound and Vibration*, **25**, 561 (1972).
27. DAVIS, R., HENSHELL, R. D. and WARBURTON, G. B., *Journal of Sound and Vibration*, **22**, 475 (1972).
28. DAWE, D. J., *Journal of Mechanical Engineering Science*, **7**, 28 (1965).
29. DEN HARTOG, J. P., *Mechanical Vibrations*, 4th Ed., McGraw-Hill, New York (1956).
30. ELISHAKOFF, I., *Journal of the Acoustical Society of America*, **57**, 361 (1975).
31. FLÜGGE, W., *Stresses in Shells*, 2nd Ed., Springer-Verlag, Berlin (1970).
32. FLÜGGE, W. and ZAJAC, E. E., *Ingenieur-Archiv (Festschrift)*, **28**, 59 (1959).
33. FRAEIJS DE VEUBEKE, B. M., *A Variational Approach to Pure Mode Excitation based on Characteristic Phase Lag Theory*, AGARD Report no. 39 (1956).
34. FRÝBA, L., *Vibrations of Solids and Structures under Moving Loads*, Noordhoff, Groningen (1972).
35. FUNG, Y. C. and BARTON, M. V., *Journal of Applied Mechanics, Trans. ASME*, **80**, 365 (1958).
36. GOLDSMITH, W., *Impact*, Arnold, London (1960).
37. HENSHELL, R. D. and ONG, J. H., *International Journal of Earthquake Engineering and Structural Dynamics*, **3**, 375 (1975).
38. HENSHELL, R. D., WALTERS, D. and WARBURTON, G. B., *Journal of Sound and Vibration*, **20**, 381 (1972).
39. HENSHELL, R. D. and WARBURTON, G. B., *International Journal for Numerical Methods in Engineering*, **1**, 47 (1969).
40. HURTY, W. C. and RUBINSTEIN, M. F., *Dynamics of Structures*, Prentice-Hall, Englewood Cliffs, N.J. (1964).
41. JACOBSEN, L. S. and AYRE, R. S., *Engineering Vibrations*, McGraw-Hill, New York (1958).
42. JAEGER, L. C., *Elementary Theory of Elastic Plates*, Pergamon Press (1964).
43. JENNINGS, P. C. and BIELAK, J., *Bulletin of the Seismological Society of America*, **63**, 9 (1973).
44. KOLSKY, H., *Stress Waves in Solids*, Oxford University Press, Oxford (1953).
45. KRAUS, H., *Thin Elastic Shells*, Wiley, New York (1967).
46. LECKIE, F. A. and LINDBERG, G. M., *Aeronautical Quarterly*, **14**, 224 (1963).
47. LEE, E. H. and SYMONDS, P. A., *Journal of Applied Mechanics, Trans. ASME*, **74**, 308 (1952).
48. LEISSA, A. W., *Vibration of Plates*, NASA SP-160 (1969).
49. LEISSA, A. W., *Vibration of Shells*, NASA SP-288 (1973).
50. LEISSA, A. W., *Journal of Sound and Vibration*, **31**, 257 (1973).
51. LIAW, C-Y. and CHOPRA, A. K., *International Journal of Earthquake Engineering and Structural Dynamics*, **3**, 33 (1974).
52. LIAW, C-Y. and CHOPRA, A. K., *International Journal of Earthquake Engineering and Structural Dynamics*, **3**, 233 (1975).

53. LIN, Y. K., *Probabilistic Theory of Structural Dynamics*, McGraw-Hill, New York (1963).
54. LOVE, A. E. H., *The Mathematical Theory of Elasticity*, 4th Ed., Cambridge University Press, Cambridge (1927).
55. LYSMER, J. and KUHLEMEYER, R. L., *Journal of the Engineering Mechanics Division, ASCE*, **95**, 859 (1969).
56. MALTBAEK, J. C., *International Journal of Mechanical Sciences*, **3**, 197 (1961).
57. MASSONNET, C., *Bulletin des Cours et des Laboratoires d'Essais des Constructions du Génie Civil et d'Hydraulique Fluviale*, **1** (1940).
58. MAUNDER, L., *Quarterly of Applied Mathematics*, **17**, 437 (1960).
59. MCLEOD, A. J. and BISHOP, R. E. D., *The Forced Vibration of Circular Flat Plates*, Mechanical Engineering Science Monograph no. 1, Institution of Mechanical Engineers, London (1965).
60. MINDLIN, R. D., *Journal of Applied Mechanics, Trans. ASME*, **67**, A69 (1945).
61. MINDLIN, R. D. and GOODMAN, L. E., *Journal of Applied Mechanics, Trans. ASME*, **72**, 377 (1950).
62. NOVOZHILOV, V. V., *The Theory of Thin Shells*, Noordhoff, Groningen (1959).
63. PARMELEE, R. A., PERELMAN, D. S., LEE, S. L. and KEER, L. M., *Journal of the Engineering Mechanics Division, ASCE*, **94**, 1295 (1968).
64. REISSNER, E., *Journal of Mathematics and Physics*, **25**, 80 (1941).
65. RICHART, F. E. Jr., HALL, J. R. Jr. and WOODS, R. D., *Vibrations of Soils and Foundations*, Prentice-Hall, Englewood Cliffs, N.J. (1970).
66. ROBSON, J. D., *An Introduction to Random Vibration*, Edinburgh University Press, Edinburgh (1963).
67. SANDERS, J. I. Jr., *An Improved First Approximation Theory for Thin Shells*, NASA TR R-24 (1959).
68. SCANLAN, R. H. and WARDLAW, R. L., Reduction of flow-induced structural vibrations, Section 2, pp. 35–63, of ASME AMD vol. 1, 1973, *Isolation of Mechanical Vibration, Impact and Noise*, edited by J. C. Snowdon and E. E. Ungar.
69. SNOWDON, J. C., *Vibration and Shock in Damped Mechanical Systems*, Wiley, New York (1968).
70. SNOWDON, J. C., *Journal of the Acoustical Society of America*, **56**, 1177 (1974).
71. SRINIVAS, S., JOGA RAO, C. V. and RAO, A. K., *Journal of Sound and Vibration*, **12**, 187 (1970).
72. THOMSON, W. T., CALKINS, T. and CARAVANI, P., *International Journal of Earthquake Engineering and Structural Dynamics*, **3**, 97 (1974).
73. TIMOSHENKO, S. P. and GOODIER, J. N., *Theory of Elasticity*, 3rd Ed., McGraw-Hill, New York (1970).
74. TIMOSHENKO, S. P. and WOINOWSKY-KRIEGER, S., *Theory of Plates and Shells*, 2nd Ed., McGraw-Hill, New York (1959).
75. TIMOSHENKO, S. P., YOUNG, D. H. and WEAVER, W. Jr., *Vibration Problems in Engineering*, 4th Ed., Wiley, New York (1974).
76. UNGAR, E. E., *Journal of Engineering for Industry, Trans. ASME*, **89B**, 626 (1967).
77. VAISH, A. K. and CHOPRA, A. K., Dynamic analysis of structure-foundation systems, *Proc. 5th World Conference on Earthquake Engineering*, Rome, 1973, Paper no. 115.
78. VELETSOS, A. S. and WEI, Y. T., *Journal of the Soil Mechanics and Foundations Division, ASCE*, **97**, 1227 (1971).
79. VLASOV, V. Z., *General Theory of Shells and its Applications in Engineering*, NASA TT F-99 (1964) (English translation).

80. VOGEL, S. M. and SKINNER, D. W., *Journal of Applied Mechanics, Trans. ASME*, **87**, 926 (1965).
81. WARBURTON, G. B., *Proceedings of the Institution of Mechanical Engineers*, **168**, 371 (1954).
82. WARBURTON, G. B., *Journal of Mechanical Engineering Science*, **7**, 399 (1965).
83. WARBURTON, G. B., *Dynamics of Shells*, University of Toronto, Department of Mechanical Engineering, Technical Publication TP 7307 (1973).
84. WARBURTON, G. B. and HIGGS, J., *Journal of Sound and Vibration*, **11**, 335 (1970).
85. WEAVER, W. Jr., BRANDOW, G. E. and HÖEG, K., *Bulletin of the Seismological Society of America*, **63**, 1041 (1973).
86. WEBSTER, J. J., *International Journal of Mechanical Sciences*, **12**, 157 (1970).
87. WILKINSON, J. H., *The Algebraic Eigenvalue Problem*, Oxford University Press, Oxford (1965).
88. WILLIAMS, D., *Aeronautical Quarterly*, **1**, 123 (1950).
89. WOINOWSKY-KRIEGER, S., *Journal of Applied Mechanics, Trans. ASME*, **72**, 35 (1950).
90. ZIENKIEWICZ, O. C., *The Finite Element Method in Engineering Science*, McGraw-Hill, London (1971).

Principal Notation

SYMBOLS which are defined in one section and used only in that section are not listed.

Matrices are printed in bold type with capital letters for square (or rectangular) matrices and lower-case letters for column matrices (vectors) and row matrices; det $|\mathbf{K}|$ signifies the determinant of the square matrix $\mathbf{K}$.

$|k|$ indicates the modulus of the quantity k.

$\overline{x(t)y(t + \tau)}$ is the mean of the product of x at time t and y at time $(t + \tau)$, averaged over the time interval $-T$ to $+T$ as $T \to \infty$. (Used in Chapters 1 and 2.)

$\bar{r}$ signifies that r is a vector quantity (used only in Appendix 4).

Re $\{\cdots\}$ and Im $\{\cdots\}$ signify the real and imaginary parts respectively of the complex quantity within the brackets.

For a displacement x, velocity $dx/dt = \dot{x}$ and acceleration $d^2x/dt^2 = \ddot{x}$.

General

c	Viscous damping constant
k	Stiffness
l	Length
m	Mass
q	Principal coordinate
t	Time
u, v, w	Displacements in X-, Y- and Z-directions, respectively
x, y, z	Cartesian coordinates

341

E	Young's modulus
G	Shear modulus
L	Length
M	Bending moment
P	Applied force
Q	Generalized force
p	Vector or column matrix of applied forces
q	Vector or column matrix of principal coordinates
C	Viscous dampin matrix
H	Hysteretic damping matrix
I	Identity matrix
K	Stiffness matrix
M	Mass matrix
0	Null matrix
$\mathfrak{F}$	Dissipation function
$\mathfrak{J}$	Impulse
$\mathfrak{S}$	Strain energy
$\mathfrak{T}$	Kinetic energy
$\mathfrak{U}$	Potential energy
$\delta\mathfrak{W}$	Virtual work
γ	Viscous damping parameter
ε	Strain
μ	Hysteretic damping parameter
v	Poisson's ratio
ρ	Density
τ	Time
ω	Frequency (circular)
Ω	Eigenvalue matrix

Chapters 1 and 2

f	Natural frequency
h	Hysteretic damping constant
$h(t)$	Response function for unit impulse

i	$= (-1)^{1/2}$
p_d	Probability density function
$H(\omega)$	Complex frequency response or receptance
Pr	Probability
$R_x(\tau)$	Autocorrelation function of x
$S_x(\omega)$	Spectral density of x
T	Period
V	Velocity
$_r\mathbf{e}$	Vector or column matrix of amplitudes $_re_j$ for mode r
$\mathbf{x}$	Vector or column matrix of displacements
$_r\mathbf{z}$	Vector or column matrix of normalized amplitudes $_rz_j$ for mode r
$\mathbf{Z}$	Modal matrix
α	Phase angle
σ_x^2	Variance of random variable x

Chapters 3 and 4

a	Length
b, d	Breadth and depth of beam
d_{xa}	Flexibility function
$g(x)$	Influence function
i	$= (-1)^{1/2}$
A	Cross-sectional area
F	Transverse force
I	Second moment of area in bending
J	Polar second moment of area
K_r	Radius of gyration
N	Axial force
R	Radius of curvature
S	Shear force
T	Torque (T = period in Section 4.7)
$\mathbf{f}$	Vector or column matrix of forces

g   Vector or column matrix of parameters Γ_j (Sections 3.7, 3.8 and 5.2)

g   Polynomial row matrix (Sections 3.9, 5.4 and 5.5)

J   Dynamic stiffness matrix

α Phase angle

θ Angular rotation

λ Frequency parameter

σ Stress

ϕ Slope of beam (Sections 3.9 and 4.5)

$\phi_r(x)$ rth mode shape in flexure (Sections 3.4 to 3.8 and 4.2)

Chapter 5

h Thickness of plate or shell

r Radial coordinate

D Flexural rigidity

R Radius of curvature

S Transverse shear force

$\mathbf{u}_e$, $\mathbf{v}_e$, $\mathbf{w}_e$ Vectors or column matrices of nodal displacements for element

D   Elastic properties matrix

θ Angular coordinate or angular rotation

σ Stress

$\phi_r(x)$, $\psi_s(y)$ Beam mode shapes (Section 5.2)

$\phi = -\partial w/\partial x$ (Section 5.5)

$\psi = \partial w/\partial y$ (Section 5.5)

δ Vector or column matrix of nodal displacements

Chapter 6

$a_0 = \omega R(\rho/G)^{1/2}$ Non-dimensional frequency factor

$b = m/\rho R^3$ Mass ratio

f_j	Non-dimensional dynamic flexibility functions
r	Radial coordinate
C	Wave velocity
H	Depth of water
K	Bulk modulus
Q	Reaction force or moment on the ground due to a foundation
R	Base radius of disc or foundation
$\mathbf{x}$	Vector or column matrix of displacements
$\mathbf{y}$	Vector or column matrix of free field displacements of the ground
β_j	$= (2j - 1)\pi/2H$
$\phi(x, y, t)$	Velocity potential for fluid

Answers to Problems

Chapter 1

1. 5·03 Hz.
2. 20 cos 31·6t mm.
3. 2·12 kN s/m; 0·0168; 0·105.
4. 2·98 mm; 31·6 rad/s; 89·0°.
5. (a) 25·0 mm; (b) 25·0[exp $(-5t)$(0·258 sin 19·36t + cos 19·36t) − cos 20·0t] mm.
6. $\omega/\omega_n = (1 - 2\gamma^2)^{-1/2}$.
7. 4·50 Hz; 4·58 Hz, 3·80 mm; < 3·78 and > 6·31 Hz; 4·64 Hz, 4·12 mm.
8. (a) 0·0909; (b) 0·0449; (c) 0·00990; (d) 0·000622.
9. 1·732P_0/k; 0·667T.
10. (a) a, $0 \le t_1/T \le 1/6$, $5/6 \le t_1/T \le 7/6, \ldots$
        $|2a \sin \pi t_1/T|$, $1/6 \le t_1/T \le 5/6$, $7/6 \le t_1/T \le 11/6, \ldots$
    (b) $2a \sin \pi t_1/T$, $0 \le t_1/T \le 1/2$
        $2a$, $t_1/T > 1/2$.
11. (a) 1·970 × 10^8 N^2, 14·03 kN; (b) 1·976 × 10^8 N^2, 14·06 kN; 0·44 per cent; 99·74 kN.

Chapter 2

1. 4·91, 12·0 Hz; 1·5, −1·0.
2. 0·004 cos 30·9t + 0·006 cos 75·6t (m); (a) 0·015 m, (b) −0·01 m.

3. (i)   (a) $x_1 = 0.0135 - 0.0108 \cos 30.9t - 0.0027 \cos 75.6t$ m,

           $x_2 = 0.0135 - 0.0162 \cos 30.9t + 0.0027 \cos 75.6t$ m.

     (b) $x_1 = 0.0270(1 - \cos 30.9t)$ m,

           $x_2 = 0.0405(1 - \cos 30.9t)$ m.

   (ii)   (a) $x_1 = 0.0135 - 0.0108 \exp (-2.38t)(\cos 30.8t + 0.0774 \sin 30.8t)$

              $- 0.0027 \exp (-14.29t)(\cos 74.2t + 0.192 \sin 74.2t)$ m.

           $x_2 = 0.0135 - 0.0162 \exp (-2.38t)(\cos 30.8t + 0.0774 \sin 30.8t)$

              $+ 0.0027 \exp (-14.29t)(\cos 74.2t + 0.192 \sin 74.2t$ m.

(b) $x_1 = 0.0270[1 - \exp(-2.38t)(\cos 30.8t + 0.0774 \sin 30.8t)]$ m,

$x_2 = 0.0405[1 - \exp(-2.38t)(\cos 30.8t + 0.0774 \sin 30.8t)]$ m.

4. (a) 1.35, 7.07, 0.762 mm; (b) 1.34, 12.02, 4.51 mm.

5. Central difference: $\Delta t = 0.025$: (i) $x_1 = 2.67$ cm, error 0.32 cm,

$x_2 = 2.73$ cm, error -0.32 cm.

(ii) $x_1 = 2.22$ cm, error 0.06 cm,

$x_2 = 2.62$ cm, error -0.04 cm.

$\Delta t = 0.0125$; (i) $x_1 = 2.44$ cm, error 0.09 cm,

$x_2 = 2.97$ cm, error -0.08 cm.

(ii) $x_1 = 2.19$ cm, error 0.03 cm,

$x_2 = 2.64$ cm, error -0.02 cm.

Newmark: $\Delta t = 0.025$: (i) $x_1 = 2.15$ cm, error -0.20 cm,

$x_2 = 3.21$ cm, error 0.16 cm.

(ii) $x_1 = 2.08$ cm, error -0.08 cm,

$x_2 = 2.74$ cm, error 0.08 cm.

$\Delta t = 0.0125$: (i) $x_1 = 2.23$ cm, error -0.12 cm,

$x_2 = 3.16$ cm, error 0.11 cm.

(ii) $x_1 = 2.13$ cm, error -0.03 cm,

$x_2 = 2.70$ cm, error 0.04 cm.

6. $2.310 \sin 10\pi t - 2.355 \sin 30.9t + 0.0018 \sin 75.6t$ m; $3.474 \sin 10\pi t - 3.533 \sin 30.9t - 0.0018 \sin 75.6t$ m.

7. 81.6, 186 Hz; -0.0623, 1.78 rad/m.

8. 33.6, 72.0 μm.

9. (i) 33.6, 72.0 μm; $-0.8°$, 0.6°; (ii) 4.32, 10.74 mm; 86.7°, 88.7°.

10. (i) $x_A = 0.0236, 0.0151, 0.00627$ m; $x_B = -0.00092, 0.00816, 0.00471$ m.

13. (a) 0.6488, 0.6782; (b) 0.6488, 0.6802; (c) 0.6480, 0.6748 mm.

Chapter 3

1. 1288, 2576, 3864 Hz.

3. 66.2, 182.5, 357.7 Hz.

5. $(\lambda l)^3 (\sin \lambda l \cosh \lambda l - \cos \lambda l \sinh \lambda l) = (2kl^3/EI) \sin \lambda l \sinh \lambda l$.

    (a) 0.416, 3.927; (b) 1.310, 3.943; (c) 3.142, 6.283.

6. Values of VEI/Pl^3: (a) (i) 0.0249, (ii) 0.0168; (b) (i) 0.0314, (ii) 0.0260.

7. V/V_{st}: (a) (i) 97·1 (1), (ii) 97·1 (1); (b) (i) 2·47 (1), (ii) 0·395 (2).
M/M_{st}: (a) (i) 113·8 (1), (ii) 113·8 (1); (b) (i) 18·2 (1), (ii) 2·90 (1).

8. (a) 0·0277Pl^3/EI; (b) 0.

9. (a) 0·341$P_0\,l^3/EI$; (b) 0·650$P_0\,l^3/EI$. $\pm 2\cdot25P_0\,l/Z$.

10. (a) 0·123$p_0\,l^4/EI$; (b) 0·252$p_0\,l^4/EI$.

11. Values of vEI/Pl^3: (a) (i) 0·235, (ii) 0·232; (b) (i) 0·862, (ii) 0·830.

13. (a) $\omega_1 = 5\cdot683(EI/ml^3)^{1/2}$.

15. (a) $\omega_1 = 0\cdot860(EA/ml)^{1/2}$.

Chapter 4

1. (a) 2·24, (b) 37·8 m/s; (c) none. 6·11 to 10·11 Hz.

3. (a) $a(1 - x/l)\sin \omega t - \dfrac{2a}{\pi}\sum_s \dfrac{1}{s}\sin\dfrac{s\pi x}{l}\left[\dfrac{\sin \omega t - (\omega_s/\omega)\sin \omega_s t}{1 - (\omega_s/\omega)^2}\right]$.

(b) $a\sin \omega t - \dfrac{4a}{\pi}\sum_{s = 1,\,3,\,5,\,\ldots}\dfrac{1}{s}\sin\dfrac{s\pi x}{l}\left[\dfrac{\sin \omega t - (\omega_s/\omega)\sin \omega_s t}{1 - (\omega_s/\omega)^2}\right]$.

(c) $\dfrac{M_0 l^2}{EI}\left[\left(\dfrac{x}{3l} - \dfrac{x^2}{2l^2} + \dfrac{x^3}{6l^3}\right) - \dfrac{2}{\pi^3}\sum_s\dfrac{1}{s^3}\sin\dfrac{s\pi x}{l}\left\{\dfrac{\sin \omega t - (\omega_s/\omega)\sin \omega_s t}{1 - (\omega_s/\omega)^2}\right\}\right]$.

4. (a) $a(1 - \cos \omega t) + a\sum_s\left[\dfrac{\phi_s(x)}{\lambda_s^4 l}\left(\dfrac{d^3\phi_s}{dx^3}\right)_{x = 0}\dfrac{\cos \omega_s t - \cos \omega t}{1 - (\omega_s/\omega)^2}\right]$

where $\phi_s(x)$ and λ_s relate to a uniform cantilever.

(b) $\dfrac{a}{2}\left[3\left(\dfrac{x}{l}\right)^2 - \left(\dfrac{x}{l}\right)^3\right](1 - \cos \omega t) - a\sum_s\left[\dfrac{\phi_s(x)}{\lambda_s^4 l}\left(\dfrac{d^3\phi_s}{dx^3}\right)_{x = l}\dfrac{\cos \omega_s t - \cos \omega t}{1 - (\omega_s/\omega)^2}\right]$

where $\phi_s(x)$ and λ_s relate to a uniform beam, clamped at $x = 0$ and simply supported at $x = l$.

5. 770 kN; −1·23 per cent.

6. 2·90 kN.

7. 5·03, 12·3, 31·4 Hz.

12. (a) $1 \le \zeta \le 3$. (b) $\dfrac{3(\zeta - 1)M_0\,t_1^2}{2\rho Al^2}$; $\dfrac{3\zeta(\zeta - 1)M_0\,t_1^2}{2\rho Al^2}$.

13. $\omega_1 = 0\cdot432\,(ED^2/\rho l^4)^{1/2}$.

Chapter 5

3. −2·18, −2·18, 0 per cent.

4. 0·252Pa^2/Eh^3.

6. $\dfrac{10\cdot21}{R^2}\left(\dfrac{D}{\rho h}\right)^{1/2}$.

7. $\omega R^2(\rho/E)^{1/2}/h = 18\cdot0$ for $w = \Gamma(r - \frac{1}{2}R)(R - r)\sin 3\theta \sin \omega t$. [Value from exact solution is $17\cdot0$ [80].]

8. $46\cdot8$, $52\cdot6$, $61\cdot7$ Hz.

9. $51\cdot2$, $58\cdot9$, $65\cdot6$ Hz; none; 8.

10. Values of ω_{jn}/ω_0 for modes $j/n = 1/1$, $2/1$, $1/2$ and $1/3$: (i) 1, $2\cdot5$, $2\cdot5$, $5\cdot0$; (ii) $1\cdot671$, $3\cdot293$, $2\cdot557$, $5\cdot007$; (iii) $3\cdot494$, $5\cdot912$, $2\cdot836$, $5\cdot045$.

Index